Middle East and North Africa

Edited by

Eckart Ehlers
Katajun Amirpur

BRILL

LEIDEN | BOSTON

Cover illustration: Ḫāne Āqāzādeh, Abarkūh, Iran. © Ben Müller, 2019.

Library of Congress Cataloging-in-Publication Data

Names: Ehlers, Eckart, 1938-editor. | Amirpur, Katajun, 1971- editor.
Title: Middle East and North Africa : climate, culture, and conflicts / edited by Eckart Ehlers and Katajun Amirpur.
Description: Leiden ; Boston, MA : Brill, 2021. | Series: Climate and culture, 2213–0519 ; volume 6 | Includes bibliographical references and index.
Identifiers: LCCN 2020046623 (print) | LCCN 2020046624 (ebook) | ISBN 9789004444454 (hardback) | ISBN 9789004444973 (ebook)
Subjects: LCSH: Middle East—Environmental conditions. | Africa, North—Environmental conditions. | Climatic changes—Social aspects—Middle East. | Climatic changes—Social aspects—Africa, North. | Environmental management—Middle East. | Environmental management--Africa, North. | Human beings—Effect of climate on—Middle East. | Human beings—Effect of climate on—Africa, North.
Classification: LCC GE160.M628 M53 2021 (print) | LCC GE160.M628 (ebook) | DDC 304.20956—DC23
LC record available at https://lccn.loc.gov/2020046623
LC ebook record available at https://lccn.loc.gov/2020046624

KULTURWISSENSCHAFTLICHES
INSTITUT ESSEN

FORSCHUNGSKOLLEG DER
UNIVERSITÄTSALLIANZ RUHR

Typeface for the Latin, Greek, and Cyrillic scripts: "Brill". See and download: brill.com/brill-typeface.

ISSN 2213-0519
ISBN 978-90-04-44445-4 (hardback)
ISBN 978-90-04-44497-3 (e-book)

This book is printed on acid-free paper and produced in a sustainable manner.

Contents

PART 3
Present

PART 4
Prospects

Foreword

Humans influence the environment and climate, with the consequences now felt across the globe. National or regional efforts to restrict or at least contain the damage are invariably insufficient: in principle, environmental and climate protection requires a global approach.

Paradoxically, the way we *perceive* environmental and climate change and respond to the harm that they cause is closely linked to local or regional patterns of perception. It is these particularistic perceptions that often lead to different, and in many instances conflicting, reactions to preventive and curative environmental and climate protection measures.

These local views are grounded not only in the different paths that socio-economic development has taken in specific regions of the world, but also in varying *cultural patterns*. Think, for example, of the vastly different ways in which current problems are perceived, or of how policy styles and politico-social environments differ. Also, the disturbance of the environment and climate causes relatively rapid social changes, in which the interpretation of symbols for the relationship between man and nature plays an important part. The history of climate and culture, patterns of perception of environmental and climate change, and an informed assessment of the future direction of environmental and climate policy in different parts of the world have to be taken into account in order to get to grips with the problem.

From a variety of angles, such as the history of ideas, historiography, the study of civilisation, and the political sciences, the monographs and edited volumes in *Climate and Culture* will all deal with the following questions:

- How do local and regional cultures perceive historical and contemporary changes in the environment and climate?
- How did and do they adjust to these changes?
- How do their various representatives and spokespeople introduce their respective views into the global debate and into emerging systems of international negotiation?

The following titles are included in the series:

VOLUME 1: *Nature, the Environment and Climate Change in East Asia*, edited by Carmen Meinert, 2013.

VOLUME 2: *Climate Change in South and Southeast Asia*, edited by Barbara Schuler, 2014.

VOLUME 3: *Cultural Dynamics of Climate Change and the Environment in Northern America*, edited by Bernd Sommer, 2015.

VOLUME 4: *Climate Change and Cultural Transition in Europe*, edited by Claus Leggewie and Franz Mauelshagen, 2017.

VOLUME 5: *Environmental Change and African Societies*, edited by Ingo Haltermann and Julia Tischler, 2019.

VOLUME 6: *Middle East and North Africa: Climate, Culture, and Conflicts*, edited by Eckart Ehlers and Katajun Amirpur, 2021.

Carmen Meinert
Claus Leggewie

Acknowledgements

The preparation and publication of this book has been attended by a series of more or less serious obstacles: a change of the editorial responsibility for this volume, regrettable cancellations of promised contributions, sometimes very long delays in the submission of papers and, consequently, the need to find new authors with recognised competence and scholarship in their specific fields of research. But "all's well that ends well"—this Shakespearean promise may therefore be an appropriate omen for this volume.

The editors of this book on *Middle East and North Africa: Climate, Culture, and Conflicts* are grateful to the series editors of the CLAC-series, Carmen Meinert and Claus Leggewie, for their trust, patience and understanding in view of the long and sometimes frustrating production process.

The foremost gratitude, however, goes to the authors of this publication. As indicated, some of them have been contributors from the very beginning of this venture, and their readiness to update earlier versions of their papers is highly appreciated. Equally appreciated are the endeavours of those colleagues who joined us at a later date and who submitted their contribution in time, postponing other and sometimes equally important obligations. We are well aware of the sometimes almost impolite pressure that we have put on some of our co-authors. All the more do we wish to acknowledge their understanding and cooperation.

This book on climate and culture in the Middle East and North Africa and the close interactions and interdependencies of both in the millennia-old natural and cultural history of this region contains a comparatively large number of maps and diagrams, of figures, tables and photos—possibly more than in other volumes of this series. This is owed to many helpful hands behind the scenes. The editors would like to thank all those colleagues who submitted ready-to-print versions of and attachments to their papers; others relied on outside help. The co-editor Eckart Ehlers is greatly indebted to Mrs. Ursula Dörken for her steady support in (re-)writing different versions of manuscripts; and to Dipl.Ing. Martin Gref, Department of Geography, University of Bonn, for his professional skill in preparing maps and figures. The co-editor Katajun Amirpur would like to thank Ingrid Overbeck for her help and support.

Yet the highest measure of our appreciation and gratitude goes to Ben Müller, Student Assistant at the Centrum for Religious Studies (CERES) at the Ruhr-Universität Bochum. His expertise and knowledge in the preparation and editing process of this book, acquired in the editing work of earlier volumes of the CLAC series, can hardly be overestimated. Without his help, advice, and patience, this book would not have been published in its present form and at this point.

Eckart Ehlers and Katajun Amirpur

Figures, Maps, and Tables

Figures

Maps

Tables

Abbreviations

AC	Alternating Current
Akk.	Akkadian
AP	Arboreal Pollen
Arab.	Arabic
ARC	Alliance of Religions and Conservation
a.s.l.	Above sea level
BCE	before the Common Era/before Christian Era
BMZ	German Federal Ministry for Economic Cooperation and Development
BP	Before Present
CCS	Carbon Capture Storage
CERES	Center for Religious Studies, Ruhr-University, Bochum
Chin.	Chinese
CLAC	*Climate and Culture*
CSP	Concentrating Solar Thermal Power Stations
CO_2	Carbon dioxide
DC	Direct Current
Deut.	*Book of Deuteronomy*
DLR	Department of Systems Analysis and Technology Assessment of the German Aerospace Center, Stuttgart, Germany (Germ. Deutsches Zentrum für Luft- und Raumfahrt)
EACI	European Agency for Competitiveness and Innovation
EU	European Union
EUMENA	Europe, Middle East, and North Africa
ECOSOC	Economic and Social Council
ESCWA	(United Nations) Economic and Social Commission for Western Asia
esp.	especially
et al.	et alii/aliae/alia
Ex.	*Book of Exodus*
Ez.	*Book of Ezekiel*
fig.	figure
FSI	Forum for Social Innovation
GA	General Assembly
GCC	Gulf Cooperation Council
Gen.	*Book of Genesis*
GERD	Grand Ethiopian Renaissance Dam
Germ.	German

GeSI	Global e-Sustainability Initiative
GHG	Greenhouse gas
HVDC	High Voltage Direct Current
Ibid.	ibidem
ICRC	International Committee of the Red Cross, Geneva
ICT	Information and Communication Technologies
i.e.	*id est*
IFEES	Islamic Foundation for Ecology and Environmental Sciences
IFS	International Foundation for Science
IHDP	International Human Dimensions Programme
IOM	International Organization for Migration
IPCC	Intergovernmental Panel on Climate Change
IS	Islamic State
ITCZ	Intertropical Convergence Zone
ITU	International Telecommunication Union
LAS	League of Arab States
LDC	Least Developed Countries
Lev.	*Book of Leviticus*
LLC	Limited Liability Company
MADE	Muslim Action for Development and Environment
MEA	Millennium Ecosystem Assessment
MENA	Middle East and North Africa
MS	Member States
MSP	Mediterranean Solar Plan
Mt.	*Gospel of Matthew*
MUI	Indonesian Ulema Council
NAO	North Atlantic Oscillation
NAP	Non Arboreal Pollen
NGO	Non-Governmental Organisation
NKGCF	German National Committee on Global Change Research
NTC	Net Transfer Capacity
NU	Nahdlatul Ulama
OECD	Organisation for Economic Co-operation and Development
Pers.	Persian
PMN	Progressive Muslims Network
PRC	People's Republic of China
Ps.	*Psalms*
PSI	Planetary Security Initiative
R&D	Research and Development

RCC	Rapid Climate Change
RED	Renewable Energy Directive
SEM	Southern European and Mediterranian
SEP	Solar Electricity Pipeline
SDEWES	"Sustainable Enhancements of Power, Cooling and Water for Hot Climates"
SOAS	The Middle East Centre, School of Oriental and African Studies
Sum.	Sumerian
SWP	Stiftung Wissenschaft und Politik
tab.	table
TOC	Total Organic Carbon
TREC	Trans-Mediterranean Renewable Energy Cooperation
UIN	Syarif Hidayatullah State Islamic University
UK	United Kingdom
ULRP	Urmia Lake Restoration Program
UN	United Nations
UNESCO	United Nations Educational, Scientific and Cultural Organization
UNDP	United Nations Development Programme
WACC	Weighted Average Cost of Capital
WBGU	Wissenschaftlicher Beirat der Bundesregierung Globale Umweltveränderungen
WEF	World Economic Forum
WIN	Wisdom in Nature
WWAP	United Nations World Water Assessment Programme
WWF	World Wide Fund for Nature
YHWH	Tetragrammaton, four-letter biblical name of the God of Israel
ZEF	Zentrum für Entwicklungsforschung

Notes on Contributors

Katajun Amirpur
is Professor of Iranian Studies at the University of Cologne. Her doctoral thesis and postdoctoral habilitation were on Iranian reformist thinkers. From 2011 to 2018, she served as Deputy Director of the Academy of World Religions at Hamburg University. One of her most recognised publications is Amirpur Katajun, *New Thinking in Islam* (Munich: C.H. Beck, 2019). Her main research activity is Shiʿi Islam.

Helmut Brückner
is a Geomorphologist. Until 2019, he was Professor for Physical Geography at the University of Cologne. His research foci are coastal geomorphology, geo-archaeology, and geochronology. His recent publications include: Brückner, Helmut et al., "Life Cycle of Estuarine Islands—From the Formation to the Landlocking of Former Islands in the Environs of Miletos and Ephesos in Eastern Asia Minor (Turkey)," *Journal of Archaeological Science: Reports*, 12 (2017): 876–894. doi.org/10.1016/j.jasrep.2016.11.024 and; Herda, Alexander et al., "From the Gulf of Latmos to Lake Bafa: On the History, Geoarchaeology, and Palynology of the Lower Maeander Valley at the Foot of the Latmos Mountains," *Hesperia: The Journal of the American School of Classical Studies at Athens* 88.1 (2019): 1–86. https://www.jstor.org/stable/10.2972/hesperia.88.1.0001?seq=4#metadata_info_tab_contents.

Eckart Ehlers
is Professor Emeritus of Geography at the University of Bonn and Senior Fellow at its Center for Development Studies (Ger. Zentrum für Entwicklungsforschung (ZEF)). His special fields of interest cover the social and economic geography of the Near and Middle East as well as the ecological aspects of global environmental change and interactions between nature and society in the Anthropocene. In these fields he has served in the past as chair of the German National Committee on Global Change Research (NKGCF), as chair of the Scientific Committee of the International Human Dimensions Programme (IHDP) on Global Environmental Change and other national and international scientific organisations like START (System for Analysis, Research and Training) in Washington/DC or the International Foundation for Science (IFS) in Stockholm. His recent publications and editorials include Ehlers, Eckart, Das *Anthropozän. Die Erde im Zeitalter des Menschen* (Darmstadt:

Wissenschaftliche Buchgesellschaft, 2008) and; Ehlers, Eckart, "Globalization, UNESCO, Urban Morphology and Local Responses," *Urban Morphology* 22 (2018): 101–108.

Max Engel
is a geomorphologist. He is head of the Lab of Geomorphology and Geoecology at Heidelberg University and specialises in coastal and dryland geomorphology, geoarchaeology, palaeoenvironmental changes, and coastal hazards. His recent publications include: Boesl, Fabian et al., "Digital Mapping of Coastal Boulders—High-Resolution Data Acquisition to Infer Past and Recent Transport Dynamics," Sedimentology (2019). doi:10.1111/sed.12578 and; Engel, Max, Fabian Boesl, and Helmut Brückner, "Migration of Barchan Dunes in Qatar—Controls of the Shamal, Teleconnections, Sea-level Changes, and Human Impact," Geosciences 8 (2018): 240. doi:10.3390/geosciences8070240.

Kerstin Fritzsche
is a political scientist and sustainability researcher. She heads the research area digitalisation at the Institute for Futures Studies and Technology Assessment (IZT) in Berlin, Germany. She is specialised in the field of the interdependencies between digital transformation and sustainable development in countries of the Global South. Her recent publications include: Fritzsche, Kerstin, Silke Niehoff, and Grischa Beier, "Industry 4.0 and Climate Change—Exploring the Science-Policy Gap," Sustainability 10.12 (2018): 4511. http://doi.org/10.3390/su10124511 and; Fritzsche, Kerstin, Stefanie Kunkel, and Marcel Matthess, "Digitalized Economies in Africa. Structural Change in a Dynamic Environment," *IASS Fact Sheet* (2020): https://doi.org/10.2312/iass.2020.050.

Ursula Kowanda-Yassin
is a teacher of Religious Education and a specialist in Islamic studies. She is a researcher on Muslims in Europe at Sigmund Freud University, Vienna, specialising in Islamic environmental education. Her recent publications include: Kowanda-Yassin, Ursula, *Öko-Dschihad. Der grüne Islam: Beginn einer globalen Umweltbewegung* (Salzburg, Wien: Residenz-Verlag, 2018) and; Kowanda-Yassin, Ursula, *Mensch und Naturverständnis in den Grundlagen des sunnitischen Islam. Ein Beitrag zum aktuellen Umweltdiskurs* (Würzburg: Ergon-Verlag, 2011).

Ephraim Meir
is Professor Emeritus of Modern Jewish Philosophy at Bar-Ilan University, Ramat Gan, Israel. He specialises in modern Jewish philosophy and interreligious

theology. His recent publications include: Meir, Ephraim, *Interreligious Theology: Its Value and Mooring in Modern Jewish Philosophy* (Berlin, Jerusalem: De Gruyter and Magnes, 2015) and; Meir, Ephraim, *Becoming Interreligious. Towards a Dialogical Theology from a Jewish Vantage Point* (Münster, New York: Waxmann, 2017).

Rosel Pientka-Hinz
is an Assyriologist. She is Adjunct Professor (Ger. *Privatdozentin*) of Ancient Near Eastern Studies at the Center for Religious Studies (CERES) at the Ruhr-University of Bochum and specialises in the cultural and religious history of the ancient Near East (above all animal symbolism). Her recent publications include: Pientka-Hinz, Rosel, "Making Contact in Mesopotamia: Powerful Kisses, Forbidden Kisses," in *Tabou et transgressions, Actes du colloque organisé par le Collège de France, Paris, les 11–12 avril 2012*, ed. Jean-Marie Durand, Michael Guichard, and Thomas Römer, 69–81 (Freiburg, Göttingen: Universitätsverlag Vandenhoeck and Ruprecht, 2015) and; Pientka-Hinz, Rosel, "Bunte Kühe? Zu den frühesten Farbbezeichnungen im Alten Orient," in *Ancient Near Eastern Studies: The Empirical Dimension of Ancient Near Eastern Studies/Die empirische Dimension altorientalischer Forschungen*, ed. Gebhard J. Selz and Klaus Wagensonner, 325–374 (Wien, Berlin: LIT, 2011).

Matthias Schmidt
is Professor of Human Geography and Resource Strategies at the University of Augsburg and specialises in political ecology, human environment geography, development, and transition research. His recent publications include: Schmidt, Matthias and Sebastian Purwins "Simultaneity as a Challenge for Development," *GeoJournal* 83 (2018): 1193–1204 and; Pearson, Olivia and Matthias Schmidt, "Repercussions of Institutional Governance Changes on Communication Practices in the Afar National Regional State, Ethiopia," *Singapore Journal of Tropical Geography* 39 (2018): 438–453.

Franz Trieb
works as Project Manager and Scientific Advisor of the International Energy-Water-Nexus Team at the Department of Systems Analysis and Technology Assessment of the German Aerospace Center (DLR) in Stuttgart, Germany. He specialises in renewable energy integration for power and seawater desalination. His recent publications include: Trieb, Franz, et al., "Rescuing the Concept of Solar Electricity Transfers from North Africa to Europe," *International Journal of Energy Sector Management* 10.3 (2016): 448-473, accessed May 15, 2017. http://www.emeraldinsight.com/1750-6220.htm and;

Trieb, Franz, Massimo Moser and Jürgen Kern, "Liquid Solar Fuel—Liquid Hydrocarbons from Solar Energy and Biomass" (presented at the conference on "Sustainable Enhancements of Power, Cooling and Water for Hot Climates (SDEWES)," Dubrovnik, Croatia, October 4-8, 2017.

Tobias von Lossow

is a political scientist and Research Fellow in the Sustainability Unit at the Clingendael, Netherlands Institute of International Relations. His research focuses on water conflicts, hydro-politics and water diplomacy in transboundary river basins in the Middle East and Africa. Moreover, he works on the role of water in armed conflicts and fragile post-conflict settings and on climate change impacts on water security. Tobias von Lossow is a lecturer at the Free University Berlin, Consortium Expert of the Planetary Security Initiative (PSI), a member of the Water and Conflict in the Middle East project of Georgetown University, Qatar (2017–2019) and member of the Water Security in the Middle East by EuroMeSCo & IPCRI (2019–2020). His recent publications include: von Lossow, Tobias, "More than Infrastructures: Water Challenges in Iraq" *Policy Brief, Clingendael & PSI* (2018) and: von Lossow, Tobias, "The Multiple Crisis. Perspectives on Water Scarcity in the Euphrates and Tigris Basin, *Orient* 58.1 (2017): 45–53.

INTRODUCTION

The MENA Region: Climate, Culture, and Conflicts in the Middle East and North Africa

Eckart Ehlers and Katajun Amirpur

1 Concepts, Contexts, and Contents of this Volume

This book, volume 6 of the *Climate and Culture* series concludes the commendable attempt by publisher and editors to provide an overview on the intricate interactions between nature and societies in their specific regional and cultural environments.[1] As such, this book is also part of an ambitious attempt to create awareness and understanding for one of the most challenging threats to humanity in the 21st century: the impacts and consequences of global environmental changes to all forms of life. In doing so, however, this series goes far beyond well-trodden paths. Unlike the almost countless numbers of scientific reports, handbooks, and academic monographs, *Climate and Culture* goes further and deeper. It takes a regionally and culturally differentiated approach to explore the fundamental ideas, basic foundations, and historical consequences of what one may call "environmentalism" in the broadest sense of the word. Such a perspective includes the sometimes dramatic reactions of their political implementation, impact, and problems. In this sense, this volume—like its predecessors—takes an approach predominantly based on humanities and social and political science to achieve a better understanding of the aforementioned challenges as they affect the MENA region.

It is obvious that the magnitude and diversity of the omnipresent ecological, socio-economic, and political problems in the Middle East and North Africa (MENA), their causes and consequences cannot be fully addressed either within one volume like this or by any single academic or scientific discipline alone. The complexity of the problems is far too great, solutions not in sight.

One further aspect should be kept in mind when evaluating the conception, contexts, and contents of this book. It is the fact that they all have to follow certain pre-set rules in format and content: Ideas—Past—Present—Prospect

1 Here and in the following texts the term 'environment' will be used in a general sense as both natural and cultural environment. For more details see Eckart Ehlers and Carl F. Gethmann, ed, *Environment across Cultures* (Berlin: Springer, 2003).

 | DOI:10.1163/9789004444973_002

are formal prescriptions that the editors had to follow. The same holds true for the requested multi-disciplinary character of the contributions to this volume in order to ensure a multi-faceted look at the close interactions and interconnectivities between climate and culture, between nature and society. Thus, this book—like others of this series—claims a form of holistic approach. What does that mean? We see it reflected in the fact that while all contributions to this book are analyses of specific aspects of the climate and culture context rooted in one discipline, each paper is also part of a greater whole, of a universal message. Jointly and individually, they carry messages and transferable insights applicable across all disciplines. They all deal with the overarching *conditio humana*, i.e. with the fact that human existence and human survival, be it as an individual, be it as a society or part of it, is always and has always been a struggle with a specific natural or social environment. Such a hypothesis and claim, surely not modest, must apply to all contributions of this book, independent of their thematic, chronological or chorological contents and focus. Whatever happened in the past has consequences for the present—whatever happened here, has consequences there. In other words: the given division into Past—Present—Prospect should be seen as a more or less arbitrary one, as a guiding principle for the conception and context of this book.

Unity and diversity are leading principles in the contextualisation of the individual contributions. Yet unity and diversity also determine the overall conception of this book as well as its contents. However, climate and culture in a vast area like the MENA region stretching from the borders of the Indian subcontinent in the east to the shores of the Atlantic Ocean in the west are in themselves so versatile and diversified that it is almost impossible to subsume them under one conceptual approach. Thus the following introductory remarks should be seen as an attempt to clarify at least some of those admittedly difficult commonalities as well as the diversity of our region which appears to be portrayed as an undifferentiated entity here. That is not the case![2] And this makes these explanations indispensible:

Unity: The problem starts with the demarcation of the term MENA. This region is notoriously hard to define or to delineate. The following section 2 will provide an overview of how geographers have tried to define this conglomerate of different natural and cultural environments as well as showing the obvious problems of such attempts. It will also discuss some of the main intellectual reservations and critiques of those manifold 'Western' attempts to

2 This statement does not contradict later attempts by different authors (see section 2 and 3 of this introduction) to create and substantiate certain, yet incomplete forms of natural and cultural uniqueness of the MENA region.

formulate and to create common understandings of what this region is about, especially the devastating critique that ties into the charge of 'Orientalism'. Not making light of these concerns, sections 2 and 3 will nonetheless highlight several commonly shared characteristics of the physical environment across this vast region. Special emphasis will be placed on climate and vegetation. Yet these physical characteristics, decisively influenced by a mosaic of various landforms and their influence on climate and orohydrography, are hardly sufficient to argue for a unifying image of our region.[3] Unfortunately, the same holds true for the anthropogenic aspects of the area reviewed in section 4. Are population and ethnicities, history or religion really suited to substantiate a form of regional identity, a sort of cultural identity? We doubt that. Thus we are left with the somewhat sobering insight that neither nature nor culture obviously justify any claim of homogeneity for the MENA region as a whole. Environmental problems of climate change and land degradation, however, the political consequences of globalisation as well as the fight over natural resources (water, land, oil etc.) may turn out to be the predominant and common characteristics. Paragraph 5 will discuss these aspects in greater detail. A final summary of the introduction may therefore come to the conclusion that disunity is the unifying principle of the present situation of this part of the world.

Diversity: In contrast to the introduction's difficult endeavour to develop a unifying concept to facilitate our understanding of the MENA region, diversity is the predominant characteristic of the following chapters. These chapters are the very essence of this publication and its *raison d'être*. They represent a diversity that is centered on its focus on climate and culture in the Middle East and North Africa. Though the ten chapters vary in their specific disciplinary approach, in their regional and/or historical foci, and in their assignment to the predetermined categories of Ideas—Past—Present—Prospects, they all address the central question of the *Climate and Culture* series in general and of this volume specifically. Therefore, we want to remind our readers once again that all papers are part of a greater whole, facets of the overarching theme: the interconnectivities and interactions between nature and society in specific regional and historical contexts on a local, regional, and even global scale. This overarching aim and goal sometimes makes it difficult to assign the individual chapters to clear and unambiguous positions in the table of contents. Even worse—or better: it is almost unavoidable that at least some chapters have immediate repercussions on and contextual connections with other chapters.

3 See above; and specifically Peter Beaumont, Gerald H. Blake, and J. Malcolm Wagstaff, *The Middle East. A Geographical Study* (London, New York, Sydney, Toronto: John Wiley and Sons, 1979).

Thus, occasional repetitions and cross-references to earlier or later chapters of the book are justified.

IDEAS: Chapters 1 and 2 deal with basic ideas of Muslim and Jewish discourses on environmental perception and desirable patterns of behavior, derived from the holy texts of both religious traditions and their interpretations. It is by no means self-evident that religious and philosophical considerations shape human behavior or have consequences for our understanding of cultural or political ideas and actions. Are Amirpur's reflections in chapter 1 not of immediate relevance for efforts to overcome the disastrous cascade of ecological and physical violence against nature and people as e.g. exemplified in figure 0.3 of our introduction to this volume or as described in later chapters on conflicts over land, water and other resources? The same holds true for Meir's historical analysis of ancient traditions as a basis for today's human attitudes *vis-à-vis* nature and the environment (chapter 2). It goes without saying that these two essays cannot replace the abundance of relevant religious, theological or philosophical discourses on the relationship between nature and humankind. Nevertheless, they are important reminders of the moral obligations of individuals and societies as creative caretakers and responsible stewards of our common Planet Earth.

PAST: The same message applies to chapters 3, 4, and 5. The papers by Brückner-Engel and Pientka-Hinz deal with seemingly remote, yet realistic scenarios and events. They touch on specific aspects of water and its great importance as the indispensible basis of all forms of life in this predominantly arid environment. Both papers also stress the Janus-faced character of this element: its constructive as well as potentially highly destructive power. Brückner and Engel (chapter 3) demonstrate impressively how and to what extent historical climate oscillations and flooding events have not only shaped and changed the fragile natural landscapes of Mesopotamia, but also influenced the rise and fall of civilisations. They also point to the impacts of such natural events on human societies interpreting them as having been triggered by mythological deities and heroes: mythologies that have spread from here into almost all ancient civilisations of the Near and Middle East, Egypt and beyond. Pientka-Hinz (chapter 4) gives a vivid account of these ancient Sumerian mythologies in which water plays a central role. The title of her paper speaks for itself. It discusses the central role of water in mythologies of creation and flood and points to the close archetypal connections between destruction and recovery, between death and life. Chapter 5 marks the bridge between past and present: qanats as a vanishing technology of water engineering and water management with a history of almost 3000 years of sustainable water use. The longevity of this remarkable technology is not only an extraordinary example of human ingenuity to cope with the adversities of a hostile environment, it is also a paradigm of human adaptation to nature through the development of adequate organizational rules and customs to ensure the sustainable use of water as a limited resource.

PRESENT: Three chapters of this book are devoted to present-day problems of the climate-culture-nexus. One may well argue that under present circumstances, this is too little. However, we want to repeat that all contributions to this book should be considered as *pars pro toto*, i.e. as offering insights transferable to specific aspects of today's problems in the MENA region. And again we wish to point to section 5 of our introduction, in which the critical interactions and interdependencies between the two cascades of nature and society (fig. 0.3) are outlined. These scenarios are valid and meaningful for many symptoms of the ecological and political crisis in our region. Chapter 6 deals with one of most dramatic ecological disasters of the Middle East: the vanishing of Lake Urmia in northwestern Iran due to the overuse of all water resources, i.e. the surface waters of the lake's feeder streams and the groundwater strata in the catchment area of the lake. Schmidt's cross-reference to the fate of Lake Aral is a good example of the transferable character of this case study which also provides a focused account of the intricate interactions between nature and local society in pursuit of ecological, socio-economic, and political solutions to fight desertification and to ensure the survival of this agriculturally important lake basin. Lake Urmia's problems reflect the incomparably greater problems of water use in the Middle East in general on a local scale. von Lossow's analysis of water scarcity in the Near and Middle East (chapter 7) is a perfect example of our hypothesis that water—like land, oil, gas etc.—is a highly contested issue fuelling political and even military confrontations. International conflicts over water have been recorded in the past, and there are looming conflict potentials in the future. Water scarcity will increase due to climate change, but also due to population growth and growing demand by changing life-styles and consumption patterns. Hydropolitics will gain momentum.[4]

4 For a somewhat more detailed discussion of the crucial water issue see e.g. Marcus D. King, ed. *Water and Conflict in the Middle East* (London: Hurst, 2019); Leif Ohlson, ed. *Hydropolitics. Conflicts over Water as a Development Constraint* (Dhaka: University Press, London, New Jersey: ZED Books, 1995); Christopher Ward and S. Ruckstuhl, ed., *Water Scarcity, Climate Change and Conflict in the Middle East* (London: Tauris, 2017); Mark Zeitoun, *Power and Water in the Middle East. The Hidden Politics of the Palestinian-Israeli Water Conflict* (London, New York: Tauris, 2008, Paperback 2011). For broader and comprehensive contexts of security threats in the MENA region in general see: Ashkok Swain and Anders Jägerskog, *Emerging*

Will Fritzsche's observations on the slowly emerging information societies in the Arab world (chapter 8) be enough, and will these developments take place soon enough? These are serious and essential questions in view of the threatening consequences of rising temperatures and decreasing precipitation in this climatically disadvantaged as well as politically and economically fragmented region.

PROSPECT: As noted previously, the MENA region is a strongly differentiated region rich in natural resources. One resource, however, is abundant throughout: sunshine or solar energy. Trieb's analysis (chapter 9) provides prospects for a solution and reasons for hope. The abundance of solar energy is not only a potential solution for the energy problems of many MENA states and territories. Its abundance is so great that even the demand of European countries can be met by its almost unlimited potential. This will, however, require an economically feasible technology which allows the transfer of solar energy from the Near and Middle East and especially from North Africa to remote destinations in Europe. Trieb's analysis from both scientific and engineering perspectives gives hope, even if this vision may not be realised for decades to come.

Another source of hope may be derived from Kowanda-Yassin's concluding considerations (chapter 10). Can a combination of environmental and religious education offer a realistic chance to change human behavior and attitudes towards nature? Is this a potential strategy to diminish the obvious threats posed by individuals, local societies, or regional politics towards MENA's fragile ecosystems? And are the experiences of one cultural area transferable to another? These and similar questions almost perfectly bracket Amirpur's initial chapter and Toynbee's visionary conclusion about the future of Mother Earth (section 6 of this introduction).

One final remark: the title of this book places a special emphasis on the term 'conflict'. Conflicts, although not specifically defined in the framework of this publication, are almost ubiquitous concomitants of many of the following essays. Without going into the wide field of an in-depth definition of conflicts and conflict theories, it must suffice here to understand conflicts in a very general manner. Firstly, in a traditional Marxist understanding as competition for limited resources; secondly, however, also as any controversy between individuals and society over different group interests including religious differences;

Security Threats in the Middle East. The Impact of Climate Change and Globalisation (Latham MD: Rowman and Littlefield, 2016); Anders Jägerskog, Michael Schulz, and Ashkok Swain, ed., *Handbook of Middle East Security* (Latham MD: Rowman and Littlefield 2019).

and finally, as individuals' and societies' diverging interests in the appropriation of natural goods and services.

2 The MENA Region: Problems of its Definition, Demarcation, and Characteristics. Western Perspectives and the Question of 'Orientalism'

The regional focus of the following essays is the Middle East and North Africa, often abbreviated as the MENA region. But what does that mean: Middle East—Near East—Orient? Not to mention all the other regional terms applied in Western languages such as English (*Near and Middle East*), French (*Proche et Moyen Orient*), or German (*Naher und Mittlerer Osten*) for this part of our planet. The terminology even found its way into languages of the MENA region itself, as into Arabic (*aš-šarq al-ausaṭ*) or Persian (*ḫāwar-e miyāne*). They all express the same thought: This region is a Western perception, a Eurocentric "construction" and vision. It still is what Pearcy called an "indefinable region".[5] Its delineations and limits changed again and again, and these changes reflect historical, colonial, and imperialistic perceptions of the MENA region.[6] In the words of Edward Said:

> The Orient was almost a European invention, and had been since antiquity a place of romance, exotic beings, haunting memories and landscapes, remarkable experiences [...].[7]

Said speaks of "imaginative geographies" and their tendencies to "orientalise the oriental" and argues that it was especially "geography [...] essentially underpinning for knowledge about the Orient [...]."[8] As a matter of fact: our

5 George E. Pearcy, *The Middle East—An Indefinable Region* (Washington: Department of State Publication, 1964), 1–12.

6 Cf. Rolf F. Krause, "Orient, Naher und Mittlerer Osten. Die Begriffe im Wandel der Zeit," *Geographische Rundschau* 45.1 (1993): 4–9 and; Georg Stöber, "Nahost—Mittelost—Vorderer Orient: Wer ist was?," in *Der Islamische Orient. Grundlagen zur Länderkunde eines Kulturraums*, ed. Eckart Ehlers and Abdoldjavad Falaturi (Köln: Studien zum Islam in Interkulturellen Wechselbeziehungen. Schriftenreihe der Islamischen Wissenschaftlichen Akademie Bd. 1, 1990); Michel E. Bonine, Abbas Amanat, and Michael E. Gasper, ed., *Is there a Middle East? The Evolution of a Geopolitical Concept* (Redwood City: Stanford University Press 2011).

7 Edward W. Said, *Orientalism. Western Conceptions of the Orient* (New York: Routledge 1978, reprint in London: Penguin Books 1991), 1.

8 Ibid., 216.

'knowledge' of the Near and Middle East as well as that of North Africa is still scanty, highly imperfect, and full of misconceptions, in spite of rapidly growing interactions with these countries due to intensive economic, and political and touristic ties to the region and in spite of the almost unlimited potential of internet and modern media in the so-called information societies. Knowledge is still very scanty. Perceptions and preoccupations prevail; images and blurry (mis)conceptions dominate the past as well at present, and very often they have been (and still are) associated with religious conflicts, not only from a cultural perspective, but also in connection with struggles over land, water and natural resources.

Our specific region, the Near/Middle East and North Africa, may well be a perfect example for the close fruitful or problematic interactions between nature and cultures in space and over time. From Greek and Roman times onwards, relationships between East and West were characterised by conflict and military warfare between Athens, Rome and the 'barbarians', i.g. the Persians, Egyptians and Nubians; the rise of Islam and its conquest of the Mediterranean World, the crusades and the quests for the 'Holy Land'; and finally European colonialism and imperialism. No doubt many of these encounters also brought about a positive exchange in philosophy and religion, in arts, science and technology, in short: in culture and civilisation. Yet it is true that from the 16th century onwards, expanding European powers destroyed indigenous cultures and traditions or transfigured them according to their liking. Above all, they began to claim colonial, economic, and military spheres of interest, fix boundaries, and shape the political maps of the world. The MENA region, i.e. the Near/ Middle East and North Africa, is no exception to this rule.[9] On the contrary: political interference has been continuous since the 18th and 19th centuries, especially after the World Wars, and it continues to this day with all the disastrous consequences for nature, people, and cultures.

Before going into details and specific background information of the MENA region and its nature, history, and culture, it may be appropriate to recall once again Edward Said's basic arguments in regard to his "Orientalism" analysis. As mentioned before, he claims that it was specifically geography's impact that shaped our Western perception of 'the Orient'. This, however, is only partly true and Said himself is well aware of it. Geography is a generalising headline, an umbrella term under which all kinds of information was summarised. It started with descriptions of occupied foreign territories, their resources, their people, cultures and customs. Travellers, traders, and political agents were the

9 Joshua C. Baylson, *Territorial Allocation by Imperial Rivalry. The Human Legacy in the Near East* (Chicago: University of Chicago, Department of Geography, 1987).

first reporters, followed by missionaries, explorers, archaeologists, historians, and those that we may call geographers in the proper sense of the word. Before all and behind most of these activities were, however, the imperial interests of European nations—the UK, France, the Soviet Union and others—to control and command territory, traffic and trade within their spheres of interest. Said's conclusion that it was "expansion, historic confrontation, sympathy, classification" that became accompanying elements of these expansions of European imperialism and its penetration of Western Asia and North Africa (not to mention worldwide similarities!) and "on whose presence the specific intellectual and institutional structures of modern orientalism depend",[10] is true. One of the central and probably the most important one was Napoleon's military campaign to Egypt in 1798. The French army of approximately 35,000 soldiers was accompanied by 167 scientists whose *Description de l'Égypte* is one of the early testimonies to Said's observation. This event also marks the beginning of a specific interest of European powers in the history, art, and culture of Middle Eastern and ancient North African/Mediterranean empires: Pharaonic Egypt, ancient Mesopotamia, Achaemenid and Sassanid Persia, Palestine and its biblical traditions, medieval Islamic caliphates, Seljuk and Ottoman Turkey and so on. It did not take very long for British and French novelists and painters, French and Italian composers or German poets to be caught up in those 'places of romance' and the images and experiences of oriental life and customs—historical as well as contemporary.

Viewed in this context, these introductory remarks should only serve to set the framework within which the following contributions by philosophers and historians, by geographers and political scientists, by anthropologists and by scholars of comparative religion should be seen. And here again, both geography and history play crucial roles, especially in regard to climate and culture. Yet the attendant problems and consequences are manifold. They begin with the previously mentioned attempts to clearly demarcate and define this region and to argue for its uniqueness and cultural peculiarity. This immediately takes us back to Said's well-argued observation that Western "sympathies" are counterbalanced by Western tendencies to "classify" and to divide an entity into its individual parts. As a matter of fact: is there a better approach to grasp an "undefinable region"? And how can we cover and characterise a region so rich in natural diversity, civilisational history, ethnic, religious, social, political, and economic variety in time and space? Thus, "sympathy and classification" are hardly avoidable in the introductory chapter to this book.

10 Said, *Orientalism*, 120.

In view of the fact that—according to Said—geography plays a major role in the intellectual annexation and interpretation of what Westerners believe to be "oriental" (but also due to the fact that one of the editors of this book is a geographer!), it may not be surprising that this introduction to the volume *Middle East and North Africa: Climate, Culture, and Conflicts* also begins with a systematic, or better, a diversifying classification of our region. In doing so we refer to two previous geographic attempts. The first one, suggested by the French geographer Xavier de Planhol,[11] developed and published before Said's elaborate criticism, represents the attempt to differentiate his world of Islam, more or less identical with the MENA region and its fringes. He divides it into seven subregions, i.e.:

(1) The Arab World: the Near East;
(2) The Arab World: North Africa;
(3) The Arab World: the Sahara;
(4) The Turkish—Iranian Realm;
(5) The Areas of Failure in Europe;
(6) The Tropical Fringes and;
(7) The Oceans of the South.

Placing special focus and emphasis on the close interrelationships between the environmental milieus and historical as well as cultural specificities, de Planhol's approach clearly goes beyond our concept of the MENA region. Nevertheless, de Planhol delivers a broad perspective of regional variations which help us to understand the remarkable diversities and conflicts of interests within a cultural area generally perceived as homogeneous, with Islam as its unifying bond.[12]

The second approach is that of the German geographer Eugen Wirth.[13] Unlike de Planhol, Wirth focuses on the MENA region as a whole. His approach is based on the by now outdated concept of *Kulturerdteile* developed by the German geographer Albert Kolb. His definition of a 'cultural world region'

11 Xavier de Planhol, *Le Monde Islamique. Essai de Géographie Religieuse* (Paris: Presses Universitaires de France, 1957/1959), English Edition: *The World of Islam*, (Ithaca: Cornell University Press, 1995) and; Xavier de Planhol, *Les Fondements Géographiques de l'Histoire de l'Islam* (Paris: Flammarion, 1968/1975), German Edition: *Kulturgeographische Grundlagen der islamischen Geschichte* (Zürich, München: Artemis Verlag: Die Bibliothek des Morgenlandes, 1975).

12 See also Xavier de Planhol, *Les Nations du Prophète. Manuel géographique de politique musulmane* (Paris: Fayard, 1993).

13 Eugen Wirth, "Einleitung: Der Orient—Versuch einer Definition und Abgrenzung," in: *Nordafrika und Vorderasien*, ed. Horst Mensching und Eugen Wirth (Frankfurt a.M.: Fischer, 1973), 15–26.

(Ger. *Kulturerdteil*) describes a space of subcontinental extent that is defined by the individual origin of its culture and its unique relations towards nature:

> Unter einem Kulturerdteil [wird] ein Raum subkontinentalen Ausmaßes verstanden, dessen Einheit auf dem individuellen Ursprung der Kultur, auf der besonderen einmaligen Verbindung der landschaftsgestaltenden Natur- und Kulturelemente, auf der eigenständigen, geistigen und gesellschaftlichen Ordnung und dem Zusammenhang des historischen Ablaufes beruht.[14]

This concept, which seems to be well suited to the overall philosophy of the *Climate and Culture* (CLAC) book series, shows certain resemblances to Alfred L. Kroeber's anthropological concepts of "culture area or cultural areas". His approach, exemplified above all by his classic study on "Cultural and Natural Areas of Native North America", tries to follow a holistic approach by contextualising the potentials of the natural environment and specific cultural developments, attitudes and traditions as the basis of specific regional identities or cultural areas.[15]

Our second approach should be seen in line with this concept. It is an attempt not to differentiate our region regionally, but to identify its commonalities, its specificities and its natural as well as cultural uniqueness. Among the many attempts to define the MENA region in this way,[16] Wirth's attempt is probably the most sophisticated and differentiated approach. He mentions six characteristics,[17] which, in his own words, characterize the very essence of our region:

> [(1)] Der Orient ist der westliche and mittlere Teil des großen altweltlichen Trockengürtels.
>
> [(2)] Der Orient ist nach heutigem Wissen die erdölreichste Großregion der Erde.
>
> [(3)] Der Orient ist kulturgeschichtlich das Ursprungszentrum sowohl der neolithischen Revolution als auch unserer Hochkulturen.

---

14 Albert Kolb, "Die Geographie und die Kulturerdteile," in *Hermann von Wissmann*-Festschrift, ed. Leidlmair A. (Tübingen: Geographisches Institut der Universität Tübingen, 1962), 46.

15 Alfred L. Kroeber, *The Nature of Culture* (Chicago: University of Chicago Press, 1952).

16 See also Eckart Ehlers et al., *Der Islamische Orient. Grundlagen zur Länderkunde eines Kulturraumes* (Köln: Studien zum Islam in interkulturellen Wechselbeziehungen. Schriftenreihe der Islamischen Wissenschaftlichen Akademie, 1990), 1–19.

17 Wirth, "Einleitung: Der Orient," 15–26.

[(4)] Der Orient ist derjenige Teil der antiken, hellenistisch-römisch beeinflussten Mittelmeerwelt, der im 7. Jahrhundert n. Chr. von den Arabern erobert wurde und heute überwiegend dem Islam zugehört.

[(5)] Im Mittelalter und in der frühen Neuzeit haben mächtige Staatensysteme mit ihrer politisch-wirtschaftlichen Organisation und ihrem kulturellen Gestaltungswillen einige Teilregionen des Orients bis heute nachhaltig geprägt.

[(6)] Die Länder des Orients sind heute rentenkapitalistisch geprägte Entwicklungsländer alter Kulturtradition.

The introductory juxtaposition of two geographical perspectives of the MENA region—testimony to Said's observation of Western attitudes in dealing with the Orient at large and specifically Western Asia and North Africa—reveals a number of insights that are of relevance for the following academic contributions to this book. First: the climate and culture of our region show a wide variety of natural and ecologically diversified environments. These have caused different cultural adaptations to nature, but also historically given rise to different forms of Islam as one of the unifying bonds of our region.[18] These diversities, however, are overlaid by a number of additional characteristics that give coherence and cohesion to the region as a whole. Climate and natural resources, ancient and medieval clashes of culture and civilisation, cross-cultural exchanges of ideas and fashions, religious conflicts, political, and socio-economic interrelationships and dependencies: they all are addressed in Wirth's attempt to cover the very essence and uniqueness of the MENA region. Therefore, these characteristics will serve as a guiding principle in the structure of the following introductory remarks, but also as the background to the following collection of articles.

3 Climate, Climate Change, and Environmental Problems in the MENA Region

As mentioned previously, climate and aridity are probably the most characteristic features of MENA's physical environment and the most distinct determinant of the region's nature.[19] It is also nature's greatest challenge to all those

---

18 de Planhol, *Le Monde Islamique* and; de Planhol, *Les Fondements Géographiques de l'Histoire de l'Islam* and; de Planhol, Xavier, *Les Nations du Prophète*.

19 As mentioned earlier (footnote 1) the term 'environment' will be used in a very general sense.

that have been living in this area stretching from the Atlantic Ocean in the West to the dry desert and semi-desert regions of the Indian subcontinent and Central Asia in the East and Northeast. The so-called dry belt of North Africa and the Near and Middle East stretches approximately 10,000 kilometres to the East. Deserts and barren lands prevail. Precipitation is extremely scarce and irregular. Vegetation is absent or confined to a few scattered shrubs or thorny bushes. This great natural uniformity is one of the many common images of the deserts of the MENA region. The following words describe the situation very appositely:

> Certain shared characteristics give a degree of unity to the region. The most fundamental is climate. The region is characterized by extremely arid summers and a winter-spring maximum of precipitation. Continental effects are so marked in the centre of the main land masses—Asia Minor, Iran, Arabia and North Africa—and the precipitation is so low that population is confined to oases or thinly scattered as nomadic groups. Human activity throughout the region is closely adapted to climatic conditions.[20]

At the same time, however, the authors emphasise a remarkable diversity within this uniformity:

> Diversity is as characteristic of the Middle East when examined at a detailed level, as unity appears to be at the regional scale. In many ways the lack of unity looms very large. In detail, a variety of terrain and climate exists in the region, and human response is not uniform.[21] (ibid., 4)

More than many words, map 0.1 reflects the ecological diversity of our region and its impact on land use and land cover. Taking up Wirth's aforementioned first and foremost characteristic of our region, we have to bear in mind that the term dry belt (Ger. *Trockengürtel*) implies more than a dearth or total absence of water. Aridity has direct consequences for the natural vegetation cover, soil development, and above all for water availability—and without water, there can be no human existence or culture!

Without going into further detail of the close interactive relationships between nature and culture, one must state that the MENA region and especially its Asian part are the perfect example of nature's stimulating impact on human development and the rise of cultures and civilisations (see section 4). But the

20 Beaumont, Blake, and Malcolm Wagstaff, *The Middle East*, 3–4.

21 Ibid., 4.

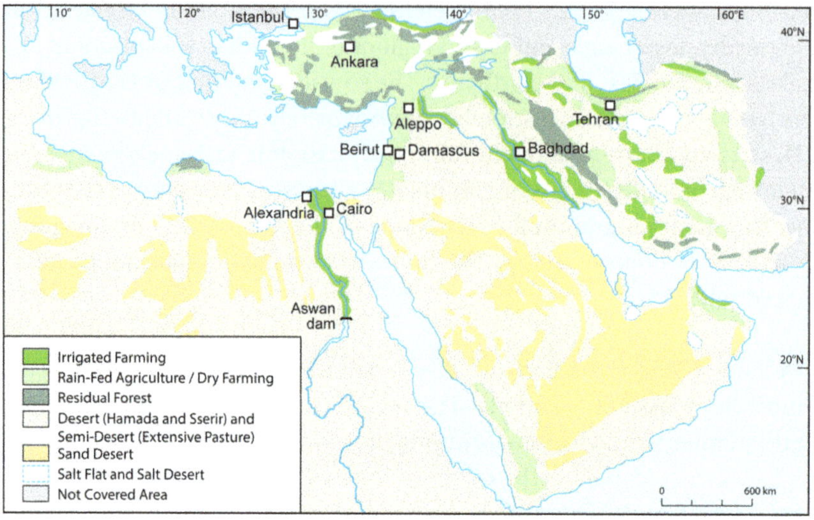

MAP 0.1 The MENA region: land use and land cover: a generalised survey. © ECKART EHLERS. COMPILED FROM DIFFERENT SOURCES

opposite also holds true: In a region where rainfall is scarce and variable, where the natural vegetation cover is thin or extremely extensive, and where soils do not offer any protection against the erosive forces of winds and storms, slight changes in the long-term temperature and precipitation regimes can have disastrous consequences (see also section 5).

With the exception of some climatically and ecologically favoured regions, e.g. the mountain regions of Pontus and Taurus in Turkey, of Lebanon, of Alborz and Zagros in Iran, parts of the Atlas mountain ranges in the Maghreb, to name just a few, the entire MENA region today suffers from aridity and the consequences of climate change. As a matter of fact, it likely is one of the parts of our globe where climate change and its negative effects will be felt most seriously, physically as well as in its socio-economic consequences. The Middle East and the Mediterranean part of the MENA region is already experiencing severe climatic constraints, and it will continue to do so to a greater extent for the foreseeable future and until the end of the 21st century. The forecast looks bleak and alarming. All models and scenarios of the IPCC (both 2007 and 2014) point in the same directions: higher temperatures and less precipitation! The assumptions vary. Mean annual temperature increases between 2°C and 4°C for the Mediterranean region, North Africa, the Near and Middle East are not unlikely. While the winter months may be warming up by 'only' 2.5°C to 3°C along the Mediterranean shores of North Africa and the Maghreb in general resp. 3°C to 4°C in the Asian part of the MENA region, predictions for the summer months are even more alarming: 3.5°C to 4°C in the West and more than 4°C in the East are possible. Climate change, however, will not only affect the global and regional temperature regimes, but also the long-term precipitation patterns. Model simulations for the IPCC region Southern Europe/ Mediterranean, which covers the North African countries bordering it between Egypt and Morocco, predict developments similar to those for MENA's Asian part, i.e. the Near and Middle East: annual precipitation losses of 15 to 20 per cent compared to today's average values. Once again, the summers are particularly crucial: climatic models and scenarios predict deficits of up to thirty per cent and more compared to the present situation, which is already under severe climatic stress. The overall regional climate change projections for the 21st century—here reduced to their main components of temperature and precipitation and based on the outcome of more than twenty global models—predict the following trends for southern Europe and the Mediterranean (SEM) and for Central Asia (CAS), covering large parts of the Middle East (see table 0.1).

TABLE 0.1 Regional averages of temperature and precipitation projections for the MENA region until the end of the 21st century*

		Temperature responses (°C)			Precipitation responses (%)			Extreme seasons (predictions) (%)			
Region	Season	Min	50 Av.	Max	Min	50 Av.	Max	T years	Warm	Wet	Dry
Southern Europe	DJF	1.7	2.6	4.6	-16	-6	6	>100	93	2	12
Mediterranean	MAM	2.0	3.2	4.5	-24	-16	-2	60	98	1	31
30N-10W to	JJA	2.7	4.1	6.5	-53	-24	-3	55	100	1	42
48N-40 E	SON	2.3	3.3	5.2	-29	-12	-2	90	100	1	21
	Annual	2.2	3.5	5.1	-27	-12	-4	45	100	0	46
Central Asia	DJF	2.2	3.2	5.2	-11	+4	+22	–	84	8	–
30N-40E to	MAM	2.3	3.9	4.9	-26	9	+3	>100	94	–	16
50N-75E	JJA	2.7	4.1	5.7	-58	13	+21	>100	100	3	20
	SON	2.5	3.8	4.9	-18	+3	+24	–	99	–	–
	Annual	2.6	3.7	5.2	-18	-3	+6	–	100	–	12

SOURCE: IPCC, 2007: CLIMATE CHANGE 2007: THE PHYSICAL SCIENCE BASIS... 2007, 854–855.

* This table is a condensation of a large data set. The above values are the minimum, maximum and average predictive values of temperature and precipitation projections from a set of 21 global models. For further explanation see IPPC 2007, 854.

General as these projections of the IPCC 2007 may be, the expected scenarios have been fully confirmed by the most recent IPCC report of 2013/2014. Their authors conclude that:

> […] in recent decades North African annual and seasonal observed trends in mean near surface temperature indicate an overall warming that is significantly beyond the range of changes due to natural […] variability.[22]

And in regard to precipitation they state:

> Over the last few decades the northern regions of North Africa (north of the Atlas Mountains and along the Mediterranean coast of Algeria and Tunisia) have experienced a strong decrease in the amount of precipitation received in winter and early spring.[23]

Transferring these observations into projections for the 21st century, the signals are alarming: reduction of rainfall over northern Africa is "very likely",[24] while temperatures will increase at the same time.

Similar developments are to be expected for the Asian parts of the MENA region. Observed data suggest that "in West Asia, upward temperature trends are notable and robust in recent decades";[25] tendencies for precipitation, on the other hand, are characterised by "a weak but non-significant downward trend".[26] As a result a growing imbalance of projected temperature increases and decreasing precipitation figures is foreseeable in Central and West Asia as well.

As previously indicated, increases in temperature and decreases in precipitation are clearly discernable already. So are their immediate environmental consequences. The number of extreme events already seems to be rising. Droughts on the one hand, severe floods and destructive inundations on the other, have clearly increased. Dust storms, heat waves, cold spells, and other weather extremes, many of them at local to regional level, are concomitant phenomena of the overall warming trend in the MENA region. Warming in combination

---

22 Isabelle Niang et al., "Africa," in *Intergovernmental Panel on Climate Change (IPCC)—Climate Change 2014: Impacts, Adaptation, and Vulnerability. Part B: Regional Aspects. Contribution of Working Group II to the Fifth Assessment Report of the Intergovernmental Panel on Climate Change* (Cambridge, New York: Cambridge University Press, 2014), 1206.

23 Niang et al., "Africa." 1209.

24 Ibid., 1210.

25 Ibid., 1333.

26 Niang et al., "Africa." 1209.

with declining precipitation has dramatic consequences for its highly sensitive terrestrial ecosystems. Thus it is not surprising that the Millennium Ecosystem Assessment (MEA) of 2005 comes to the conclusion that the dryland regions of our globe and especially their hyperarid, arid and semiarid parts (covering almost one third of the terrestrial surface of the Earth) are more sensitive to climate change than other ecosystems and biomes—even leaving aside the fact that loss of biodiversity and degradation of the few remaining ecosystem services and desertification are already irrevocable.[27] These observations are even more alarming since they are not only a consequence of purely natural phenomena, but severely aggravated by human interference in the fragile semi-desert and highly vulnerable grassland/steppe ecosystems of the MENA region. The hyperarid, arid, and semi-arid drylands, covering more than 47 million square kilometres or one third of the terrestrial surface of the Earth with the MENA region at its old-world centre, "where plant production is limited by water availability [...] and where annual precipitation is less than two thirds of potential evapotranspiration",[28] will suffer additionally since "fresh water availability [...] is projected to be further reduced"[29] in the future.

Given these observed and observable trends, climate change and its environmental consequences are alarming. Middle Eastern and North African countries and regions are facing dim climatic futures in the 21st century. If the changes projected by the IPCC 2013/2014 in its medium to high-confidence scenarios according to which the temperature increases in large parts of Africa "will rise faster than the global land average, particularly in the more arid regions",[30] are correct, there will not only be a massive impact on terrestrial and the few available freshwater ecosystems and their services. More importantly, there will be socio-economic consequences. Water shortage, droughts and famines, migration, ethnic clashes and warfare over land, water, and shrinking resources are very likely to impact not only vulnerable groups and societies, but may even end up creating millions of environmental refugees and mass migrations—the beginnings of which we are already experiencing today (see section 5).

27 Millennium Ecosystem Assessment (MEA), *Ecosystem and Human Well-being: Synthesis* (Washington, D.C.: Island Press, 2005).

28 MEA, *Ecosystem*, 29.

29 Ibid.

30 Niang et al., "Africa." 1202.

4 Culture and Civilisation: The Great Past, Controversial Present, and Insecure Future of the MENA Region

In line with Wirth's attempts to characterise the specifics of the *Kulturerdteil Orient* as a cultural identity, we can apply three of his criteria to a discussion of the MENA region and its outstanding importance for the history, culture, and civilisation for humanity and humanism in general. These are

(1) its role and function as the origin of the Neolithic Revolution and as a cradle of cultures and civilisations;

(2) its role as a melting pot of the ancient Mediterranean cultures of Western Asia, North Africa and Southern Europe and their mutually beneficial cultural cross-fertilisation and;

(3) its role as the meeting point and centre of three monotheistic religions, of the Arab conquest and of Islamisation.

Wirth's emphasis of "rent capitalism" as a specific form of economic activity also deserves mention here (see also Chapter Five).

Culture and civilisation, their historical achievements and their present danger, represent the human dimension of today's debate on climate and global environmental change. As such they not only fit the basic 'Climate and Culture' approach of this series, but they address culture-specific interpretations of the intricate interrelations between nature and society at a local and regional scale. Aside from the indicated consequences of a harsh and often hostile natural environment, there is population growth, ethnic or religious conflicts, socio-economic disparities: if these conflict potentials collide or overlap, societies or even states may be destabilised and become highly vulnerable. In section 5, such scenarios will be discussed in more detail.

While the present problems of the MENA region clearly are the decisive determinants of today's discourses on the political and/or socio-economic future of this region, its great history and its contributions to the intellectual, cultural, and religious developments of humanity in general must be appropriately acknowledged. Such acknowledgement is necessary and absolutely justified because these testimonies to MENA's significance for the development of human civilisation are under severe stress today through archaeological vandalism, through terrorism by religious fanatics, and through the weakness of the failed states that are harbouring these world heritage sites. What about Niniveh or Palmyra, Mosul or Aleppo? In any case and in almost all regards, both nature and society are inseparably interwoven in those achievements and failures, be it in form of material culture or of ideas and metaphysical belief systems. This book contains two contributions to these contexts both from mythological-religious perspectives (Ephraim Meir and Rosel Pientka-Hinz)

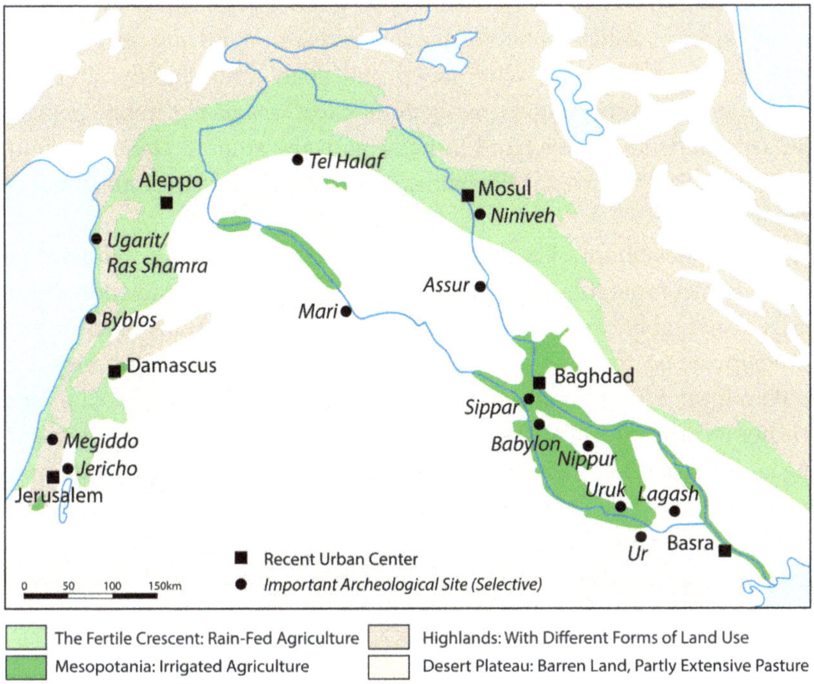

MAP 0.2 The Fertile Crecent and Mesopotamia. The maps displays selected sites of the Neolithic Revolution and Hydraulic Civilisations.
© ECKART EHLERS. COMPILED FROM DIFFERENT SOURCES.

and on the basis of archaeological and geological evidence (Max Engel and Helmut Brückner).

The role and significance of the ancient Middle East and North Africa for the cultural and civilisational evolution of humankind is without parallel. As previously mentioned, this region is the birthplace of the Neolithic Revolution.[31] The Fertile Crescent, i.e. the natural steppes and grasslands of the western fringe of the Zagros Mountains, the southern forelands of the Taurus system and the hilly promontories of the eastern Lebanon ranges are the natural steppe regions with temporary, yet sufficient average rainfall and grasslands that allowed hunters and gatherers to settle down and found the first permanent settlements of mankind ten to twelve thousand years ago (map. 0.2). The abandonment of a migratory lifestyle and the gradual sedentarisation process are substantiated by countless archaeological excavations and sites, especially in modern-day Iran, Iraq, Syria, Jordan, Libanon, Palestine and Israel. It was accompanied and shaped by the domestication of plants and animals,

31 James Mellaart, The *Neolithic of the Near East* (London: Thames and Hudson, 1975).

i.e. the selective collection of specific steppe grasses (esp. the wild forms of wheat and barley: *triticum monococcum, tr. dicoccum—hordeum sp.*) and their ultimate human cultivation. Wild sheep and goats, and slightly later also wild cattle, seem to represent the early domestications of big steppe game.[32] Human sedentarisation goes hand in hand with the smooth transition from villages to urban structures, the diversification of society and the emergence of social hierarchies.[33]

A second achievement of early Near and Middle Eastern cultural development and its incomparable contribution to the history and civilisation of humanity is inseparably connected with the hydraulic civilisations of the MENA region: Egypt along the Nile and Mesopotamia between the Euphrates and Tigris. Pharaonic Luxor, Amarna or Memphis, Sumer, Assur or Babylon and many other locations recall city states, larger territories and even empires embracing large tracts of the whole MENA region and beyond. From the 4th millennium BCE onwards, culture and civilisation exploded and created what may be called a multi-faceted intellectual and cultural revolution. The development and growth of these advanced civilisations was accompanied by remarkable innovations and progress in all spheres of life with a lasting impact to this day. Politically, the rise and fall of these empires and their dynasties was linked with conflict over territories and their demarcation and over trade routes, and it was pursued by 'sophisticated' weapons and warfare. Controversies over differing belief systems were supported by powerful and numerically extensive priesthoods. These priests and a rapidly expanding caste of 'civil servants' were part of an equally powerful administrative bureaucracy, the expression of remarkable social hierarchies and stratifications in the populations of these city and temple states and their territories. While the oldest urban settlements like Jericho in present Palestine or Çatalhöyük in southern Anatolia, dating back to the 9th resp. 8th/7th millennium BCE, can be considered exceptional Neolithic forerunners of later urbanism, those of the 4th and 3rd millennium onwards are characterised by an explosion of differing urban forms and functions. Temple cities and capitals become the hubs of political, military, and religious centres with a broad range of cultural achievements. These include the development of writing in the form of pictograms, cuneiform, and hieroglyphs,

32 Wilhelm Rudorf, *Zur Geschichte und Geographie alteuropäischer Kulturpflanzen. Eine Übersicht aufgrund archäologischer Quellen* (Berlin, Hamburg: Verlag Paul Parey, 1969); Peter J. Ucko and G.W. Dimpleby, *The Domestication and Exploitation of Plants and Animals. Proceedings of a Meeting of the Research Seminar in Archaeology and Related Subjects Held at the Institute of Archaeology* (London: London University, 1969).

33 Charles Burney, *From Village to Empire. An Introduction to Near Eastern Archaeology* (Oxford: Phaidon Press Ltd., 1975); Hamblin, Dora J. et al., *The First Cities* (New York: The Life Books, 1973).

the codification of laws and rules (Codex Ḫammurabi, mid-18th century BCE), the preservation of mythological and religious texts and epics (e.g. the epic of Gilgamesh on clay tablets, about 2000 BCE, or the cosmogenic documents of ancient Egypt)—not to forget the very practical developments in astronomy, mathematics, and other scientific observations that are so important for administrative or economic purposes and for the calendarisation of the year. *Pars pro toto*, figure 0.1 illustrates the ingenuity and progress of the cultural and civilisational achievements throughout the ancient MENA region. The reproduction of an ancient clay tablet dating to about 1500 BCE and showing a city plan of Nippur combines all characteristics that stand for the cultural progress of the ancient empires: writing, mathematics, hydraulic technology, mythology, religion, and so on.[34]

The third and probably most enduring importance of the MENA region lies in its lasting impact on world history through to today's religious discourses and political actions/activities, by its role and function as the central region of a profound turning point in the history of mankind. As we have seen, all ancient cultures and civilisations of the MENA region were characterised by polytheism. Egypt and Sumer, Akkad, Assur and Babylon, Mitanni, Elam or the Hittites, people and tribes of the Mediterranean coast between early Greece, Asia Minor and the lands of Palestine: they all worshipped deities and idols of different kinds and for different purposesand transmitted their mythological narratives and memories.[35] In the middle of the last millennium before Christ, however, the foreshadowing of a profound intellectual revolution, a "Great Transformation"[36] started to dawn, a real "Götterdämmerung" in the proper sense of the word. And the MENA region was the most influential arena of this development. What happened?

Polytheism and its mythological traditions dominated the spiritual life of the early advanced civilisations and was integral to daily life in almost all cultures of the ancient Near and Middle East, of Egypt and pre-classical Greece. Then, however, new and truly revolutionary ways of thinking gained ground: mythologies that had endowed life with meaning through traditional

34 A modern, up-to-date and very comprehensive survey is e.g. Stephen Bourke, *The Middle East. The Cradle of Civilization* (London: Thames and Hudson, 2008), revised and updated edition 2018 (Quarto Publishing).

35 Karen Armstrong, *A Short History of Myth* (Edinburgh: Canongate, 2005); Stephanie Daley, *Myths from Mesopotamia. Creation, the Flood, Gilgamesh, and Others. With Introduction and Notes* (Oxford: Oxford University Press, 2008).

36 Karen Armstrong, *The Great Transformation—The Beginning of Our Religious Traditions* (New York: Alfred A. Knopf, 2006); Mircea Eliade, *Histoire des croyances et des idées religieuses* 4 vols. (Payot, 1978/1992), esp. volume 1, *De l'âge de pierre aux Mystères d'Eleusis*.

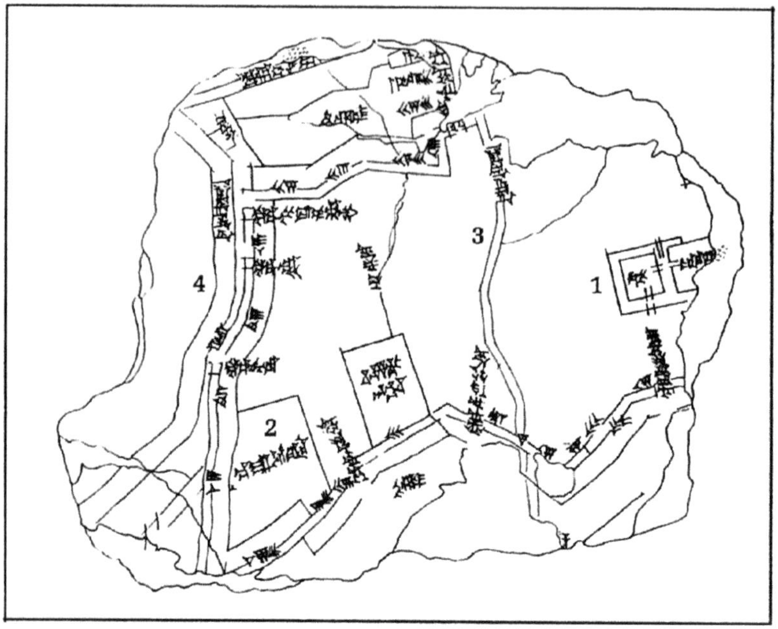

FIGURE 0.1 Clay tablet showing a plan of Nippur (ca. 1500 BCE) with details of urban culture: temples (1), parks and gardens (2), central canal (3), the Euphrates (4), walls and city gates.

narratives and rituals (especially in Pharaonic Egypt and Babylon) lost their importance. The ancient pantheons of gods and goddesses and godlike heroes and heroines faded. They became slowly yet increasingly part of a "cultural memory".[37] Instead of mythological history and polytheistic worship of local deities and objects of ritual veneration, new ways of religious feelings and thinking emerged as preconditions of this transformation. Egyptologists point to Pharao Akhenaten, Amarna and Heliopolis as an early, yet short-lived attempt to replace the immense diversity of deities by the conception of a single God. It would take many more centuries and developments in different locations until the idea of a monotheistic alternative gained momentum. The MENA region was one of the most important centres of religious innovation. Historians, cultural anthropologists, philosophers or historians of religion tend to agree that the first millennium BCE and especially the time between its 9th and 3rd centuries can be understood as an "Axial Age" (German *Achsenzeit*). This term was coined by the German philosopher Karl Jaspers in 1947 in response to the catastrophe of the Nazi era and its role for human culture and

37 Jan Assmann, *Das kulturelle Gedächtnis. Schrift. Erziehung und politische Identität in frühen Hochkulturen* (München: C.H. Beck, 1992).

all civilisational achievements of mankind. But how does this tie into our book and the MENA region? The reason is that the *axial age* and its historically outstanding significance for the development of the human spirit and humanity's moral, religious, and intellectual embeddedness into a new relationship between mankind and the spiritual world is based on a new foundation: monotheism. And, as indicated, the MENA region was likely the most important single centre of this profound development. Assmann's recent book *Achsenzeit* offers what may be the most concise and precise description of this transformation. He writes:

> Die empirische Dimension des Achsenzeit-Mythos besteht zunächst in nichts anderem als der erwähnten, zweifellos faszinierenden Beobachtung, dass in drei der maßgeblichen Kulturkreise der damaligen Welt, China, Indien/Persien und Vorderer Orient/östliche Mittelmeerwelt, gleichzeitig und unabhängig voneinander Männer auftraten, deren Ideen/Schriften, die Welt veränderten. Die geschichtsphilosophischen Dimensionen kommen mit dem Versuch ins Spiel, die Gemeinsamkeiten dieser Gründerfiguren herauszuarbeiten und die Gleichzeitigkeit ihres Auftretens zu erklären.[38]

The striking contemporaneousness of new religious and philosophical pioneers like Confucius (Chin. *Kongzi* 孔子) and Laozi (老子) in China, Siddhārtha Gautama, i.e. the Buddha Śakyamuni in India, the semi-legendary figure of Zarathustra in Iran, the prophets of ancient Israel like Moses, Isaiah, and Jeremiah or the philosophers of Greece like Anaximander, Socrates, Plato has instigated a broad theological and philosophical discourse which continues until today. And this is part of the MENA region's great past, of its controversial present, and obviously also of its politically insecure future. The Near and Middle East as the cradle of three monotheistic religions Judaism, Christianity, and Islam, is an element of the cultural memory of this part of our world we must not neglect. The cultural memory of the region and its inhabitants articulates itself especially in the 'Holy Land' and in the city of Jerusalem, a holy city for all three Abrahamic religions.

The indisputable fact that Islam is one of the most central constituent elements of the MENA region agrees well with Wirth's afore-mentioned characterisation of it. Islam, the youngest of the great monotheistic 'world religions', has been advocating responsible stewardship of nature and the environment since its inception in the 7th century. Islamic ideas of environmentalism are

38 Jan Assmann, *Achsenzeit. Eine Archäologie der Moderne* (München: C.H. Beck, 2018), 22–23.

deeply rooted in its holy book, the Qur'ān, and in the traditions of its prophet Muḥammad, collectively termed the *aḥadīṯ*. Muslims learn from the Qur'ān that God created the universe with all its atoms and molecules and that the laws of creation include the elements of order, balance, and proportion. E.g., in sura *al-Furqān* [The Criterion] it is said:

> *allaḏi la-hu mulku as-samawāti wa 'l-'arḍi wa lam yataḫiḏ waladan wa lam yakun la-hu, šarīkun fi al-mulki wa ḫalaqa kulla šay'in faqaddarahu taqadiran* [He to whom belongs the dominion of the heavens and the earth and who has not taken a son and has not had a partner in dominion and has created each thing and determined it most exactly].[39]

And in sura *al-Yūnus* [Jonas] it is said:

> *huwa allaḏi ǧa'ala aš-šamsa ḍiyā'an wa 'l-qamara nūran wa qaddarahu manāzila li-ta'lamū 'adada as-sinīna wa 'l-ḥisāba ma ḫalaqa allahu ḏalika 'illā bi-'l-ḫaqiyufaṣṣilu al-'ayāti li-qaumin ya'lamūna* [He Who made the sun to give radiance and the moon to shine and determined for it phases so that you may know the number of years and account of time. Allah did not create this except in truth. He makes the signs clear for a people who know].[40]

The Qur'ān and the *aḥadīṯ* are the sources of authentic teachings on preserving the environment, but it was only in the 1960s that academic scholars engaged with the topic in an academic context of Islamic environmental theology called *ecotheology*:[41] The question of ecological balance was placed in a religious context, based on a cosmological view of the world, with nature, humanity, God or heaven, and Earth forming a natural and well-balanced order. Seyyed Hossein Nasr, the author of *Man and Nature: The Spiritual Crisis of Modern Man* published in 1968, believes that human beings have been denying this order since the time of the European Enlightenment, replacing it with an anthropocentric order that places human beings at the centre and gives them free rein to exploit nature in the absence of a connection to any higher order. Nasr sees the environmental crisis as a spiritual crisis. He does not want to see Islam exhausting itself in the observance of rites and rituals, but instead embrace a personal responsibility for the world, one that goes

---

39 Qur'ān 25:2.

40 Qur'ān 10:5.

41 Monika Zbidi, *The Call to Eco-Jihad*. https://en.qantara.de/content/islamic-environmentalism-the-call-to-eco-jihad.

beyond denominational boundaries. This is something that each person has to work on for him- or herself. Environmental and climate protection will thus become matters of spiritual rather than just fashionable significance.[42] Fuad Kamal promotes the idea that science and Islam can collaborate to invest social activities with a sense of direction.[43] He claims that much of the foundation of modern science is built on Muslim roots. Islamic science, however, has not been a cold pursuit devoid of any ethical considerations but a search for God's signs, limitless bounty, and mercy in nature. Therefore, nature and the environment have always played an important part in the lives of Muslims. Muslims understand that God has not created all this for nothing. In fact, Muslims have been commanded to find the wonderful signs of God around them so that cause their awe of their *Rabb* (Cherisher and Sustainer) will increase: Behold! In the creation of the heavens and the earth, and the alternation of night and day—there are indeed signs of understanding men who celebrate the praises of God, standing, sitting, and lying down on their sides, and contemplate the (wonders of) creation in the heavens and the earth (with the thought): "Our Lord! Not for nothing have You created (all) this! Glory to You! Give us salvation from the penalty of the Fire."[44] In Islamic belief, business is seen as part of wider society, a social role is part of its identity. There is no clear dividing line between the economic incentives for prosperity and the public interest and social values, both are connected under the umbrella of *tauḥīd*[45] which is the submission and unity of all creations to the oneness of The Creator. Individual and common values should operate under the same revered rules and directives.[46] These revered rules include justice and equality for all of God's creations, which includes the natural environment and all beings in it. This central notion is superseded by the broader principle that all pursuit of economic gain is subjected to the rules of the permissible (Arab. *ḥalāl*) and forbidden (Arab. *ḥarām*).[47] A growing Muslim movement, led by thinkers like Nasr and Kamal, defined as the highest priority the need to address the intractable problems that the human race is creating by overexploiting and degrading the planet. Salman Zafar, a representative from EcoMENA,

---

42 Fazlun M. Khalid, "Islam and the Environment," in *Encyclopedia of Global Environmental Change*, ed. Ted Munn (Chichester: John Wiey & Sons, 2002), 1–8.

43 Fuad Kamal, *Easily Understand Islam* (Desert Well Network LLC, 2006).

44 Qur'ān 3:190–191.

45 Hassan H. El-Yacoubi, and J. El-Yacoubi, *Islam: The Language of Oneness: Tawhid, the Islamic Paradigm* (Colorado: Boulder, 1994).

46 Khaliq Ahmad and Kazeem Ogunsola, "An Empirical Assessment of Islamic Leadership Principles," *International Journal of Commerce and Management* 21.3 (2011): 292.

47 John Zinkin, "Islam and CSR: A Study of the Compatibility between the Tenets of Islam, the UN Global Compact and the Development of Social, Human and Natural Capital," *Corporate Social Responsibility and Environmental Management* 14.4 (2007): 206–218.

a Qatar-based organisation providing a strong platform for youngsters in the region to express their concern for contemporary environmental issues, positioned himself on Eco-Islam by stating:

> Islamic beliefs, traditions and values provide an effective and comprehensive solution to the current environmental challenges faced by the human race. Islam has a rich tradition of highlighting the importance of environmental protection and conservation of natural resources. According to Islamic law, the basic elements of nature—land, water, fire, forest, and light—belong to all living things, not just human beings. The Holy Qur'ān and Sunnah are a guiding light to promote sustainable development in Islamic countries as well as around the world. Allah [...] commands human beings to avoid doing mischief and wasting resources as these acts cause degradation of the environment. The privilege to exploit natural resources was given to the mankind on a guardianship basis, which implies the right to use another person's property on the promise that it will not be damaged or destroyed.

This example is not meant to demonstrate that all Muslims embrace the notion of Eco-Islam. But there is value in adopting environmental ethics in Muslim-majority markets because of the evident disconnect between ethical teachings and practice which early proponents of Eco-Islam voiced in the Assisi Declarations:

> We often say that Islam is a complete way of life, by which it is meant that our ethical system provides the bearings for all our actions. Yet our actions often undermine the very values we cherish. Often while working as scientists or technologists, economists or politicians, we act contrary to the environmental dictates of Islam.[48]

In summary, geography and climate, history and cultural achievements, religion and its societal impacts—they all are causes of both a great past of the MENA region, but also of its present problems and of its sometimes very dim outlook for the future. Especially when and where conflict over limited natural resources (land, water, oil, gas etc.) is overlain by religious or political antagonisms, even military conflicts seem almost unavoidable. Climate change,

48 Abdullah O. Nasseef, *Alliance of Religions and Conservation. The Assisi Declarations: Messages on Humanity and Nature from Buddhism, Christianity, Hinduism, Islam & Judaism.* www.arcworld.org/downloads/THE%20ASSISI%20DECLARATIONS.pdf.

environmental conflicts, and the impact of globalisation contribute to these uncertainties and become part of the conflicts between climate and culture, nature and society.

In this context, it is important to remember Islam's specific and repeated references to the divine creation of the Earth and all its natural beauties and offerings to mankind. The fact that the Qur'ān and the *aḥadīṯ* advocate human responsibility for nature as part of a well-balanced order between nature and its stewardship is one side of the coin. We have seen above that it has gained ecological and political importance in recent decades and is part of an Islamic discourse on the responsibilities of an enlightened community of believers for the preservation of God's creation today. On the other hand, political Islam and Islamist movements have become a major regional force, especially since the end of European Imperialism and World War II, and its proponents embrace a significantly different agenda and different priorities.

5 Middle East and North Africa: A Complex Crisis Region, Stress of Global Environmental Change and Globalisation, and the Search for New Equilibria

Climate and climate change are ubiquitous environmental problems of global dimensions. Their impact varies from region to region in terms of the form and intensity of its consequences.[49] Variation and changes in temperature and precipitation, sea level rise, loss of biodiversity, draughts and floods, storms and other extreme events: these are just a few components of what can be summarised as 'global environmental change'. It goes without saying that regions with *a priori* natural handicaps like the arid Middle East and/or North Africa are especially exposed to those changes.

Climate change and its consequences, detected as a primarily meteorological phenomenon from the 1980s onwards, has developed into one of the most pressing political problems of today. Testimony to this fact is the installation of a specific UN body, the Intergovernmental Panel on Climate Change (IPCC), observing and evaluating climate change on a global scale and its regional impacts as well as assessing expected future developments. National research

49 Rüdiger Glaser, *Global Change. Das neue Gesicht der Erde* (Darmstadt: Primus Verlag, 2014); Jeanne X. Kasperson, Roger E. Kasperson, and B.L. Turner II, ed, *Regions at Risk. Comparisons of Threatened Environments* (Tokyo, New York, Paris: United Nations University Press: UNU Studies on Critical Environmental Risks, 1995).

programmes with special foci on political issues like vulnerability research, human and societal adaptation, and mitigation strategies are associated with these international scientific activities and research endeavours. An outstanding example of such analyses are the reports by the Wissenschaftliche Beirat der Bundesregierung Globale Umweltveränderungen (WBGU). Its report for the year 2007 is devoted to a careful examination of climate change as a global security risk,[50] that for the year 2011 advocates a new *contrat social* for a great global transformation towards ecological sustainability and socioeconomic and political stability,[51] and the latest one to date addresses the global phenomenon and the transformative power of urbanisation.[52]

As mentioned earlier (see section 2 and 3) it would have been tempting to go into more detail regarding the almost countless conflict potentials of the MENA region. The editors of this volume are well aware of these lacunae. Especially the history of ancient and medieval Egypt would have added interesting details concerning the development of its hydraulic civilisation and its comparison with that of Mesopotamia. However, several attempts to close this gap proved fruitless for reasons beyond their influence. It must therefore suffice to point to some earlier publications analysing the close interaction between nature and society, climate and culture in the history of ancient Egypt as well as its impact on the development of its hydraulic civilisation.[53]

Climate and culture, nature and society: as much as these seemingly incompatible juxtapositions have shaped the history of mankind and its ideas, and have produced outstanding and incomparable contributions to human culture and civilisation, they are today the causes of conflicts and crises of global dimensions. Global environmental change triggered by climate change along with population growth and the effects of globalisation are at the bottom of these conflicts and crises. And since these causes go hand in hand and interact with each other, one may argue that climate change, human security and

---

50 Wissenschaftlicher Beirat der Bundesregierung Globale Umweltveränderungen (WBGU), *Sicherheitsrisiko Klimawandel* (Berlin, Heidelberg: Springer, 2008).

51 Wissenschaftlicher Beirat der Bundesregierung Globale Umweltveränderungen (WBGU), *Welt im Wandel. Gesellschaftsvertrag für eine Große Transformation* (Berlin: WBGU, 2011).

52 Wissenschaftlicher Beirat der Bundesregierung Globale Umweltveränderungen (WBGU), *Der Umzug der Menschheit: Die transformative Kraft der Städte* (Berlin: WBGU, 2016).

53 Karl W. Butzer, *Early Hydraulic Civilization in Egypt. A Study in Cultural Ecology* (Chicago: University of Chicago Press, 1976); Wolfgang Helck, *Wirtschaftsgeschichte des Alten Ägypten im 3. und 2. Jahrtausend v. Chr.* (Leiden, Köln: Brill, 1975); Wolfgang Schenkel, *Die Bewässerungsrevolution im Alten Ägypten* (Mainz: Philipp von Zabern: Deutsches Archäologisches Institut, Abteilung Kairo, 1978); W.A. Ward, "Relations Between Egypt and Mesopotamia from Prehistoric Times to the End of the Middle Kingdom," *Journal of the Economic and Social History of the Orient* 7 (1964): 1–45, 121–135.

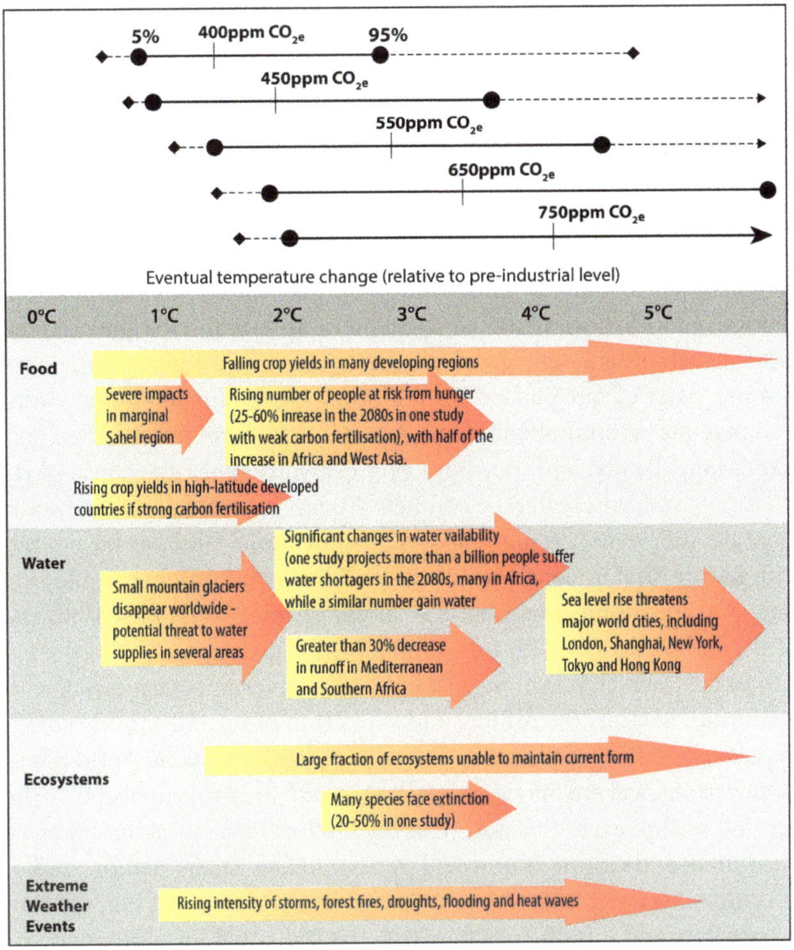

FIGURE 0.2 Global warming scenarios and their ecological consequences. Modified after Nicholas Stern 2007.

violent conflicts go hand in hand.[54] Nicholas Stern has comprehensively compiled some of the effects and consequences of climate change on human societies and economies, many of which apply to the MENA region.[55]

While the consequences of global warming Stern identifies (fig. 0.2) speak for themselves, the interactions between climate change and human vulnerability resp. the individual or collective human responses to nature's threats

54 WBGU, *Sicherheitsrisiko Klimawandel*.

55 Nicholas Stern, *The Economics of Climate Change. The Stern Review* (Cambridge: Cambridge University Press, 2007).

are much more complicated. They differ from region to region, depending on political, socioeconomic, and ideological framework conditions. The MENA region with its great diversity is no exception to this rule. On the contrary: Any attempt to develop a comprehensive and all-embracing scheme in which the interconnectivities and mutual interdependencies of nature and society within this region and their impact on conflicts and disasters are comprehensively addressed is difficult to achieve. Thus, the following attempt does not claim completeness or perfection.

As indicated in the first and third section of this introduction, climate, climate change, and environmental consequences are ubiquitous phenomena in the MENA region. In some regards they may even be more dramatic here than in other parts of our planet. On the other hand, climate and environmental change are natural phenomena. As such, they have always been part of any landscape history, and they have very likely also been the cause of repeated military and humanitarian conflicts. Today, however, the situation is fundamentally different: it is not nature alone any more. Humans have interfered with nature, first unconsciously, today with scientific and technological know-how and increasingly without any moral or ethical scruples. With the dawn of the industrial era, with the coming of the steam engine and the exploitation of coal (which is nothing but fossilised subterraneous forests), with industrialisation at large, the drilling and extraction of oil and gas (also originally biomass), and the increasing demands of a rapidly growing world population, a new geological era may be in *statu nascendi*: the Anthropocene![56] The birthdate and birthplace of this new relationship between nature and society, between climate and culture is now widely accepted as England at the end of the 18th century. Its irrefutable indicator is the steady increase of CO_2 concentrations on Earth, and it is obvious now that the Industrial Revolution and its consequences, summarised socially, economically, ecologically, and politically under the term "The Great Transformation"[57] have abolished the strict distinction between physical and social causes of climate and environmental changes to Earth's system. On the contrary: modern research on climate and climate change considers both nature and society as equivalent causors of change, its triggers and problems. Humankind has become a geological factor. Nature and society, climate and culture are no longer separate triggers of global environmental changes or environmentally caused conflicts. No, they are closely

---

56 Paul J. Crutzen,"Geology of Mankind," *Nature* 415 (2002): 23; Eckart Ehlers, *Das Anthropozän. Die Erde im Zeitalter des Menschen* (Darmstadt: Wissenschaftliche Buchgesellschaft, 2008).

57 Karl Polanyi, *The Great Transformation. The Political and Economic Origins of our Time* (New York: Farrar and Rinehart, 1944); WBGU, *Welt im Wandel*.

interconnected and very often mutually dependent. The Industrial Revolution and the Great Transformation from 1769 onwards have changed the rules of the game in the MENA region as much as anywhere.

Returning to our topic, i.e. climate and culture in the Middle/Near East and North Africa, we have to observe that climate continues to be a decisive and increasingly effective and aggressive driver of change. On the other hand, our region is—probably more than many others—the scene of political and cultural disputes and conflicts, quite a few of them originating in physical constraints like droughts or freshwater scarcity. Whenever and wherever such climatic or environmental stressors happen in fragile political and socio-economic situations or when they collide with ethnic or religious conflicts and clashes, climate and culture may lead to violence and even military warfare.

Figure 0.3 tries to summarise the mutually interdependent scenarios of physical and cultural threats in today's MENA region. Two observations can easily be made: First, both climate and culture have their own cascade of causes and effects. The climate cascade ends up in the deterioration of natural environments, their degradation and loss of sustainable self-regulation. As a result nature loses its ecological welfare functions for society and people.[58] The same holds true for the cascade of the human dimension. Triggered by demographic factors, especially by population growth in the 20th century—quantitatively in terms of numbers, qualitatively in terms of increasing demand for goods and services—the human demand side creates a cascade of socio-economic, political, and cultural consequences ending in crisis and conflict scenarios. Secondly—and this is a basically new situation in the relationship between climate and culture: the two cascades merge and create synergies that are disastrous for both sides. Since the beginning of the Industrial Revolution, humanity has changed its status from being a victim of nature to being its commander and manipulator. And what is the outcome?

Climate and culture, nature and society have entered a new relationship in a region with a highly fragile natural environment in combination with equally unstable political cultures and structures. This new relationship is a heavy burden on the present and foreseeable future of the MENA region. Fig. 0.3 complements the theoretical discussion of both natural and cultural/political drivers of conflict with a number of practical examples of current reality. While the climatic/natural sequence in itself is now heavily influenced, if not determined, by human intervention, the cultural cascade—and that explicitly includes the political, socio-economic, sociopolitical, technological and/

58 MEA, *Ecosystem.*

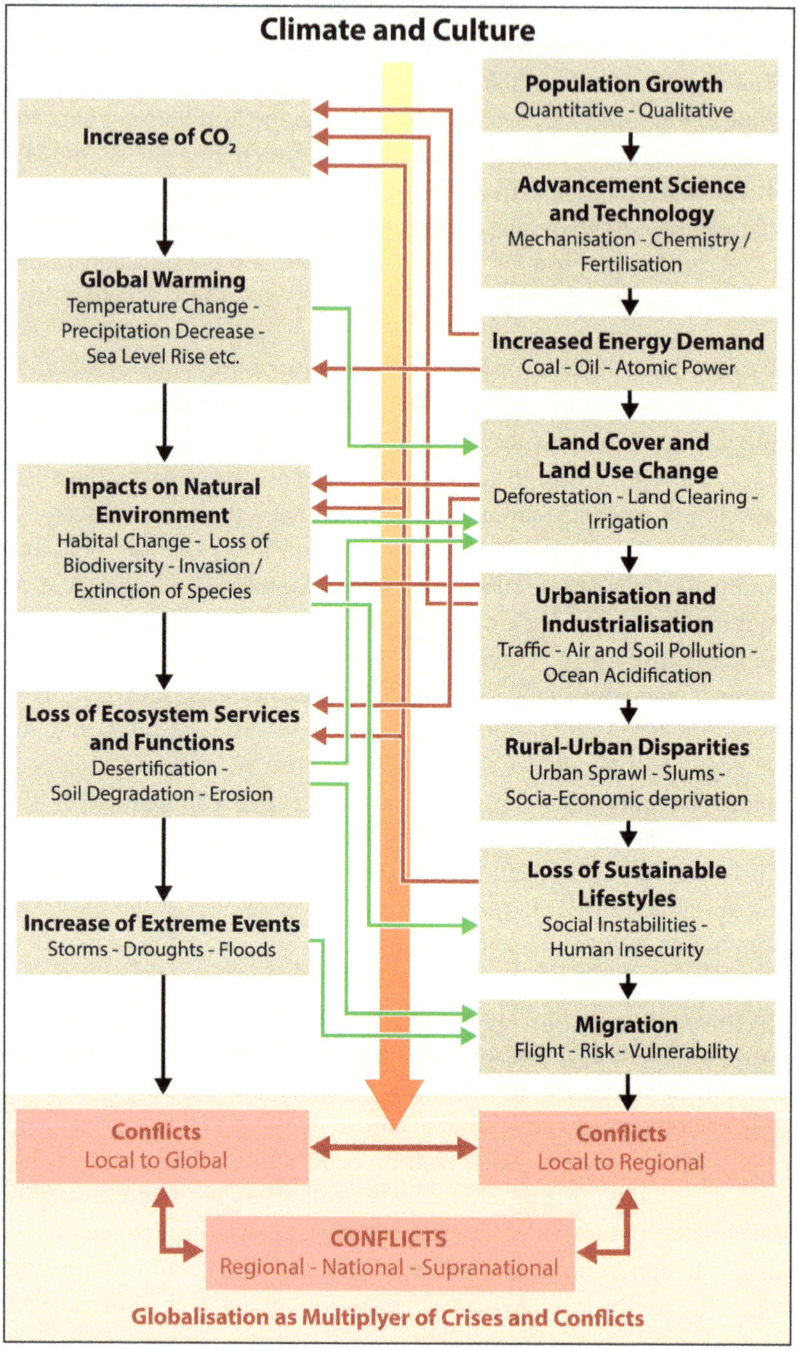

FIGURE 0.3 Climatic and socio-economic consequences of global warming and population growth: interactions between climate and culture and potential conflicts.

or spiritual aspects of the human dimension and of human culture—is also closely combined with its climatic counterpart.

Population growth and the advancement of science and technology trigger rapidly increasing human intervention in nature, and nature strikes back. Both fig. 0.2 and fig. 0.3 try to demonstrate some of these interrelationships in general terms (fig. 0.2) resp. specifically for the MENA region (fig. 0.3). Incomplete as these attempts surely are, they nevertheless reveal some of the region's specific environmental and cultural crisis symptoms. The climatic cascade, triggered by natural and man-made global warming, depletes the already harsh environment to such a degree that socio-economic disruption (e.g. poor harvests, increase of food prices), water shortage, ethnic or religious clashes (e.g. regarding land and water access) or political confrontations (from local via regional to national levels) may serve to accelerate conflicts and crises. The MENA region of today is full of tragic examples for these contexts and quite a few of them end in destabilisation of states and societies (failed states) and/or international conflicts and warfare. Fig. 0.3 mentions some of these conflicts and clashes in the Near and Middle East and North Africa at the interface of natural and cultural/political agents. This book also offers striking examples of such conflict potentials (see e.g. Chapter 5, 6, and 7).

The climatic cascade triggered by the steadily increasing concentrations of CO_2 and other greenhouse gases exerts its influence on the nature of the deserts and the steppe regions as well as the forests and fertile stretches of cultivated lands throughout the MENA region through the above-mentioned changes of temperature and precipitation. These cause slight and slow changes to the natural habitat. Extreme events like storms, droughts and floods promote the deflation and erosion of soils, the destruction of vegetation cover and desertification—not, however, without the human contribution of e.g. turning natural pastures into agricultural lands, destroying the natural vegetation cover and causing deflation of soil and desertification. The cultural cascade, basically triggered by population growth and its growing demand for people's well-being and poverty reduction (food, health, shelter, security etc.), has its own agenda. Many reactions, however, greatly aggravate the negative development of the climate cascade. Increased energy demand, deforestation and land clearance, chemical fertilisers and subsequent groundwater pollution: these are just a few human contributions to climate and environmental changes. But there is, of course, a specifically anthropogenic side to the cultural cascade. Especially at its lower end, the indicated conflict potentials appear not so much influenced by climate change or environmental hazards as conditioned by the globalisation processes in which the MENA region is heavily involved. Urbanisation, industrialisation, militarisation: these are not only starting points of specifically cultural developments, but indirect causes of consequences that leave their footprint on global warming and its effects as well as on the natural environment. The rich oil and gas producing states of the Middle East and North Africa are major polluters of the atmosphere; their extravagant lifestyle, their wasteful use of water (through desalinisation plants along their coasts), and their unprecedented urban building activities contrast sharply with their poorer neighbours. Rural-urban and rich-poor dichotomies are determinants of migration at the national and international level. These migrations cause additional conflict, especially when and where they are associated with environmental problems (droughts, floods) or triggered by humanitarian catastrophes (famines) or ethnic and religious clashes.

In the long run, both the climatic and the cultural cascades intermingle. The final results are often enough conflicts which may lead to the destabilisation of societies and states. Many examples of this are found in the Middle East and North Africa in the form of 'failed states': Afghanistan, Yemen, Iraq, Syria or Libya are probably the most notorious cases.

Again, it would have been tempting to expand our two examples of water scarcity (Chapters 6 and 7) by additional analyses of climate and culture-dependent conflict scenarios. These are abundant in the MENA region: clashes over territories and natural resources, especially oil and gas, but also over trade routes as well as rapid population growth, urbanisation and urban sprawl, loss of fertile land, unemployment, and both national and international migration are just a few of the many causes and consequences of the intricate interdependencies between nature and culture. They will surely increase in the future due to the undeniable effects of climate change, land degradation, and of the turbulences of globalisation.

6 Conclusion and Outlook

Climate and culture in the Near and Middle East and in North Africa: our introductory survey of several thousand years of natural and cultural history and present-day problems in this region has carefully indicated close interrelationships and sometimes even obvious conflicts between nature and society. Some of the following essays reveal causes and consequences of such encounters in much more detail.

Our attempt to understand and interpret the historical and cultural achievements in the millennia-old history of the MENA-region brings us back to our initial references to cultural theory and to the interpretative potentials of the *Kulturkreislehre*. In this context and at this point we can refer to Arnold

Toynbee's monumental twelve-volume compendium *A Study of History* (1934–1961). This work, devoted to an analysis of the rise and fall of great empires, covers the history of humanity under the specific aspects of the challenges they face and the responses they develop. Challenges are, among others, history itself, climate and geography. Toynbee's final 1976 publication, a shortened and summary review of his opus magnum, bears the descriptive yet highly appropriate title *Mankind and Mother Earth. A Narrative History of the World.* Out of its 82 concise chapters, more than thirty deal explicitly with historical and geographic elements of the MENA-region, starting with the challenging conquest of the amphibious lowlands of Mesopotamia, their importance for the rise of the early Sumerian civilisation and its mutual cross-fertilising interactions.

Arnold Toynbee's impressive look at the mechanisms and courses of the evolution, rise and fall of civilisations may seem somewhat historical itself today. In his almost philosophical world-view, societies' creative adjustment to environment, their dependence on religious tradition and their importance for cultural development play decisive roles. Toynbee's insights into the close interrelationships between nature and society, however, were conceived and published before the modern Great Transformation took place and a new world order came into existence. Nowadays, challenges to and responses by individual states, cultures and/or civilisations are no longer isolated, local or regional events. Instead: conflicts, threats and dangers have reached global dimensions. Planet Earth has shrunk to a global village and is now characterised by a hitherto unknown intricate simultaneousness of time and space on a global scale. What happens here and now has immediate effects, and responses are instantaneous: political conflicts, international trade, oil production and prices, banking business, information flows, climatic events, sea-level change, environmental degradation, population growth and overpopulation, human migration and so on. Many of these phenomena are closely interconnected and inseparable from each other, no longer isolated but closely dependent.

Arnold Toynbee, often considered one of the last, if not the last universal historian, may also have the last word in this introductory essay. It relates closely to the rise and fall of Middle Eastern and North African cultures and civilisations. The final paragraphs of his "Mankind and Mother Earth", written in 1973 and almost prophetic, are obviously more valid today than 45 years ago. They read as follows:

> Man possesses consciousness, he distinguishes between good and evil, and in his acts he makes choices. In the ethical field, in which Man's choices are either for evil or for good, his choices produce a moral credit-debit account. We do not know whether this account is closed at the

> death of each short-lived human being [...] For the network of relations between incarnate human beings that constitutes human society, the account is still open and will remain open so long as mankind allows the biosphere to remain inhabitable. Will mankind murder Mother Earth or will he redeem her? He could murder her by misusing his increasing technological potency. Alternatively he could redeem her by overcoming the suicidal, aggressive greed that, in all living creatures, including Man himself, has been the price of the Great Mother's gift of life. This is the enigmatic question which now confronts Man.[59]

59 Arnold Toynbee, *Mankind and Mother Earth. A Narrative History of the World* (Oxford; Oxford University Press, 1976), 585–586.

PART 1

Ideas

CHAPTER 1

And We Shall Save the Earth: Muslim Environmental Stewards

Katajun Amirpur

Abstract

Theology in our times bears a great responsibility due to the deep and rapid changes in people's lives, in particular in dealing with the growing of ecological crisis. This essay provides an overview of Muslim ethics and attitudes towards Islam and ecology. Muslim scholars who suggest a new theology, 'eco-theology', which honours the Earth and worldly life as well as enriching people's lives with faith and divinity are explored. Islamic modules of the relationship between God and humans as Creator and creature are presented. In addition, this essay deals with the question how Muslims implement Islamic environmental ethics in sustainability mobilisation and activism in Western as well as Muslim-majority countries.

1 Introduction

Muslim environmentalists use Islam to argue and legitimise their position, increasing awareness of the strong link between faith and ecology. They are extending the discourse to the ecological question by constructing an alternative environmental imaginary based on the sovereignty of God and the responsible trusteeship of humankind. Scholars applied central ethical precepts detailed in the Qurʾān and Sunnah to construct a theocentric ecological philosophy of Islam.[1] The so-called eco-ethic put forward below centres upon six key principles which foreground the fundamental aspects of the environmental message of Islam. It is not intended to weigh the cogency of their arguments, i.e. the question whether or not environmentalism is Islamic. No judgement on their position is passed. Instead, the aim of this essay is to attempt to understand

1 Seyyed H. Nasr, "Islam and the Environmental Crisis," *al'Ilm* 17 (1997): 1–23; Seyyed P. Ouis, "Islamic Ecotheology Based on the Qur'ān," *Islamic Studies* 37.2 (1998): 151–181; Fazlun Khalid, "Islam and the Environment—Ethics and Practice: An Assessment," *Religion Compass* 4.11 (2010) 707–716.

 | DOI:10.1163/9789004444973_003

how Muslim environmentalists are able to develop a strong argument relating their activism to their cultural background at both the theoretical and the practical level.

This connection is already made explicit in the names of many initiatives: "Green Khutba Campaign," "Green Guide to Hajj," "Muslim Green Guide to Reducing Climate Change," "Greening Ramadan," or "Clean Medina Campaign".[2] Evidently, the activists choosing these names see the connection between Islam and environmentalism as central. In recent years, this strand of environmentalism has come to be referred to as "Green Islam" or "Eco-Islam".

The term aims to express the idea that Muslims value their natural environment as God's creation and should therefore regard sustainable living as a religious duty inherent in Islam. Its proponents believe that creation in its totality exists to praise God and view all parts of the Earth as 'divine signs'. This means that God is omnipresent and the protection of the natural environment is necessary for God's sake alone. Further, nature is understood as a totality of mutually complementing elements.

A sustainable lifestyle in the Islamic sense practically means: using fair-trade products, growing fruit and vegetables or purchasing local produce, saving water in ritual ablutions, and many other points that will be addressed in the following.

2 Origins

Environmentalism has been featuring as an important topic for Muslims since the 20th century. They encountered it in part as an intellectual challenge in the then current claim that all three monotheistic religions were centrally to blame for the emerging environmental crisis by focussing on humanity and neglecting nature.

In 1967, the historian Lynn White Junior had presented a controversial thesis regarding the ecological crisis: He claimed that it had its roots in monotheistic faith. The responses by religious leaders and institutions to this accusation laid the groundwork for what would become a new eco-theological discourse. Though White mainly focused on Christianity and Judaism in his publications,

2 "Green Khutba Campaign," "Green Guide to Hajj," "Muslim Green Guide to Reducing Climate Change," "Greening Ramadan," "Clean Medina Campaign"—the names of these initiatives should indicate the connection of Islam and nature. The labels "Green Islam" or "Eco-Islam" have gained ground to call this contemporary current in recent years.

his claims came at a time of increasing sensitivity to ecological concerns and also encouraged Muslims to engage with the issue more thoroughly.

A further reason why environmentalism became an increasing concern to many Muslims was the way in which people in majority-Muslim countries were exposed to the consequences of ecological devastation. They saw water sources polluted, forests cut down and soil eroding as industrialisation advanced. Turning to the Qur'ān to solve this predicament was an intuitive solution; Many Muslims experience Islam as a guiding light in all aspects of life.

3 The Qur'ān and Environmentalism

The need for environmental protection and a sustainable lifestyle can be derived from numerous individual Qur'ānic passages. In addition, the understanding of the world that is put forward in the Qur'ān as a whole does not allow for the idea of inanimate matter. The entire world consists of living matter and everything in it relates immediately to God. This assumption leads to the theological question to what extent humanity is allowed to interfere with this living matter and potentially destroy its relationship to God.

By the world view it proposes, the Qur'ān also offers numerous starting points to develop a comprehensive ecological ethics and theology of sustainability. It contains fundamental ethical principles interpreted concretely based on the actions of the Prophet. Surah 6:141, e.g., states that God "does not like those who commit excess."[3]

This verse is further illustrated through the behaviour of the Prophet. At one time, he admonished one of his companions for pouring away water left over after washing himself. The Prophet told him to return the leftover water to the river so people downstream would be able to use it. What he had done, the Prophet said, was wasteful. There are many traditions of the Prophet Muhammad cautioning against wasting or polluting water. Similar examples calling for a sparing use of natural resources and humane treatment of animals are also found in the recorded words and deeds of early Muslims such as Imam Ali. These, too, served to guide the actions of future generations.

These indications that the Prophet Muhammad followed an environmentally friendly lifestyle are especially important because he is considered the example for all Muslims to emulate. The accounts of his life tell us that he was careful in his use of natural resources. The message is that even where the

3 See Qur'ān 6:141.

rivers are flowing, where resources are readily available, humanity must always use them in measure.

Islamic environmental ethics are based on the Qur'ān and the *aḥādīṯ*, the records of Muhammad's words and deeds. Careful stewardship of nature and creation, in particular of the flora, fauna, and water, is a duty incumbent on all Muslims in this interpretation of the Qur'ān and *sunnah*. Water is accorded particular significance in Islam as it is considered both the source and basis of all life as well as being important in the ritual ablutions required of all Muslims.

Plants are also accorded great significance in Islam. They nourish humans and animals[4] and also provide humanity with the oxygen they require to live. Humans are thus called upon to honour them and acknowledge their worth. "The Shariah recognises animals as beings whom humans are allowed to make use of, but who are owed good treatment in return."[5]

Planting a tree is considered a particularly meritorious act. A well-known *ḥadīṯ* claims that every Muslim who planted a tree will be rewarded in the afterlife for every animal or human eating of its fruits.

Thus, each part of nature fulfils a specific purpose in creation beyond the praise of God. All are important for the functioning of the world, and all things in nature are thus interrelated and depend on one another. Yet there is a second, indirect way in which exegetes derive an Islamic view of humanity's relation to the natural environment from the Qur'ān and the *sunnah*. It can be argued that the far better conduct of animals with regard to nature is used to hold up a mirror to humanity's faults:

According to the Qur'ān, the Prophet Solomon had a close relationship with horses. He also spoke with birds and once stopped a crowd of people to free a path for a swarm of ants. Other foundational texts of Islam share this theme. The *Epistles of the Brethren of Purity* contains a chapter titled *The Case of Animals versus Man before the King of the Jinn* where animals and humans compete for divine recognition. When the humans claim the ability to do something, the animals retort that they, too, are able to do this, and do it better. God's judgement invariably favours the animals.

Humans are repeatedly criticised for their treatment of God's creation which is compared unfavourably to the conduct of animals. The central points of criticism are their arrogance, carelessness, and excess. At the end of the story, we learn what distinguishes humans from animals: The ability to enter into a reflected relationship with God. While animals' relationship to God is

4 Qur'ān 80:24–32.

5 Ursula Kowanda-Yassin, *Mensch und Naturverständnis im sunnitischen Islam: Ein Beitrag zum aktuellen Umweltdiskurs* (Würzburg: Ergon-Verlag, 2011), 173.

intuitive and they are naturally able to enter into it, humans can enter into such a relationship using their reason. They are exhorted to follow the example of the animals in their responsible treatment of their natural environment.

Further, Muslims can base their conception of a proper relationship between humans and the natural environment on more universal principles. To check human conduct and maintain proper limitations, we frequently find six religious principles quoted alongside modest restraint and justice (Arab. *ʿadl*). These are:

3.1 *The Principle of Unity (Arab.* tauḥīd*)*

The principle of *tauḥīd* applies to Islamic ecological theology at three levels. Firstly, it refers to the monotheistic unity of God in contrast to polytheism (Arab. *širk*) and the idolatry of pre-Islamic times, secondly to the divine oneness in contrast to the Christian concept of the Trinity which the Qurʾān criticises repeatedly. Thirdly, the principle of *tauḥīd* expresses the unity of God with all of creation.

3.2 *The Principle of Creation (Arab.* fitra*)*

In Islamic environmental discourse, *fitra* refers to the original state of Creation or the original nature of things. This includes the original state of humanity existing in harmony with nature. Humanity's need and duty to protect the natural environment follow from it.

The argument is thus made that humans used to live in a natural state of *fitra*, unconsciously respecting the boundaries set for them by unwritten natural law. This state ended with the advent of industrialisation. Though humans had the same positive and negative attributes in the past as they do now, their tendency towards good or bad actions had been limited by the natural order.

Previous civilisations did not leave their permanent footprints in the shape of persistent pollutants, toxic residues, and nuclear waste. This shows that humanity's burden of responsibility is greater today than it was since we now have to contend with the ability to actually destroy our natural environment in many ways. Thus, Islamic environmentalism, strives to restore *fitra* and protect the Earth.

3.3 *The Principle of Stewardship (Arab.* ḫilāfa*)*

Humans act as stewards or representatives (Arab. *ḫalīfa*) of God on Earth. Thus, humanity is responsible for the natural environment before God who entrusted his creation to them and placed them in its service.

Humans are not the owners or overlords of the Earth; that position is reserved for God. Nonetheless, they have an important role in the order of creation. The Islamic environmental movement calls on humanity to responsibly fulfil its duty as stewards of nature rather than continue to subject it to their will.

3.4 *The Principle of Responsibility (Arab.* amāna)

This principle is closely linked with that of stewardship (*ḫilāfa*). Responsibility (*amāna*) applies to the fulfilment of human duties in all aspects of life. This responsibility is inherent in the position of stewardship which humanity was offered by God and accepted. It is typically supported with a Qur'ānic passage stating that God offered this responsibility to the heavens, the Earth, and the mountains, but all refused to assume it for fear of the burden it meant.

Humanity, though, agreed to assume the responsibility.[6] In a sense, *amāna* represents a limitation on the authority of stewardship and a moral burden. Islamic environmental ethics see humanity's exalted position based not on greater power or the rule over the rest of creation, but in the fact that humanity alone is answerable to God for them.

3.5 *The Principle of Servitude*

This principle refers to the status of all humanity as servants of God that complements the principles of stewardship and responsibility. Their role as servants limits the power of humanity.

3.6 *The Principle of Balance*

The Arabic term can refer to balance in both physical and figurative way as well as to a set of scales, and Islamic environmental theology also links it to ecological balance or a middle path. This principle calls for the preservation or restoration of balance on Earth both with regard to the harmony of nature and to justice and morality in human interaction. God created the Earth and all upon it perfectly, faultlessly and in balance, and the task of humanity is to maintain it in that state.[7]

4 Leading Muslim Scholars and Activists on the Environment

Muslim ecological theologists, some of whom we will look at in more detail now, cite problems such as global climate change, earthquakes, or rising sea

6 See Qur'ān 33:72.

7 See Monika Zbidi, "*The Call to Eco-Jihad. The Islamic Environmental Movement,*" Art&Thought, 99 (2013): 8–9.

levels as evidence that the divine balance of Earth was disturbed. To Muslim ecological theologists, this means that humans, as servants of God, are bound to obey laws including care for their natural ecosystem and proper use of its resources.

Ziauddin Sardar[8] writes:

> Combine the concepts of *tawhid, khilafah, amana, halal*, and *haram* with words of justice (*adl*) and moderation, temperance, balance, equilibrium, harmony and the concept of *ihtihsan* (preference for the better) and *istislah* (public welfare) and one has the most sophisticated framework for an environmental ethic that one can possibly desire.[9]

Current Islamic theology mainly discusses two distinct approaches. One is that of the founding father of Islamic environmentalism, the Iranian-born philosopher Seyyed Hossein Nasr whose 1967 book *Man and Nature. The Spiritual Crisis of Modern Man* was greatly ahead of its time.[10] Born in 1939, Seyyid Hossein Nasr, like some Christian theologians, views ecological balance as a religious issue. He bases this view on a cosmology in which nature, humanity, and God, heaven and Earth, initially existed in a well-balanced order. He believes that humanity has been denying this cosmic order, making the environmental crisis a spiritual one. This diagnosis is made with a view to getting Muslims back onto their original, rightly guided path.

European Enlightenment, Nasr argues, had replaced theocentric order with an anthropocentric cosmological model. Placing humanity at the centre of the universe, this system lacks any connection to a higher authority, allowing humans free rein to exploit their natural environment.

---

8 Cultural critic Ziauddin Sardar was born in Dipalpur, Pakistan in 1951, but he grew up in London. He is Professor of Law and Society, Middlesex University. He is the Chair of the Muslim Institute, a fellowship society that promotes knowledge and debate, which publishes the quarterly 'Critical Muslim'. Ziauddin Sardar wrote over 45 books, guest blogged in the Guardian and presented a number of programs for the BBC and Channel 4. His research interests went from 'The Future of Muslim Civilisation' over 'Why Do People Hate America?' to 'The A to Z of Postmodern Life' and his professional life includes periods of research work for the Hajj Research Centre in the seventies and an advisory position in the cabinet of Anwar Ibrahim, the former Deputy Prime Minister of Malaysia in the late 80s.

9 Ziauddin Sardar, "Towards an Islamic Theory of the Environment," in *Islamic Futures: The Shape of Ideas to Come*, ed. Ziauddin Sardar (London, New York, 1985), 228.

10 Seyyed H. Nasr, *The Encounter of Man and Nature. The Spiritual Crisis of Modern Man* (Chicago: ABC International Group, 1967).

Nasr assumes humans to both be limitlessly egotistical and greedy, but also to strive for transcendence. Without any cosmological anchoring, they are incapable of attaining true happiness and will seek an outlet for their desire in ever greater consumption and more sophisticated technology.[11] Even when removed from God, humans by nature strive for transcendence and hunger to experience the infinite. Where this is no longer possible through religion, their striving for transcendence necessarily and vainly leads to perpetual excess in consumption. A reasonable handling of nature is only possible when those who can be awakened must be made to open their eyes and realize that the modern world is walking on the edge of a precipice and needs only to take another forward step to face its perdition.[12]

Nasr views the environmental crisis as fundamentally spiritual. He thus postulates the existence of a natural balance between humanity, nature, and creation. The roots of the ecological crisis to him lie in a spiritual crisis—the increasing tendency to view humanity as distinct from God and nature. The solution lies in reorientation and a renewed connection with God. Since God is manifested in nature, this would also entail a re-spiritualisation of the natural world and thus a return to greater mindfulness in its treatment. If there is a return to the religious view of nature, particularly, by Sufis, human beings may come to understand that the natural world is more than just an entity to be utilised. On Nasr's word, "we need a deep transformation of our understanding of nature and environment which is His creation."[13] This is both a very conservative attitude and a very modern one, especially considering the numerous calls on consumers to change their lifestyles.

An entirely different approach is developed by Farid Esack. He locates humans in a wider system of relationships with themselves, each other, with plants, animals, creation as a whole and, not least, with God. Farid Esack recognizes the fragmented nature of identity in the modern context, acknowledging that "we all comprise multiple identities, depending on where we come from, what we believe in, where we are and whom we are interacting with."[14] These relationships must be characterised by dignity and justice. When the Qurʾān states that plants and animals, too, form communities, that all things are animate and capable of praying to God, humanity must accept that this fact needs to be accounted for in the way they realise their relationships. Esack, a South

11 Nasr, "*The Encounter of Man and Nature,* 9.

12 Nasr, "Islam and the Environmental Crisis," 104.

13 Seyyed H. Nasr, "The Islamic Perspective on the Environmental Crisis: Seyyed Hossein Nasr in Conversation with Muzaffar Iqbal," in: *Islam and Science* 5.1 (2007): 81.

14 Farid Esack, *On Being a Muslim* (Oxford: Oneworld Publications, 2002), 137.

African, also refers to the liberation movement which is significant to his approach to ecological Islam. Esack characterizes Islam as a path "committed to social justice, to individual liberty and the quest for the Transcendent who is beyond all institutional religious and dogmatic constructions; an Islam that challenges us to examine our faith in personally and socially relevant terms."[15] In his view, coloured by liberation theology, it is necessary to rebalance some oppressive relationships. The relationship that humans have with God is reflected in the degree to which their relationships with other beings and nature as a whole are emancipatory and just. The key concept in this environmental theology is justice.

Farid Esack is part of the Progressive Muslims Network (PMN), a group of Islamic theologians who subscribe progressive Islam. In an internet manifesto, these theologians profess:

> Third, humankind is located within their dual position of being simultaneously autonomous beings with full agency and as returning to God. Agency implies power over one's life and one's status as a returnee to God implies both a sacredness beyond one's commodity value as well as defining the limits of that agency. In other words, our struggle to experience a personally and socially meaningful Islam is rooted in praxis geared towards creating a more humane society as part of a sustainable eco-system in the service of the Transcendent.[16]

The Turkish philosopher Ibrahim Özdemir argues that Western science has re-calibrated human understandings of nature. Western scientific methods have objectified nature, whereas medieval and Renaissance traditions viewed nature as being vital and animated with sacred life force.[17] Özdemir regards Qur'ānic teachings on nature as an appropriate vehicle for challenging materialistic conceptions of nature as they reveal nature as imbued with sentience and vitality. He considers nature as the ultimate proof of Divine existence. His ideas reflect *Ṣūfī* metaphysics in this respect. Human beings witnessing the

---

15 Esack, *On Being a Muslim*, 2.

16 Progressive Muslims Network (PMN), "Progressive Islam: A Definition and Declaration," last accessed via the Wayback Machine at Archive.org, March 25, 2019. http://web.archive.org/web/20020403150253/www.progressivemuslims.com/index2.html.

17 Ibrahim Özdemir, *Science and the Environment: Is Science Responsible for Environmental Crisis?* (JESS/CSBD, 1996), 35–46.

awesome majesty of nature should submit to the Divine.[18] His ideas here are akin to Nasr's thinking.

Fazlun Khalid[19] is counted among the 15 leading environmental theologians worldwide as well as appearing on the "Independent on Sunday" list of the top 100 environmentalists in the UK in 2008, and is also listed amongst the "500 Most Influential Muslims in the World" by the Royal Islamic Strategic Studies.[20] Khalid published numerous works on environmental ethics in the context of Islam. He was employed at the British Home Office, but decided to leave his job in politics to dedicate himself to religion and the environment. His seminal experience was being recruited as a mediator for the Race Relations Board in 1968. He learned quickly that problems concerning race, poverty, and overindebtedness in developing countries shared a common origin in political inequality.

At a WWF conference in the early 1980s, he was asked what Islam had to say on environmental issues. Lacking an immediate answer, he followed up the question with Qur'ānic scholars and Islam experts. None of their answers satisfied him as he found them all inadequate to the challenges of these days. In the end, he resigned from his Home Office position and began studying the basis of environmentalism in Islam at the University of Birmingham. In 1994, he started the Islamic Foundation for Ecology and Environmental Sciences (IFEES)[21] which developed and designed multifaceted activities.

Khalid's motivation for his activism is genuinely religious. He states that he draws strength from the fact that he, as a faithful Muslim, placed his forehead against the earth five times a day to thank God for his creation. After all, it is said in the Qur'ān that "the creation of the heavens and earth is greater than the creation of mankind".[22]

Religious scholars in Indonesia share the conviction that Islam has much to say about environmentalism. Media Zainul Bahri, Professor of Religious Studies at the Syarif Hidayatullah State Islamic University (UIN) Jakarta, says that: "there are many threats in the Quran regarding environmental issues. God

18 Ibrahim Özdemir, "Towards an Understanding of Environmental Ethics from a Qur'anic Perspective," in *Islam and Ecology*, ed. Richard C. Foltz, Frederick M. Denny and Azizan Baharuddin (Cambridge: Harvard University Press, 2003), 3–38.

19 For more information on Fazlun Khalid see: www.ifees.org.uk/the-founder.

20 Cardiff University: Centre for the study of Islam in UK, last accessed March 25, 2019. http://sites.cardiff.ac.uk/islamukcentre/virtual-centre/public-lectures/2013-public-lectures/fazlun-khalid/.

21 Interview Fazlun Khalid with Deutsche Welle by Franziska Badenschier on April 2, 2013. https://p.dw.com/p/188GU.

22 See Qur'ān 40:57.

blames people if they cause environmental damage".[23] He believes that these messages are often overlooked deliberately as many people believe that there were no environmental concerns in the Prophet's days. "Theological teaching since the 1950s has mainly focused on humanitarian concerns, and continues to do so", Bahri states.[24]

5 Transforming Islamic Eco-Theology: Initiatives, Projects, and Blogs of Islamic-inspired Environmentalism

The Islamic side of Eco-Islam is mostly seen in the occasions its proponents choose for their environmentalist activism. IFEES works beyond the national borders of Britain and is widely regarded as the ancestor of all Islamic environmentalist organisations. It is active in both local and international projects and well-connected among other such organisations that emerged across Britain in recent years. Among other international activities, it supports exchange programmes with students of religious schools in Indonesia.[25]

Its greatest international success was achieved on the island of Zanzibar where it was called in by CARE International in 1999. Local fishing communities were increasingly unable to support themselves as fish stocks dwindled due to overexploitation. Some took to dynamite fishing among the coral reefs to bring home a catch. IFEES offered the fishers of Zanzibar a two-day Qur'ān workshop and managed to persuade then within 48 hours to stop dynamite-fishing—something other environmental charities had been unable to manage in four years. The fishers were convinced because Khalid deployed arguments from the Qur'ān to dissuade them from continuing the destructive practice. He quotes one participant who told him: "We must obey the laws of the Creator because they are part of Creation, but we need not obey the laws of the government."[26]

Another high-profile action of the IFEES was the 2008 "Clean Medina Campaign" in Birmingham. Its slogan was: "It's a campaign! It's a film! It's Jihad!"

---

23 Media Z. Bahri, Interview given to Rizki Nugraha and Ayu Purwaningsih/Deutsche Welle on June 14, 2018, last accessed January 28, 2019. http://www.dw.com/en/indonesia-combines-islam-with-environmental-activism/a-44220277.

24 Ibid.

25 Rosemary Hancock, *Islamic Environmentalism: Activism in the United States and Great Britain* (New York: Routledge, 2018).

26 Interview given by Fazlun Khalid to Franziska Badenschier/Deutsche Welle.

Muslims recorded themselves as they swept and cleaned streets together and tried to motivate their coreligionists in other cities to follow their example.[27]

Many other initiatives followed the example of IFEES, one is mentioned by Martina Sabra in her report on a rural school 150 kilometres southeast of Dar al Baidaa (former Casablanca). In a lesson on environmental protection, the teacher explained that trees and plants were deserving of respect not only for the shade they gave us, but also because they were part of God's creation. To clarify this, he quoted the words of the Prophet Muhammad. The children listened attentively.[28]

Ayman Mazyek, founding member of the Islamic aid organisation Green Helmets, also sees cooperation with Islamic structures as normal:

> Of course you need close contact with Islamic dignitaries on the ground. They are highly respected people there [...] and it would certainly be helpful if aid workers had more details of the local form of Islam in the country. But in the end, technical qualifications and human qualities count for more with us than religious conviction or knowledge of religious details.[29]

The more central a role Islam plays in development cooperation, the more important a detailed understanding of it should be fostered. Muslims across the world are actively voicing their concern for the environment and harnessing the educational landscape of Islam as one of the primary conduits for environmental action. Nonetheless, we must consider this: an unquestioning support of these religious groups as the more 'authentic' or 'credible' partners runs the risk of legitimising totalitarian Islamic identity discourses and furthering antidemocratic currents. Still, it appears that people on Zanzibar and children in Morocco are readier to be convinced by religious arguments than by rational reasoning.

This experience is confirmed by a project in Indonesia. During Ramadan of 2018, Indonesia's most important Muslim assembly, the Indonesian Ulema Council (MUI), launched an awareness campaign jointly with Greenpeace and the Indonesian Ministry for Forestry and the Environment to reduce plastic pollution. During Ramadan, Indonesians tend to become more consumerist:

---

27 Natan Levy, Harfiyah Haleem, and David Shreeve, *Sharing Eden: Green Teachings from Jews, Christians and Muslim* (Leicestershire: Kube Publishing, 2012), 27.

28 Qantara, "Environmental Protection through the Koran?," last accessed January 28, 2019. https://en.qantara.de/content/islam-and-development-cooperation-environmental-protection-through-the-koran.

29 Ibid.

Many people break their fast in restaurants or order takeaway food, increasing the amount of waste. As they meet in mosques in the evening to collectively break their fast, many bring food in disposable plastic dishes. The organisations advocated the adoption of reusable carrier bags to reduce the consumption of single-use plastic bags in the country. The government and the clergy organised in the country's largest Muslim organisations, the Nahdlatul Ulama (NU) and Muhammadiyah, hoped to directly influence the consumption patterns of roughly one hundred million adherents.[30]

To start the campaign "#PantangPlastik" (#AntiPlastic), Greenpeace held an "eco-iftar" in South Jakarta. Greenpeace states that it recognises the importance of working with Islamic organisations in reaching out to the population at large. Imams can become central figures in environmental protection as many rural people, unlike most urbanites, will listen to what their clergy say.[31]

This is where we must also ask whether religiously motivated environmentalism is any more effective than other kinds. It is, after all, not the first time Indonesia bet on religion to help the environment. In 2014, the MUI issued a fatwa banning the hunting of endangered animal species.[32] The IFEES also supported the School4Trees Programme for forest protection.[33]

In Germany, projects to foster awareness of sustainable living and environmentalism specifically aiming at Muslims are no longer a great exception. The journalist Ulrike Hummel reported on the project "*Faire Moschee*" (fair mosque) in which twenty young Muslims from Dortmund developed the concept of an environmentally sustainable, socially equitable mosque community which the Moscheeverein Wali e.V. now implements.[34] The project began with a symposium on fair trade. The initiators decided to offer organically farmed tea and vegetarian food and use porcelain cups and plates rather than

30 Emerging Earth Community, "Indonesia Combines Islam with Environmental Activism," last accessed March 26, 2019. www.emergingearthcommunity.org/innerfeed?source_site=fore&pag_title=Indonesia+combines+Islam+with+environmental+activism&article_id=6&feed_type=news&news_url=news/item/indonesia-combines-islam-with-environmental-activism/.

31 The Jakarta Post of June 5, 2018, accessed March 26, 2019. www.thejakartapost.com/news/2018/06/05/ greenpeace-indonesia-holds-eco-iftar-in-pondok-indah.html.

32 Lembaga Pemuliaan Lingkungang Hidup Dan Sumber Daya Alam: Majelis Ulama Indonesia, "Wildlife Trafficking Fatwa," accessed March 26, 2019. http://mui-lplhsda.org/wildlifetraffickingfatwa/.

33 Ifees. "EcoIslam Mosques as Green Project Exemplars," accessed March 26, 2019. http://www.ifees.org.uk/mosques-as-green-project-exemplars/.

34 Ulrike Hummel, *Die grüne Welle*, 2014, accessed January 29, 2019. http://de.qantara.de/inhalt/islam-und-umweltschutz-die-gruene-welle.

disposable plastic ones. In addition, the association now sells fair trade halal argan oil products.[35]

Project leaders Hafssa El Hasbouni from the *Forum für Soziale Innovation* (Forum for Social Innovation, FSI) states that:

> The meeting in Solingen has shown that above all many young people want to join in. They feel that environmentalism and Islam are urgent issues and can't wait to become active.[36]

Many young Muslims grew up immersed in debates about environmental degradation and climate change and are now developing an Islamic perspective on this. They fundamentally want to respect God's creation which to them means a sustainable Islamic lifestyle.

The pilot project is trying to develop the foundations for a future "fair mosque" as a model for other communities. "The concept we are developing and will test at the Wali Mosque can later be used as a model for other mosque communities," says the 24-year-old Hafssa El Hasbouni.[37] They are also considering a certification scheme for the more distant future, "that is still far away and will depend on many factors."[38]

What then do young people think a "fair mosque" looks like? Is it about solar minarets, following the Green Hajj Guide, or simply reducing meat consumption? It is about none of these things, at least, not entirely. They regard all of it as not enough. What this generation focuses on above all is the social dimension of the mosque community. "I would draw a fundamental distinction between 'hardware' and 'software'," says Hafssa El Hasbouni. Hardware refers to things like using recycled paper or shopping for fair-trade products—a responsible approach to natural resources. "What many people brought up, though, is what I call the software dimension, fair treatment among ourselves, for each other, and especially for women."[39]

Mosques should not discriminate against women and should welcome them both with or without a headscarf. At the Dortmund Moscheegemeinde Wali e.V., men and women already pray together in the same room—still an exception in German mosques. The Moscheegemeinde Wali e.V. seems eminently well suited to adopt this approach. Its head Abdelhay Fadil has been a

35 Stiftung Umwelt und Entwicklung NRW, "Faire Moschee," accessed March 26, 2019. www.sue-nrw.de/2016/02/29/faire-moschee/www.sue-nrw.de/2016/02/29/faire-moschee/.

36 Hummel, *Die grüne Welle*.

37 Ibid.

38 Ibid.

39 Ibid.

fair trade activist for years. The community founded a women's cooperative of 63 farm women in Morocco to produce argan oil, assuring the Berber women of a fair and secure income. The focus lies on development aid. Before, the women were unable to make a living selling the argan fruit they grew. Now, they can produce the oil artisanallyand receive fair compensation for this high-end product.[40]

In 2008, Abdelhay Fadil founded the company *Arganpur* to market the organic cosmetics produced from the oil. By now, the trained biochemist and his staff are handling the production themselves. A further cooperative has now been founded to produce rosewater and rose oil—an idea that will also be integrated into the "fair mosque" concept. This expansion of the fair trade principle is very interesting: Fairness is to apply at all levels. Justice must be realised everywhere. Hence the plan of the initiative: A future "fair mosque" will be characterised by fair pay scales, financial transparency, flat hierarchies, and a frequent rotation of German-speaking preachers for Friday prayers. Closing the community to the outside and excluding non-Muslims is absolutely unacceptable. The group demands openness and tolerance. The latter should not least be expressed in not excluding gay and lesbian Muslims within the community.

The idea that respect for creation is deeply rooted in Islam and that this requires an ecologically conscious lifestyle is also espoused by the group of young Muslims in Germany who founded the association Hima in 2011. *Hima*, a term drawn from Islamic teaching that can be translated roughly as "nature preserve", is intended as a participatory forum for Muslims who are interested in environmental issues. Their primary purpose is raising awareness, trying to draw Muslims to ecological consciousness through education and concrete, practical projects.

The experiences of *Hima* will also be drawn on to develop the "fair mosque" project. One of the first activities of the forum was a series of nationwide "Fair Trade Brunches" as a venue for discussing the responsible use of natural resources from an Islamic perspective.[41] By now, *Hima* has become part of the Europe-wide "Green up My Community"-project. The campaign planned to showcase twenty environmentally friendly mosques throughout the continent by the end of 2014.[42] The function of *Hima* is mostly consultative. Due to its

40 Ulrike Hummel, *Die grüne Welle*.

41 Hummel, *Die grüne Welle*.

42 Hima, "Willkommen bei Hima," accessed March 26, 2019. www.hima-umweltschutz.de/geschichte.

strong connections with non-Muslim environmentalist organisations, it can draw on a great depth of experience.

The internet is increasingly important both in spreading ideas and solutions and in helping to connect Muslim activists. Along with information webpages, there are dedicated blogs and social media accounts and groups managed by people from various countries to discuss environmental issues and exchange articles, links, and videos. A.M. Schwencke adapted the words of Olivier Roy in speaking of a "globalized Eco-Islam".[43]

The blog www.theecomuslim.com by British Muslim Zaufishan Iqbal covers topics ranging from environmentally friendly *ḥalāl* foods including recipes through ethical fashion and eco-mosques to book reviews on ecological issues and both local and international news. The author titled her blog "the eco-jihadTM Enviro-news, halal living and eco-lifestyle from UK". In adopting the word "eco-jihad", she plays on Western associations with the term to tie it into the struggle for a healthy environment.

6 Declaration and Guides: The Islamic Declaration on Climate Change—The Green Guide to Hajj—Muslim Green Guide to Reducing Climate Change

Another example of activism in the name of Islam is provided by the Jordanian Queen Noor al-Hussein. Writing about a declaration published in the course of a two-day symposium held in Istanbul on Islam and climate change,[44] she clarifies why Muslims must take responsibility in advocating for the welfare of our planet:

> The Islamic Declaration on Climate Change, endorsed in August by Islamic scholars from around the world, calls on countries to phase out greenhouse gas emissions and switch to 100% renewable energy. […]
>
> First, the declaration calls on policymakers responsible for crafting the comprehensive climate agreement to be adopted in Paris to come to 'an equitable and binding conclusion'. The agreement should set clear targets and establish ways to monitor them. Additionally, prosperous countries and oil-producing states should phase out their carbon-dioxide

43 Annemarie Schwencke, *Globalized Eco-Islam. A Survey of Global Islamic Environmentalism* (Leiden: Leiden University, 2012), 4.

44 Ifees Ecoislam, "Islamic Declaration on Global Climate Change," accessed April 14, 2019. http://www.ifees.org.uk.

> emissions no later than the middle of the century; turn away from 'unethical profit from the environment'; and invest in a green economy.
>
> Second, the declaration asks people and leaders from all countries to commit to 100% renewable energy and a zero-emissions strategy as soon as possible, and to recognise that unlimited economic growth is not a viable option. Moreover, adaptation should be a high priority, particularly for the most vulnerable groups. Notably, the business sector is asked to take a more active role to reduce its carbon footprint, commit to 100% renewable energy and zero emissions, shift investments into renewable energy, adopt more sustainable business models, and assist in the divestment from fossil fuels.[45]

It is notable that the declaration, appealing to "all Muslims everywhere in the world" bases itself on Qur'ānic quotes. Here, too, the power of the text to legitimise is evident. A mindful treatment of creation, the appeal states, is a fundamental part of the Islamic message and humanity is responsible for not wasting the gifts of God.

Verse 24:45 is quoted describing God's creation of all living beings from the water, along with *ḥadīṯ* reinforcing the idea that humans are "stewards of the earth":

> The Earth is green and beautiful, and Allah has appointed you his stewards over it. The whole Earth has been created a place of worship, pure and clean. Whoever plants a tree and diligently looks after it until it matures and bears fruit is rewarded. If a Muslim plants a tree or sowed a field and humans and beasts and birds eat from it, all of it is love on his part.[46]

The declaration also refers to other religions, casting its message of environmentalism as a universal one. The Qur'ānic message, it states, reinforces what other faiths also teach: greater wisdom and beneficence in handling the planet and taking action on behalf of other beings.

Queen Noor closes:

> The Islamic Declaration on Climate Change, like Pope Francis's recent encyclical Laudato Si', is a call to humanity, regardless of faith, to work together to protect the planet upon which we depend. I hope that they

---

45 http://www.ifees.org.uk/wp-content/uploads/2018/02/Islamic_Declaration_V4.pdf.

46 Qantara, "Islam, Glaube und Klimawandel," accessed January 26, 2019. https://de.qantara.de/inhalt/oeko-islam-islam-glaube-und-klimawandel.

> will indeed provide an impetus for shifts in policies, allowing for deeper and broader reductions in CO_2 emissions. [...] We are all humans who live on the Earth and from it. We have all been entrusted to care for the planet. We can and must do this by forging bold climate and protective ocean agreements together—regardless of belief, circumstance, or status. Science tells us we have to act; our faith and humanity compel us to do so.[47]

The British Alliance of Religions and Conservation (ARC) commissioned the "Green Guide" to Hajj with Global One 2015 and EcoMuslim. Published in 2011, this guide aims to change behaviour especially during the Hajj itself, when pilgrims are in a sacred state that forbids them from hunting, killing animals, or harming and destroying trees and other plants. The Hajj to Mecca and the highest holidays of the Islamic year, the Feast of the Sacrifice (Arab. *ʿīd al-adha*) marking the end of the Hajj and the end of Ramadan (*ʿīd al-fitr*) offer occasions for Islamic environmental organisations to appeal to the ecological conscience of the faithful in religious terms.

The pilgrimage to Mecca is considered as one of the Five Pillars of Islam (Arab. *ʾarkānu ʾl-ʾislāmi ʾl-ḫamsatu*) and draws an estimated 2.5 million pilgrims who return home resolved to fulfil their religious duties all the more assiduously. This event is eminently suited to reach the hearts of the pilgrims and move them to embrace a more sustainable and environmentally friendly lifestyle. The preface to the guide was written by the Egyptian Great Mufti Ali Gomaa, and the book offers suggestions how to use only environmentally friendly products and services during the Hajj, to reduce consumption and waste, and to continue a more sustainable lifestyle after the end of the pilgrimage.

Pilgrims are called upon to forgo air travel or, where that is not possible, to donate to environmental rehabilitation projects that offset the carbon pollution caused. The "Green Guide" further asks them to go without plastic bottles entirely, to use public transport such as the Mecca Metro, and to make the pilgrimage once in their lives only, as is a Muslim's duty.

A further influential text, published as early as 2008 by Life Makers UK (founded by Amr Khaled) and the Birmingham IFEES, is the *Muslim Green Guide to Reducing Climate Change*, a practical guidebook for Muslims to

47 World Economic Forum, "How Islam can Help Combat Climate Change," accessed January 26, 2019. https://www.weforum.org/agenda/2015/09/how-islam-can-help-combat-climate-change/.

combat climate change in their daily lives.[48] Through this publication, the foundation gave its ideas of environmental Islam great media visibility. Its professionally designed newsletter "EcoIslam" presents an impressive synthesis of Islamic and environmental discourses. It offers a mix of background information on global contexts and concrete measures to take in everyday life. Advice on energy-efficient cooking and laundry, recycling, and travel by bicycle or public transport is combined with Qur'ānic quotations to demonstrate why Muslims must take action to protect the environment and that this is rooted in their faith.[49]

7 Muslims on Environmentalism in Public Discourse

So far, Muslims have not been very visible in public discourse on environmental issues. As yet, there have been no comprehensive studies on environmental consciousness among Muslims. They have not been identified as a relevant group in this context, or even considered unreachable for such issues. That is most likely because Muslims generally are not strongly represented in public discourse or visible organisations except occasionally as representatives of 'migrants'. Religious attitudes are viewed as a private matter, something not to be discussed in public in order to avoid compromising oneself. The issue is effectively taboo. Most Muslims are not even aware of the connection between environmentalism and religion, even when they see and perceive it.

8 Conclusion

The ideal of a Muslim environmental ethic contributing to a global movement to care for the planet is what Nasr, Özdemir, Khalid and all the other activists are promoting, each in their own way. These Muslim thinkers advocate a methodology shaping a contextual and practical theology towards a process that can create a sustainable global society.

As Nasr suggest, Islam is not limited to observing ritual. It implies a personal responsibility for the world that goes well beyond sectarian boundaries.

48 Environment and Ecology, "The Green Guide for Muslims," accessed March 26, 2019. http://environment-ecology.com/religion-and-ecology/76-the-green-guide-for-muslims.html.

49 Ifees Ecoislam, "Eco-Islam," accessed March 26, 2019. http://www.ifees.org.uk/category/eco-islam/.

When considering what a Muslim contribution to the intensely debated issue of environmental protection could look like once this connection is universally acknowledged, the idea of an 'Eco-Islam' is capable of providing impulses to engage with environmental questions and to reflect individual lifestyles and habits.

Environmentalism and climate activism are raised beyond the level of changing fashions to spiritual concerns. Large numbers of people who might otherwise not be open to these issues could be reached through mosque communities and Islamic organisations. Environmental discourses could be anchored here and networks with other activist groups and environmental organisations created. This would represent a great leap for the environmental cause.

We are already seeing early successes. There are solar-powered mosques, fair-trade products and green guides to the Hajj. Eco-Islam is gaining some traction.

The Muslim environmentalist movement has ambitious and universal goals. This is theologically consistent: If environmental policies are justified by respect for God's creation, then that respect must equally apply in gender relations. This is a beautiful example of holistic thinking and a promising approach for raising awareness on gender issues through the underlying discourse of justice that pervades environmental thought.

CHAPTER 2

Wonder, Gratitude, and Justice as Ingredients for a Jewish Eco-Theology

Ephraim Meir

Abstract

This chapter approaches ecology from the standpoint of Jewish religious values and Jewish philosophy. The philosophies of Abraham Joshua Heschel, Martin Buber, Joseph Ber Soloveitchik, and Abraham Yitzhak ha-Cohen Kook contain components that support a Jewish eco-theology. Additionally, the Bible, rabbinical legislation, prayer, and eco-kashrut provide us with ethical sensitivities and religious insights that bring heaven and Earth, God, and nature together.

1 Introduction

As a complex and multifaceted phenomenon, Judaism is characterised by a vast plurality of opinions and orientations. At first sight, the prevalent thought in Judaism on culture and ethics as vanquishing nature and opposed to it has little to contribute to the environmental crisis.[1] However, the idea that God is the creator of the world constitutes the kernel of Jewish environmentalism. If God created the world and continues to do so, nature and all the creatures therein have to be respected and preserved. Moreover, the Jewish view of nature as God's creation obliges humans to uphold a more equal distribution of water and food in favor of the wellbeing of all. With a divine creator and nature as a divine gift, it becomes conceivable to bring Judaism and ecology closely together.

Modern values such as democracy, ecological care or gender equality are not directly discussed in the sacred scriptures and rabbinic literature. Yet it is not too difficult to find inspiration for an environmental ethics in traditional Jewish sources. A creative interpretation and rereading of the sources

1 Paradigmatic for a radical critique of environmentalism which sacralises nature is Steven Schwarzschild, "The Unnatural Jew," *Environmental Ethics* 6 (1984): 347–362.

 | DOI:10.1163/9789004444973_004

is conducive to the protection of the environment and to a righteous use of natural resources.

Texts, customs, rituals, memory, and tradition have shaped a Jewish way of living that could develop a unique access to nature. Ecological issues concern the world, of which we are a part and which we need in order to exist. In the following, I will deal with ecological issues as perceived from a Jewish point of view. Generally speaking, environmentalism does not rank high in Jewish studies. It is not a central theme, although some excellent work has been done.[2] It is not the aim of this essay to formulate a comprehensive analysis of ecological issues in Jewish life and thought.[3] Its limited goal lies rather in the discussion of relevant elements in Jewish religious life and Jewish philosophy that contribute to ecological care and to the construction of an eco-theology. I approach ecology from the standpoint of Jewish religious values and Jewish philosophy.

2 Jewish Philosophy and Environmentalism

2.1 *Wondering*

The Jewish philosopher Abraham Joshua Heschel (1907–1972) contributed to Jewish environmentalism by writing on "radical amazement".[4] He deemed wonder a miracle. We are amazed "at the unexpectedness of being as such".[5] In Heschel's philosophy, the world is not there only to be measured, weighed, exploited, and mastered. It alludes, instead, to something higher. Yet we have become used to a scientific approach to the world in which there is no wonder anymore. A star-studded heaven no longer causes us to wonder since we

---

2 Hava Tirosh-Samuelson, "Jewish Environmentalism: Bridging Scholarship, Faith and Activism," in *Jewish Thought and Jewish Belief*, ed. Daniel Lasker (Beer Sheva: Ben Gurion University Press, 2012), 65–117 offers a good overview of the entire field. For a bibliography essay on Judaism and the environment, see Tirosh-Samuelson's bibliography essay on Judaism and the environment written for *Oxford Bibliographies Online*. The essay gives the reader an overview of the entire field.

3 Jewish thinkers such as Samson Raphael Hirsch, Isaac Breuer, Aharon David Gordon, Zalman Schachter-Shalomi, Arthur Waskow, and Arthur Green have contributed to a Jewish view on the environment. See Hava Tirosh-Samuelson, "Judaism and the Science of Ecology," in *The Routledge Companion to Religion and Science*, ed. James W. Haag, George R. Peterson, Michael L. Spezio (London: Routledge, 2011), 345–355.

4 Abraham J. Heschel, *Man Is Not Alone: A Philosophy of Religion* (New York: Farrar, Straus and Giroux, 1951), 11–17; Abraham J. Heschel, *God in Search of Man: A Philosophy of Judaism* (New York: Farrar, Straus and Giroux, 1976), 45–46.

5 Heschel, *Man Is Not Alone*, 12.

know more or less what is 'up there'. We look at nature through the lens of natural laws. We are now accustomed to regularity in nature governed by scientific laws.

In Heschel's depth-theology, nature is allusive and indicative, and "[l]ife without wonder is not worth living".[6] Heschel wondered about "the miracles which are daily with us", as in the Jewish prayer. He recognised the greatness of things familiar to us.[7] One may marvel at the view of the sunrise or a sunset. It is a wonder that occurs every day. True, it happens daily with great regularity, but Heschel's point is that we cannot get used to such a spectacle. A simple small flower may cause us to wonder. It is the product of many occurrences: it would not exist without air, rain, earth, and sun, and most of all, it is a divine gift. Heschel gives the example of a glass of water, which is perceived as one of God's continuous wonders.[8] This is the reason why a religious Jew, before drinking a glass of water, says the blessing "Blessed are You God, King of the world, by whose word all things come into being". Other blessings are said before eating bread or fruit, before drinking wine, on enjoying seasonal fruit for the first time, on seeing a rainbow or an ocean, etc. In the month of Nisan, when seeing a tree in blossom one says: "Blessed are You, Lord our God, king of the world, who reenacts the work of creation". Before smelling an aromatic plant, one says: "Blessed are You, Lord our God, king of the world, who creates fragrant plants". All these blessings are but the expression of a Jew's appreciation of nature as a divine gift.

The majesty of a tree caused Heschel to put his hat on. He related to trees with a feeling of nearness and reverence. He called an oak tree "grandpa" and, defined himself as a "grandchild" in its presence.[9] He taught people to exclaim in admiration: "How great are your works, God, how deep are your thoughts [*ma gadlu ma'asekha adonai meod amqu machshevotekha*[10]]".

Heschel told his audience that not everything has been said about reality with knowledge alone. Reality has a dimension of mystery which knowledge cannot lay bare: "The tree of knowledge and the tree of life have their roots

---

6 Heschel, *God in Search*, 46.

7 Ibid., 43.

8 Heschel refers to Ex. 15:11: "Who is like you, Lord, among the gods? Who is like you, majestic in holiness, sublime in glorious deeds, doing wonders?" He remarks that in this part of the Song of the Red Sea it is written "doing" wonders and not "did" wonders. He further quotes Rabbi Joshua ben Levi, who said: "The earning of bread is a greater wonder than the division of the Red Sea." (Midrash *Genesis Rabba* 20:22); Heschel, *God in Search*, 49, 53, n. 14.

9 Edward Kaplan, and Samuel H. Dresner, *Abraham Joshua Heschel—Prophetic Witness* (New Haven and London: Yale University Press, 1998), 88.

10 Ps. 92:5.

in the same soil. But, playing with winds and beams, the tree of knowledge often grows brilliant, sapless leaves instead of fruits. Let the leaves wither, but the sap should not dry up".[11] Mental life is therefore not restricted to perception and analytical reasoning. Before the ineffable we wonder.[12] The world is therefore to be approached not only in descriptive, but in indicative terms: the world is an allusion.[13] The self—part of nature—is "something transcendent in disguise".[14] Heschel writes: "Life is something that visits my body, a transcendent loan".[15] Therefore, we have to reeducate ourselves by learning again to wonder. In the daily prayer *shmone esreh* (the Amida prayer which contains 18 blessings) we say:

> We thank You [...] for the wonders that are daily with us and for your marvels and favors in all times, in the evening, in the morning and at midday [*modim anachnu lakh* [...] *al nisekha she-bekhol yom imanu ve-al niflotekha ve-tovotekha she-bekhol et erev va-boker ve-tsohoraim*].

For Heschel, the world is full to the brim with allusions to the transcendent reality. It is "replete with an infinite significance".[16] In this way, he offers a view upon nature that is radically different from the cold razor blade of science. One gives expression to this view, for instance, by singing or by writing poetry. The Bible, Heschel writes, "was not written *more geometrico* but in the language of poets".[17]

Heschel perceived God's presence in the natural world. The natural world possesses a spiritual dimension, and a religious person has to be educated to perceive this transcendent dimension in the natural world. Heschel did not sacralise nature. Nature in itself, he believed, was not divine, but rather had a higher meaning: it referred to God. In this way, Heschel's theology is essential if one wants to construct a full-fledged eco-theology, in which nature is not merely an object of manipulation.

Heschel was inspired by the Bible and the Jewish prayer that refers to nature as praising God.[18] In Ps. 148, the sun and moon, stars, earth and water, fire,

---

11 Heschel, *Man is not alone*, 14.
12 Ibid., 16–17.
13 Ibid., 19–23.
14 Ibid., 47.
15 Ibid., 48.
16 Ibid., 31.
17 Ibid., 37.
18 Evonne Marzouk, "Lessons from Nature: What our *Tefilla* Teaches Us about Our World," in *Compendium of Sources in Halacha and the Environment*, ed. Shenson, Ora, and Spetgant, Shai (Jerusalem: Canfei Nesharim, 2005), 54–58.

winds, trees, animals, and birds: all sing praises of God. In the Shabbat morning liturgy it says: "The soul of every living being (*nishmat kol chai*) shall bless your name, Lord our God, the spirit of all flesh shall always glorify and exalt your remembrance, our King."[19] Together with nature, human beings should praise God. Further in the Jewish liturgy, before *Shema* ("Listen Israel, the Lord is our God, the Lord is one"; Deut. 6:4), God cares for his continuous creation:

> He who illuminates the earth and those who dwell upon it, with compassion; and in his goodness renews daily, perpetually, the work of creation [*u-betuvo mechadesh bekhol yom tamid ma'ase bereshit*]. How great are your works, Lord, you make them all with wisdom, the world is full of your possessions.

In the Jewish religious experience, without God's daily work one could not live for a single moment. God gives abundantly. He lets the sun rise and lets rain pour upon all his creatures. A day of rain is deemed to be greater than the resurrection of the dead because the resurrection of the dead only concerns the righteous whereas rain falls upon both the righteous and the wicked.[20]

Following Heschel's sensitivity to the religious dimension in nature, spiritual thinkers of our time such as Zalman Schachter-Shalomi, Arthur Green, and Arthur Waskow developed a Jewish eco-theology.[21] With his non-dualistic, pantheistic thought, Schachter-Shalomi wanted to instill in his readers an earth-consciousness which he called a "Gayan consciousness": all was related to all, as if the entire universe was one great body in which human beings are only one part. Not respecting nature would mean a lack of respect for oneself. Arthur Green is another Jewish thinker who developed a mystical panentheism: God is present in all things as the One who turns into the many. Arthur Waskow, finally, is an eminent Jewish environmentalist—not only a scholar, but also a social activist.

2.2 *The Spirit Is "Between"*

Another seminal thinker who inspires a Jewish ecology is Martin Mordechai Buber (1886–1965). In his dialogical philosophy, spirit "is not in the 'I' but between 'I and you'. It is not like the blood that circulates in you but like the air in which you breathe".[22] In Buber's relational thinking, the 'I' may develop a

---

19 *Siddur Rinat Yisrael. Nusah Askhkenaz*, ed. Shlomo Tal (Jerusalem: Moreshet, 1983), 246.

20 Comp. Mt.5:45.

21 For a detailed discussion of these thinkers, see Tirosh-Samuelson, "Jewish Environmentalism," 95–108.

22 Martin Buber, *I and Thou* (New York: Charles Scribner's Sons, 1970), 89.

dialogical relationship toward nature. There is relatedness and interconnectedness between human beings and nature. 'I-you' may turn, for example, to a tree as a 'you'. In his holistic view, Buber sees relation as reciprocity. It is possible to contemplate a tree as a picture or a movement. One also can assign a tree to a species that follows the law of nature, or to count trees:

> Throughout all of this the tree remains my object and has its place and its time span, its kind and condition. But it can also happen, if will and grace are joined, that as I contemplate the tree I am drawn into a relation, and the tree ceases to be an It. The power of exclusiveness has seized me.[23]

Buber's attempt to overcome the subject-object dichotomy in his philosophy of the 'between' could serve as a remedy for a biased, hierarchical view of nature as merely an object to be exploited and mastered. It could heal our earth and our position towards nature. His famous contrast between 'I-it' and 'I-you' is helpful in overcoming our alienation from nature:

> Whoever says You does not have something for his object. For wherever there is something there is also another something; every It borders on other Its; It is only by virtue of bordering on others. But where You is said there is not something. You has no borders. Whoever says You does not have something; he has nothing. But he stands in relation.[24]

2.3 *Two Kinds of Man*

A third Jewish thinker whose thought is helpful to grow our environmental sensibilities is Rabbi Joseph Ber Soloveitchik (1903–1993). The distinguished Orthodox Rabbi called "the Rav" differentiates between the type of man depicted in Gen. 1 and the one of Gen. 2.[25] Adam the first, the man of the first creation story, is created in God's image and commanded to dominate the world. Adam the second is created from dust and his mandate is to cultivate and protect the garden (Gen. 2:15: "To till it and guard it" [*le-ovdah u-le-shomrah*]). Adam the first is creative and active; he constructs and transforms. Adam the second is contemplative and protective, and wonders at the miracle of nature.

---

23 Buber, *I and Thou*, 57–58.

24 Ibid., 55.

25 According to the documentary theory, the creation story of Gen. 1 belongs to the priestly tradition (P), whereas the second creation story of Gen. 2:4–24 belongs to the Jahwist (J). The second creation story, more specifically the divine command to "till and protect" the Garden of Eden (Gen. 2:15), has inspired Jewish environmentalists, who developed the idea of care for and custodianship of the earth.

The two types of Adam depicted by Soloveitchik are present in every man. They also appear in Ps. 8:

> When I look at your heaven, the work of your fingers, the moon and the stars, which you have established: what is man that you remember him, the son of man that you visit him? Yet, you made him a little lower than the angels, and you crowned him with glory and honor. You let him dominate over the works of your hands, you put all under his feet.

Adam the second is present in the first couple of verses (vv.4–5), Adam the first in the following verses (vv.6–7). Adam the second wonders at the beautiful nature made by God who takes care for every living being.[26] Marveling at the diversity in nature, he exclaims: "How numerous are your works. In wisdom you made them all. The earth is full of your possessions" (Ps. 104:24). Adam the first, on the contrary, is master of all animals (Ps. 8:8–9). He hears the words: "Fill the world and conquer it" (*u-milu et ha-arets ve-kivshuhah*; Gen. 1:28). Adam the first and Adam the second sing in choir: "YHWH, our Lord, how splendid is your name over all the earth" (Ps. 8:10). No wonder that Ps. 8:10 found its prominent place in the ritual of *Rosh ha-Shana*, the festival of New Year, whose central theme is the creation of the world.

The divine command to "till and guard" the Garden of Eden contributes to a Jewish environmental ethics.[27] Man has custodianship of the world. In Midrash *Qohelet Rabba* God took Adam around the trees of the Garden of Eden and said to him:

> Look at my works, how beautiful and praiseworthy they are. Everything that I created, I created for you; be careful not to damage and destroy my world, for if you damage it, there is no one to repair it after you.[28]

The delicate equilibrium between Adam the first and Adam the second prevents us from overconsumption and excessive exploitation. It shows the harmony between the active human being who dominates the powers of nature and the receptive human being, conscious of the precious gift of nature. It demands commitment to justice for all through a fair distribution of natural

---

26 See Ps. 104:27: "They all wait for you that you give them food in due time".

27 Ilana Stein, "Le'ovda Uleshomra: Judaism and the Environmental Ethic," in *Compendium of Sources in Halacha and the Environment*, ed. Ora Shenson and Shai Spetgant (Jerusalem: Canfei Nesharim, 2005), 16–20.

28 Midrash *Qohelet Rabba* 7:13 (*re'eh*).

resources as well as commitment to the preservation of nature. In view of the loss of biodiversity, the air and water pollution, and the excessive exploitation of natural resources, the two creation stories offer a sound basis for an environmental ethics and ecological awareness in which there is a balance between using and conserving.

Reflecting on the two creation stories, Reform Rabbi Daniel Fink remarks that in Gen. 1, the Earth is God's and man is created last. He refers to the Sages for whom the human being is the pinnacle of creation, but created after all the other creatures. Man is commanded to master the world, but he has ethical stewardship and should not do what he arbitrarily wants. In Gen. 2, *adam*, the earthling, is formed from *adama*, earth. He is commanded to work in the Garden of Eden and to preserve it, to guard God's creation. Shabbat, Fink notes, is the end, the goal of creation (Gen. 2:1–3): it is not only an interruption of work and nature is not merely there to be exploited.[29]

As noted at the start of this article, the relevance of creation for a Jewish ecology is paramount. Since God created the world for all, we have to care for others and not solely for our own interests in our use of nature. All are children of God who created the world for everyone. In this basic covenantal relationship between God and man, all have the right to live a healthy life, to have access to basic natural resources. Overproduction and pollution of air, rivers, lakes, and oceans go against the idea of God's creation.

Every Shabbat, every 7th day, we are commanded to leave nature as it is, without intervening, without work. for six days we intervene in nature, but the goal is to let nature praise God and to praise God with nature on the seventh. In this sense, Jewish life occurs between six days of work and one day of rest: this allows for intervention and active use of natural resources without forgetting the sabbatical aim when there is enough food for everyone.

2.4 *Rabbi Abraham Yitzhak ha-Cohen Kook*

Rabbi Avraham Yitzhak ha-Cohen Kook, the first chief Rabbi of Israel (1865–1935), is a fourth well known thinker whose approach to nature can inspire Jewish environmentalists. His holistic worldview is relevant for Jews who 'go green'. There is a famous story about Rabbi Kook who once walked in the fields with Rabbi Aryeh Levin, the *tsaddiq* (righteous man) of Jerusalem. The latter picked a flower, but was admonished by Rabbi Kook, who was careful not "to

29 Daniel B. Fink, "Judaism and Ecology: A Theology of Creation." The article is published on the internet at The Forum for Religion and Ecology at Yale "Judaism and Ecology: A Theology of Creation," accessed June 11, 2018. *fore.yale.edu/religion/judaism/*.

pluck a blade of grass or a flower needlessly, when it had the ability to grow or blossom". He continued:

> You know the teaching of our Sages that not a single blade of grass grows here on earth that does not have an angel above it, commanding it to grow. Every sprout and leaf says something meaningful. Every stone whispers some hidden message in the silence. Every creation sings its song.[30]

It would be interesting to compare Kook's mystic sensibility for every living thing with the practices and rituals of Jainism which extends *ahiṃsā* (non-violence) to the prohibition of harming any lifeforms and of Mahayana Buddhism, where *karuṇā* (compassion) is one of the two qualities of a bodhisattva—in addition to *prajñā* (wisdom)—who dedicates him- or herself to support all living beings out of love and compassion and to help them to attain *bodhi* (enlightenment). Within the Jewish life experience, the Hassidic idea that God is present in everything (*leit atar panuy mine*)[31] has the potential to lead to a positive and appreciative approach to nature.[32]

A great sensibility for the beauty and importance of nature in the land of Israel is palpable in the Israeli song *shirat ha-asabim* which was performed by the famous Israeli singer Naomi Shemer, who based her lyrical text on the words of Rabbi Nachman of Bratslav:

> Do know that each and every shepherd has his own melody. Do know that each and every grass has its own song. And from the song of the grasses a melody of the shepherd is made. How beautiful, how beautiful and pleasant to hear their song. It is very good to pray among them and to serve the Creator in joy. And from the song of the grasses the heart is filled and longs. And when the heart is filled by the song and yearns for the land of Israel then a great light is drawn forth from the land's holiness upon it. And from the song of the grasses a melody of the heart is made.

---

30 Simcha Raz, *A Tzaddik in our Time. The Life of Rabbi Aryeh Levin* (Jerusalem: Feldheim, 2008), 108–109.

31 The expression appears already in *Tiqqune ha-Zohar* 122:2.

32 See, however, Shaul Magid, "Nature, Exile, and Disability in R. Nahman of Bratslav"s 'The Seven Beggars'," in *Judaism and Ecology: The Created World and Revealed World*, ed. Hava Tirosh-Samuelson (Cambridge, MA: Harvard University Press, 2002), 333–368, who argues that Rabbi Nachman of Bratslav had a pejorative view of nature as basically imperfect.

3 Nature in Jewish Religious Life

3.1 *Ecological Paradise*

It has been claimed that Ezekiel's vision (Ez. 47:1–12) serves as a remedy for the industrial- or consumer-centered ecological destruction that leads to shortage and conflict. The solution to the problem of ecology therefore lies in reshaping one's attitude towards nature.[33] In this view, man as God's partner is responsible for the entire creation.

Ezekiel's vision may indeed be interpreted as a vision of an ecological paradise that comes into being from the water that flows from under the threshold of the temple. Of course, the water is Torah, the divine instruction, as is clear from Ps. 1:3: the man who has delight in the Torah is compared with a tree planted by streams of water, that brings forth its fruit in its season and whose leaf shall not wither. In Ezekiel's vision, the water grows to a large river. On the banks of the stream, many trees grow. When the water enters the sea, the sea will be healed and there will be a multitude of fish in it.

The rabbinical tradition that elaborates Biblical legislation can further contribute to a responsible attitude towards nature. The laws on Shabbat and on the Sabbatical year, the laws on domestic animals and the environment, and tithing for the poor have emphasised the duties of human beings towards nature which belongs to God.

3.2 *Prohibition of Wanton Destructiveness: Bal Tashchit*[34]

Deut. 20:19 forbids the destruction of fruit trees in case of the siege of a city. Fruit trees are meant to provide food for human beings. They are not to be cut down for sieges: "When you besiege a city a long time in making war against it to take it, you shall not destroy (*lo tashchit*) its trees by wielding an axe against them; for is the tree of the field a man, that it should be besieged by you [*ki ha-adam ets ha-sadeh lavo mipanekha bamatsor*]?"

From here it follows that, by extension, it is forbidden to destroy and to waste. Maimonides notes that the Torah prohibited all wanton destructiveness.[35] It is forbidden to uproot fruit trees in times of war. A fortiori, say the Rabbis, one may not cut down fruit trees in times of peace. This principle is further understood in a way that one may not use furniture for a fire, waste oil for a lamp, kill

---

33 Yehudah Levi, "Ecological Problems: Living on Future Generations' Account," in *Compendium of Sources in Halacha and the Environment*, ed. Ora Shenson and Shai Spetgant (Jerusalem: Canfei Nesharim, 2005), 27.

34 Levi, "Ecological Problems," 23–24.

35 *Hilkhot Malakhim* 6, *Halakhot* 8 and 10.

animals without reason, demolish a house in which one may live, etc. One may not spill water or waste food. A Jew should respect nature and not demolish the environment on which he depends. In the words of *Sifre Devarim* reflecting on the above passage from Deut. 20:19: "The life of a person is from the tree [*chayyav shel adam enam ela min ha-ilan*]."

It follows further that destruction out of anger is forbidden. The Babylonian Talmud 105b preserves a saying of Rabbi Shimeon, son of Elazar, "who said in the name of Hilfa son of Agra who transmitted in the name of Rabbi Yohanan, son of Nuri: the one who tears his clothes in anger, breaks a vessel in anger, or scatters his money in anger, should be in your eyes as an idolater [*ke-oved avoda zara*]". In the same vein, one shall not let one's emotions prevail and mourn excessively for the deceased by tearing a garment too much and not preserving it (Baba Kama 91b).[36]

The biblical prohibition of cutting down a fruit tree has become the basis for an entire ecological programme.[37] Yet, the prohibition is not absolute. If a tree damages the property of a neighbor, one may chop it down (Baba Batra 26a). And if wood is needed in order to make a fire for a sick person (Shabbat 129a), it is permitted to cut down trees. Nature is not only there to be preserved, it is also there in order to facilitate social justice and for the welfare of all.

In Israel, where I live, water is precious. In the daily Amidah prayer (the 18 blessings) we pray for rain. One shall be careful not to waste water. If one can recycle water, one should do so. The Zohar discusses commandments that relate to recycling. After the Festival of Tabernacles, the *etrog* (citrus) is used as *besamim* (the fragrant spices for the ceremony of marking the end of Shabbat) and the *lulav* (palm tree) is used to fuel the fire in which all the leaven is burnt before Pesach. By extension, one should recycle water in order to wash a car, water flowers, etc.[38]

3.3 *Eco-Kashrut*

The rules on kosher food are complex. One of the great principles is the prohibition to consume the blood of a bird or animal (Lev. 7:26; 17: 10–14) because the soul of any flesh is its blood; *ki nefesh kol basar damo hu* (Lev. 17:14). When shedding the blood of a bird or an animal, one shall cover the blood with dust

36 Ellen Cohn, "Growing an Environmental Ethic: The Conceptual Roots of Bal Tashchit," in *Compendium of Sources in Halacha and the Environment*, ed. Shenson, Ora, and Spetgant, Shai (Jerusalem: Canfei Nesharim, 2005), 41.

37 Tirosh-Samuelson, "Jewish Environmentalism," 99.

38 Shmuel Simenowitz, "Water Conservation and Halacha: An Unorthodox Approach," in *Compendium of Sources in Halacha and the Environment*, ed. Ora Shenson and Shai Spetgant (Jerusalem: Canfei Nesharim, 2005), 52.

(Lev. 17:13). These rules prevented Israel from becoming indifferent towards the death of animals or birds.

The rules of kashrut become even more complex if one broadens kashrut to eco-kashrut.[39] It was Rabbi Zalman Schachter-Shalomi who invented the term eco-kashrut. This kind of kashrut takes into account ethics by checking, for instance, the working conditions of employees in industrial agriculture.

I mentioned the principle of *bal tashchit* which forbids destruction and waste. Another principle is that of *tsa'ar ba'ale hayyim*, the prohibition to cause animals to suffer.[40] This principle is applied in eco-kashrut. It is forbidden to eat flesh from living animals (Deut. 12:23). In Deut. 22:6–7 we read:

> If a bird's nest happens to be before you on the way, in any tree, or on the ground, whether it is filled with young ones or eggs, and the mother bird is sitting upon the young or upon the eggs, you should not take the mother bird together with the young [*lo tiqach ha-em al ha-banim*]. Surely let the mother go and take the young [*shalach teshalach et ha-em ve-et ha-banim tiqach lekha*], so it will be good for you and you will prolong your days.

According to halakhic law, it is forbidden to take eggs from a bird's nest without chasing away the brooding bird. Further, one has to leave two eggs in the nest so that they do not realise eggs were taken. Related to this prohibition is another prohibition against slaughtering an animal and its young in one day because it is a merciless and cruel act. A cow, for instance, should not see the slaughtering of the calf.

In eco-kashrut, one applies the principle of *tsa'ar ba'ale chayyim* to the food industry. Thus one has to buy only products that are the result of good, ethical treatment of birds and animals. In many factory farms, however, animals do not enjoy good conditions: they receive unhealthy food and are prophylactically given antibiotics in order to prevent infections. Based upon Deut. 11:15, "And I will provide grass in your field for your cattle and, then, you will eat and be satisfied [*ve-natati esev be-sadekha li-behemtekha ve-akhalta ve-savata*]", it is forbidden to eat before feeding your animals.

The principle of *tsa'ar ba'ale chayyim* is also applied in kosher slaughtering (*shechita*), which minimises the animals' suffering. Kosher slaughtering is

---

39 See Tamarah Benima, *Joodser dan dit krijgt u het niet. De levenskunst van de Joodse Beschaving* (Amsterdam: Prometheus, 2015), 137–140.

40 Atara Weisberger, "An Exploration of Tza'ar Ba'alei Chaim," in *Compendium of Sources in Halacha and the Environment*, ed. Ora Shensonand Shai Spetgant (Jerusalem: Canfei Nesharim, 2005), 30–33.

based upon the prohibition to eat blood, which is the soul of the animal. One has to be sensitive to the animals, which have a *nefesh chaya*, a living soul. Vegetarianism is a further step towards mending the world.[41] According to the Halakha, it is forbidden to cause suffering to the animal. In Deut. 25:4 we read: "Do not muzzle an ox when it is treading the corn" (Deut. 25:4). The prohibition to plow a field with an ox and a donkey yoked together also follows from the prohibition of causing pain to animals (Deut. 22:10). An ox and a donkey do not work well together. Today, one has to take care that animals do not suffer in the stable or during transport.

Another principle, that of the *osheq* (oppressor), is important for a broader concept of kashrut. Halakha forbids employers to exploit employees. In eco-kashrut, social and economic justice plays a central role.[42]

Midrash *Shemot Rabba* 2:2 relates that Moses tended Yitro's flock when a little kid went astray. Moses ran after it and found it drinking water from a pool. Moses said that he did not know that this was the reason she ran way. He took the kid on his shoulder and brought it back to the flock. Therefore, God said to Moses: "Because you are merciful in leading a flock, so you will surely shepherd my flock, Israel".

3.4 *Public Domain and Climate Protection*

The Babylonian treatise Baba Kama 50b relates a beautiful short story that teaches the inescapable responsibility for the public domain: "It happened that a certain person was removing stones from his ground [*me-reshuto*] onto public ground [*li-reshut ha-rabbim*] when a pious man found him and said, 'Fool, why are you removing stones from ground which is not yours [*me-reshut she-ena shelkha*] to ground which is yours [*li-reshut shelkha*]?' The man laughed at him. Some time later he was forced to sell his field, and when he was walking on that public ground [*be-oto reshut ha-rabbim*], he stumbled over those stones. He then said, 'How well did that pious man say to me, Why are you removing stones from ground which is not yours [*me-reshut she-ena shelkha*] to ground which is yours [*li-reshut selkha*]?'"

This amazing story teaches that one is responsible for the public domain as much as for one's private domain. In a way, the pious man who reprimands the perpetrator puts into question the very distinction between the private and the public domain. In fact, the private ground is not private since one is not

41 See Richard Schwartz, *Judaism and Vegetarianism* (New York: Lantern, 2012).

42 See Arthur Waskow, "What Is Eco-Kosher?," in *This Sacred Earth: Religion, Nature, Environment*, ed. Gottlieb, Roger S. (New York: Routledge, 1996), 297–300.

allowed to do on it whatever one wants, and the public domain is one's personal responsibility.

According to this brief, powerful Talmudic story, it is forbidden to treat public ground as if it were one's private domain. The public domain is for the common good. This way of looking at what is common to all has consequences, especially for our responsibility for climate change. It is not only smoking, making excessive noise or spreading bad smells that hurt the public domain. Creating green belts around cities and protecting the ozone layer follow from the laws pertaining to the public domain which are succinctly formulated as "Do not remove stones from ground that is yours to public ground" (*lo yesalleq adam me-reshuto li-reshut ha-rabbim*). Ownership is not absolute. The public domain and its resources belong to all. To care for what belongs to all is a personal responsibility. In religious terminology: God created the world and we are the stewards of His creation.

3.5 *Shabbat*

From a Jewish perspective, God created the world with a view to the Shabbat. In Gen. 2:1–3 we read:

> And the heaven and the earth were completed and all their array. And God completed by the seventh day his work He did, and He rested on the seventh day from all his work He did. And God blessed the seventh day and He hallowed it, for thereon He abstained (*shabbat*) from all his work that God created to do (*la-asot*).

According to this biblical passage which every religious Jew recites in the *Qiddush* (hallowing) of Shabbat eve, the universe has a purpose. After the creation of space, time is hallowed: Shabbat is the crown on the divine creation. In a Jewish reading of the text, the words "to do" (*la-asot*) mean "to do henceforth." In this reading, Shabbat is the celebration of God's creation whose purpose lies in human freedom realised by not intervening in nature. Imposing limits on our use of nature creates freedom.

In Hechel's beautiful booklet on Shabbat,[43] the seventh day as a reminder of creation is detached from the world of space. Unlike the various festivals which are determined by natural events, Shabbat is not regulated by the position of the moon. It is an eternity in time. On Shabbat, not nature, but time is hallowed. It is an invitation to become sensitive again to the mystery in creation.

---

43 Abraham J. Heschel, *The Sabbath* (New York: Farrar, Straus and Giroux, 1988).

In *imitatio Dei*, Israel is commanded to hallow the seventh day as a token (*ot*) of the "eternal covenant" (*berit olam*) between God and Israel (Ex. 31:13 and 16–17; comp. Ex. 20:7–11). Like God, who rested from all the work He did and who hallowed this day, Israel shall desist from work on Shabbat. Sabbatical freedom as freedom from work and intervention in nature is meant to create freedom for all. Human dignity does not only stem from work and interference in nature. A large part of the world has adopted the Jewish view on the day of rest on which all are equally free by abstaining from intervening in nature.

3.6 *God and Nature*

Spinoza famously equated "God" (*elohim*) and "nature" (*ha-teva*).[44] He saw matter and spirit as two aspects of the one and only substance: *Deus sive natura*. There are pantheistic and panentheist views in Judaism which bring nature and God together. A more classical Jewish view on nature distinguishes between a transcendent God and nature: God can be found in nature, but He is also beyond it; nature belongs to God (*la-Shem ha-arets u-melo'ah*; Ps. 24:1). Whatever one thinks about the precise relationship between God and nature, a Jewish eco-theology adheres to the command to approach nature from the perspective of social justice. Nature is there for all human beings and, therefore, the equal distribution of food and water is a duty. The traditional Jewish approach to nature as creation, as a gift of God (*ve-ha-arets natan li-bne adam*; Ps. 115:16), is one of responsibility for the Earth. It is a remedy for materialism, consumerism, and the lack of sensitivity for sentient beings. The poor and the underserved are the privileged categories of the love of God who created nature for all of us. A Jewish spirituality is a spirituality that knows about the ladder of Jacob which was "set up on the earth, and the top of it reached to heaven; and behold the angels of God ascending and descending on it" (Gen. 28:12). It concerns body and soul and links heaven and Earth, God and nature.

4 Outlook

Modern Jewish philosophers provide us with the tools to develop environmental sensibilities. Moreover, the Jewish tradition offers thoughts and laws that regulate man's behavior towards nature. A Jewish eco-theology is, therefore, nurtured by tradition and philosophy. Such an eco-theology cannot function

---

44 Moshe Idel, "Deus Sive Natura: The Metamorphosis of a Dictum from Maimonides to Spinoza," in *Maimonides and the Sciences*, ed Robert S. Cohen and Hillel Levine (Dordrecht: Kluwer, 2000), 87–100.

without special care for climate. It is well known that fossil fuel burning, massive deforestation, and massive industrialisation contribute to global warming. Global warming leads to climate change, extreme weather conditions, and the thinning of glaciers. Factories pollute the environment and endanger and harm the health of people.

Global warming challenges us. We are responsible for the well-being of future generations. One does not have to be a Jewish eco-theologian in order to be conscious of man's responsibility for the environment. Eco-theology offers the spiritual resources to provide motivation to safeguard our planet and its inhabitants.

From a Jewish perspective, humans are stewards of the Earth, responsible for God's creation. They are called to repair the world: to protect the natural environment from pollution, to care for the ozone layer which defends us against the damaging rays of the sun and to safeguard the Earth and the people upon it.

The saying "For the earth is Mine" (Lev. 25:23) implies that our exploitation of the Earth cannot be absolute, since one has to take care of God's creation, i.e., of the common good. The rights of the individual are restricted when it comes to the protection of the public domain. A climate-conscious behavior, love of nature, and the struggle against ecological evils are the appropriate reaction to the divine gift of Earth and of all life. In a Sabbatical existence, one is invited to rediscover the mystery in creation which is a gift for all and in which all human beings can breathe, recreate themselves, and be there for others.

PART 2

Past

CHAPTER 3

Holocene Climate Variability of Mesopotamia and its Impact on the History of Civilisation

Max Engel and Helmut Brückner

Abstract

Mesopotamia, known as the cradle of civilisation, gave rise to the first complex, urban-type societies with sophisticated political hierarchies. Its rich history full of important cultural achievements was accompanied by fundamental environmental changes over the Holocene. While geo-bio-archives from the broader region reflect slightly varying climate histories, there is a clear consensus on a more humid climate regime during the Early Holocene, triggering early rain-fed agricultural practices in Northern Mesopotamia that present the foundation of initial urbanisation. In the southern basin, declining rainfall and higher competition for natural resources at a somewhat later stage in combination with the development of irrigation techniques and the transgression of the Arabian Gulf seem to have contributed to the formation of complex societies at sites such as Eridu, Ur, and Uruk, where landscape dynamics are well-preserved in the stratigraphic record. Against the background of long-term climate trends, it seems that Rapid Climate Change events—short-term climatic anomalies such as identified around 8200, 5200, or 4200 BP—have also taken their toll on Mesopotamians.

Many links between changes in climate and landscape, and socio-technical adaptation based on interdisciplinary research seem straightforward, especially where confirmation exists through cuneiform texts or archaeological evidence. The gap in chronological resolution between rather precise information on historical social development on the one hand and on climatic changes with a much higher uncertainty on the other hand may render the 'fit' between records elusive and advocates caution in any attempt at environmental determinism when trying to explain cultural history.

1 Introduction

Ancient Mesopotamia, homeland of the oldest complex, urban-type societies and the nucleus of early civilisation, covers most of the catchment of the two great rivers Euphrates and Tigris and extends from southeastern Turkey

 | DOI:10.1163/9789004444973_005

through Syria and the territory of modern Iraq down to the shores of the Arabian Gulf (map 3.1). The first mention of Mesopotamia as a geographic entity occurred in the 2nd century CE, appearing in Arrian of Nicomedia's *Anabasis of Alexander* where it is associated with the part of Syria which is located between the two rivers. It is further stated that, even though the administrative unit was first established by Alexander, the name itself derives from the inhabitants and goes back at least to the 4th century BCE before it was translated from Aramaic into Greek.[1]

1.1 *Landscapes of Mesopotamia*

The Euphrates River, formed by the confluence of the Kara Su and Murat rivers north of Lake Van (site 4 in map 3.1), is the longest stream in southwestern Asia (2,700 kilometres), fed by winter rainfall in the southeastern Taurus Mountains. The second longest river in this part of the world is the Tigris River (1,840 kilometres), which also has its source in the southeasternmost Taurus and receives contributions by several tributaries joining from the east along its way.[2] While the Tigris occupies an entrenched valley and is limited to a single stream, the Euphrates tends to separate into a number of channels.[3] There are two major flooding peaks over the year, i.e. November to March resulting from immediate surface runoff, and April to May mainly related to snowmelt.[4] Interannual discharge variability is high, ranging from 1.9×10^{10} to 6.3×10^{10} cubic metres (Tigris) and 1.0×10^{10} to 3.6×10^{10} cubic metres (Euphrates), measured during relatively unaltered streamflow settings between 1924 and 1946 (fig. 3.1a).[5]

Geographically, Mesopotamia is separated into five main physiographic regions: (1) The northern and northeastern mountain region, (2) the foothills, (3) the western desert, (4) the central *Ǧazīra* region, and (5) the southern

1 Jacob J. Finkelstein, "Mesopotamia," *Journal of Near Eastern Studies* 21 (1962): 73. doi: 10.1086/371676.

2 Richard W. Berry, Gerald P. Brophy, and Adnan Naqash, "Mineralogy of the Suspended Sediment in the Tigris, Euphrates, and Shatt-al-Arab Rivers of Iraq, and the Recent History of the Mesopotamian Plain," *Journal of Sedimentary Research* 40 (1970): 131. doi:10.1306/74D71F05-2B21-11D7-8648000102C1865D; Heidi M. Cullen, and Peter B. deMenocal, "North Atlantic Influence on Tigris—Euphrates Streamflow," *International Journal of Climatology* 20 (2000): 856. doi:10.1002/1097-0088(20000630)20:8<853::AID-JOC497>3.0.CO;2-M.

3 Galina S. Morozova, "A Review of Holocene Avulsions of the Tigris and Euphrates Rivers and Possible Effects on the Evolution of Civilizations in Lower Mesopotamia," *Geoarchaeology* 20 (2005): 406. doi:10.1002/gea.20057.

4 Cullen, and deMenocal, "Influence on Tigris—Euphrates Streamflow," 856.

5 Wafiq H. Al-Khashab, *The Water Budget of the Tigris and Euphrates Basin* (Chicago: University of Chicago Press, 1958), 48.

lowlands.[6] These regions are usually grouped to form Upper (~1–4, where 2 and 4 form the core region) and Lower (~5) Mesopotamia. Upper Mesopotamia mostly comprises steppe vegetation and desert shrub rangelands and was mainly characterised by extensive rain-fed farming in ancient times.[7] The northern and northeastern mountains comprise the northern part of the northwest-southeast-trending folded Upper Palaeozoic and Mesozoic Zagros Mountains and the southern part of the east-west-oriented Northern Taurus fold belt, reaching elevations of up to 3,500 metres above mean sea level (a.s.l.) (map 3.1). They are separated from the central *Ǧazīra* region by the gently folded foothill zone of Upper Miocene to Pliocene coarse debris.[8] The *Ǧazīra* region consists of a low, undulating plateau dropping westwards from five hundred metres at Jabal Sinjār to less than one hundred metres a.s.l. (map 3.1). Surface deposits are of Miocene age and comprise sands, silts, clays, limestones, and gypsum, the latter concentrated in numerous salt pans.[9]

The alluvial and deltaic landscapes of Lower Mesopotamia cover much of the Mesopotamian depression, a flexure zone created by the upward-thrusting Zagros Mountains to the northeast and framed by the Arabian Shield to the southwest.[10] They are dominated by active and abandoned, low-gradient alluvial plains, braided, anastomosing, and meandering river channel patterns, levees rising up to three to four metres above the surrounding flood basins, as well as numerous crevasse splays (alluvial fans formed at a levee break).[11] Arid

---

6 Al-Khashab, *Water Budget of the Tigris and Euphrate Basins*, 4.

7 Tony J. Wilkinson, "Regional Approaches to Mesopotamian Archaeology: The Contribution of Archaeological Surveys," *Journal of Archaeological Research* 8 (2000): 222. doi:10.1023/A:1009487620969; Tony J. Wilkinson et al., "Contextualizing Early Urbanization: Settlement Cores, Early States and Agro-pastoral Strategies in the Fertile Crescent During the Fourth and Third Millennia BC," *Journal of World Prehistory* 27 (2014): 95. doi:10.1007/s10963-014-9072-2.

8 Berry, Brophy, and Naqash, "Mineralogy of rivers of Iraq," 131.

9 Derk C.P. Thalen, *Ecology and Utilization of Desert Shrub Rangelands in Iraq* (The Hague: Dr. W. Junk Publishers, 1979), 73–74. doi:10.1007/978-94-009-9622-9.

10 Elazar Uchupi, Stephen. A. Swift, and David A. Ross, "Late Quaternary Stratigraphy, Paleoclimate and Neotectonism of the Persian (Arabian) Gulf Region," *Marine Geology* 160 (1999): 1. doi:10.1016/S0025-3227(99)00011-0; Jennifer R. Pournelle, "Physical Geography," in *The Sumerian World*, ed. Harriet Crawford (London: Routledge, 2013), 13. doi: 10.4324/9780203096604.ch1.

11 Kristiaan Verhoeven, "Geomorphological Research in the Mesopotamian Flood Plain," in *Changing Watercourses in Babylonia. Towards a Reconstruction of the Ancient Environment in Lower Mesopotamia*, vol. 1, ed. Hermann Gasche and Michel Tanret (Chicago: University of Chicago Press, 1998), 160–168; Wilkinson, "Regional Approaches to Mesopotamian Archaeology," 222; Morozova, "Holocene Avulsions," 405.

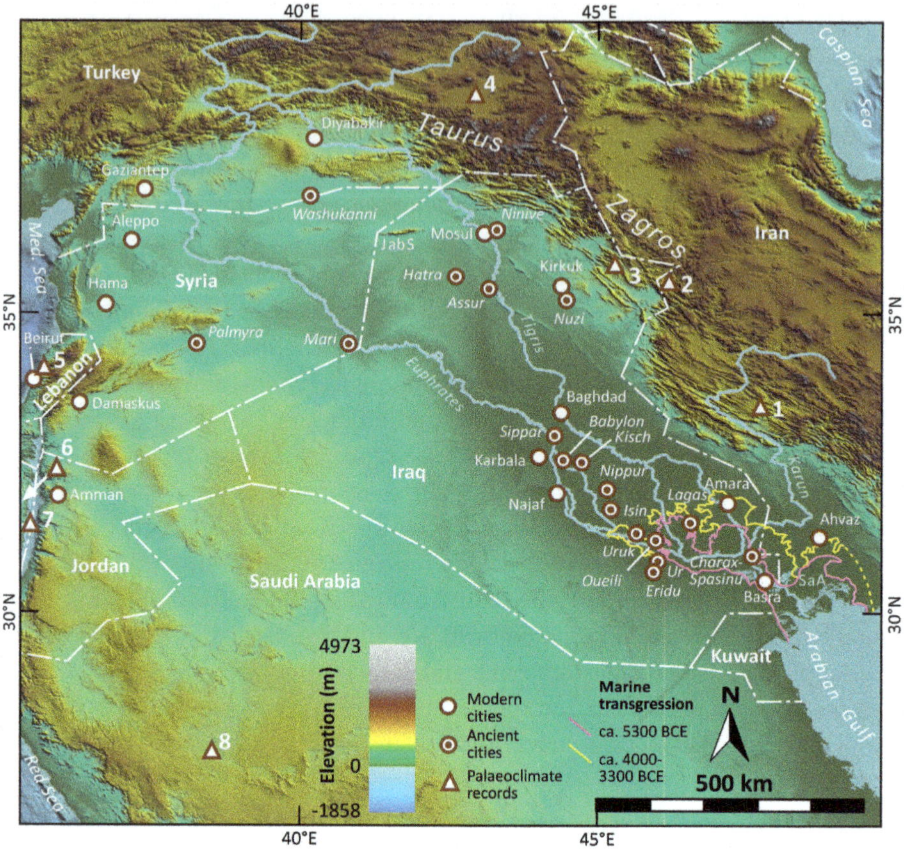

MAP 3.1 Overview of Mesopotamia and surrounding regions showing important ancient and modern cities, as well as sites of relevant palaeoclimate records. On the Mesopotamian delta plain, reconstructed shorelines of the Mid-Holocene transgression (5300 BCE and 4000–3300 BCE)[a] are shown. The map is based on the GEBCO_2014[b] grid and SRTM30[c] data provided by the US Geological Survey. Palaeoclimate records: 1=Lake Mirabad (Iran); 2=Lake Zeribar (Iran); 3=Geijkar Cave (Iraq); 4=Lake Van (Turkey); 5=Jeita Cave (Lebanon); 6=Soreq Cave (Israel); 7=Dead Sea (Israel); 8=Tayma (Saudi Arabia). SaA=Shatt al-Arab. JabS=Jabal Sinjār.

Notes:

a Jennifer R. Pournelle, "Marshland of Cities: Deltaic Landscapes and the Evolution of Early Mesopotamian Civilization" (PhD diss., University of California, San Diego, 2003), 123–124. doi:10.6067/XCV8J67GK3.

b Pauline Weatherall et al., "A New Digital Bathymetric Model of the World's Oceans," *Earth and Space Science* 2 (2015): 331–345. doi:10.1002/2015EA000107.

c Tom G. Farr et al., "The Shuttle Radar Topography Mission," *Reviews of Geophysics* 45 (2007): RG2004. doi:10.1029/2005RG000183.

landforms such as sand dunes, deflation basins and salt pans are expanding[12] where areas have been cut off from alluvial influence through the migration of channels (avulsion processes) induced by flash floods and long-term channel deposition over the last centuries and millennia,[13] and by an increase in irrigation measures further up the rivers.[14] The southern part of Lower Mesopotamia is occupied by large-scale wetlands (*ahwar*) at the confluence of the Euphrates and the Tigris, which, however, have been extensively drained and reduced to approximately ten per cent of their original size.[15] The delta of the Shatt al-Arab River consists of very young, Late Holocene deposits,[16] and its progradation is the result of relative sea-level fall in the Arabian Gulf after ca. 6000–5000 BP[17] in combination with high sediment loads, in particular supplied by the Karun River with its catchment reaching up into the Zagros Mountains[18] (map 3.1).

1.2 *Cornerstones of Mesopotamian History*

These differences in physical settings are to some extent reflected by asynchronous settlement patterns in both parts of Mesopotamia. Even though strong

---

12 Morozova, "Holocene Avulsions," 405.

13 Vanessa M.A. Heyvaert and Cecile Baeteman, "A Middle to Late Holocene Avulsion History of the Euphrates River: A Case Study from Tell ed-Dēr, Iraq, Lower Mesopotamia," *Quaternary Science Reviews* 27 (2008): 2401. doi:10.1016/j.quascirev.2008.08.024; Jaafar Jotheri, Mark B. Allen, and Tony J. Wilkinson, "Holocene Avulsions of the Euphrates River in the Najaf Area of Western Mesopotamia: Impacts on Human Settlement Patterns," *Geoarchaeology* 31 (2016): 183ff. doi:10.1002/gea.21548.

14 Adnan A.M. Aqrawi, "Stratigraphic Signatures of Climatic Change during the Holocene Evolution of the Tigris-Euphrates Delta, Lower Mesopotamia," *Global and Planetary Change* 28 (2001): 269. doi:10.1016/S0921-8181(00)00078-3.

15 Aqrawi, "Holocene Evolution of the Tigris-Euphrates Delta," 268–269; Morozova, "Holocene Avulsions," 405; Mohamed A. Hamdan et al., "Vegetation Response to Re-flooding in the Mesopotamian Wetlands, Southern Iraq," *Wetlands* 30 (2010): 178. doi:10.1007/s13157-010-0035-9.

16 Uchupi, "Stratigraphy, Paleoclimate and Neotectonism," 12.

17 Kurt Lambeck, "Shoreline Reconstructions for the Persian Gulf since the Last Glacial Maximum," *Earth and Planetary Science Letters* 142 (1996): 47. doi:10.1016/0012-821X(96)00069-6; Max Engel and Helmut Brückner, "The South Qatar Survey Project (SQSP)—Preliminary Findings on Holocene Coastal Changes and Geoarchaeological Archives," *Zeitschrift für Orient-Archäologie* 7 (2014): 299.

18 Aqrawi, "Holocene Evolution of the Tigris-Euphrates Delta," 279; Vanessa M.A. Heyvaert and Cecile Baeteman, "Holocene Sedimentary Evolution and Palaeocoastlines of the Lower Khuzestan Plain (Southwest Iran)," *Marine Geology* 242 (2007): 102–105. doi:10.1016/j.margeo.2007.01.008. For further reading, the most comprehensive overview of the fluvial landscapes of the Mesopotamian lowlands is provided by Verhoeven, "Changing Watercourses in Babylonia," 160–181.

ties existed throughout history between the upper and the lower basin, a more or less distinct cultural boundary can be drawn north of present-day Baghdad along Jabal Sinjār and related anticlinal structures.[19] The more humid north provided the opportunity for rain-fed agriculture and shows a higher density of Early Ubaid sites (fig. 3.2); however, lower counts in the south may also be related to high rates of alluvial deposition and burying of sites.[20] Settlement trends in the north during the Chalcolithic show intraregional variability followed by a supraregional trend in centralisation and increased population density in the hinterland during the Early Bronze Age. However, agro-pastoralism may still have played a significant role across both phases, in particular in more arid areas of the east and southeast.[21] It appears that population growth first emerged in the steppe environments outside the alluvial plain, but rapidly reached Lower Mesopotamia, where a settlement hierarchy seems to have developed throughout the Ubaid period into the 4th millennium BCE. At that time, urban-type structures have been identified in Lower Mesopotamia by some authors,[22] as well as a similar degree of urbanisation in Upper Mesopotamia from ca. 2600 BCE. Diverging opinions exist on the Early Dynastic II/III times and later, where Wilkinson places the peak state of urban-type centralisation in both parts of Mesopotamia lasting until the termination of the Old Babylonian period.[23] Brinkman, however, emphasises a quasi-continuous reduction of the settled area and an increase in the number of rural settlements (<10 ha) already taking place at that time[24] which, in other sources, is often linked only with the mid-2nd millennium BCE and later phases. Politically induced internal migration and dispersal during the Middle and Late Assyrian period resulted in further decentralisation towards the 9th–7th centuries BCE. Population then

19 Margarete van Ess, Helmut Becker, and Jörg Fassbinder, "Uruk—Verortung in Raum und Zeit," in *Uruk—5000 Jahre Megacity. Begleitband zur Ausstellung „Uruk—5000 Jahre Megacity" im Pergamonmuseum. Curt-Engelhorn-Stiftung für die Reiss-Engelhorn-Museen Mannheim, dem Deutschen Archäologischen Institut—Orient-Abteilung, der Deutschen Orient-Gesellschaft e. V. und dem Vorderasiatischen Museum—Staatliche Museen zu Berlin*, ed. Nicole Crüsemann et al., (Petersberg: Imhof Verlag, 2013), 42.

20 Wilkinson, "Regional Approaches to Mesopotamian Archaeology," 243–244.

21 Wilkinson, "Contextualizing Early Urbanization," 57ff., 73.

22 Douglas J. Kennett and James P. Kennett, "Influence of Holocene Marine Transgression and Climate Change on Cultural Evolution in Southern Mesopotamia," in *Climate Change and Cultural Dynamics: A Global Perspective on Mid-Holocene Transitions*, ed. David G. Anderson, Kirk A. Maasch, and Daniel H. Sandweiss, 230 (San Diego: Academic Press, 2007). doi:10.1016/B978-012088390-5.50012-1.

23 Wilkinson, "Regional Approaches to Mesopotamian Archaeology," 243–247.

24 John A. Brinkman, "Settlement Surveys and Documentary Evidence: Regional Variation and Secular Trend in Mesopotamian Demography," *Journal of Near Eastern Studies* 43 (1984): 172–174. doi:10.1086/373078.

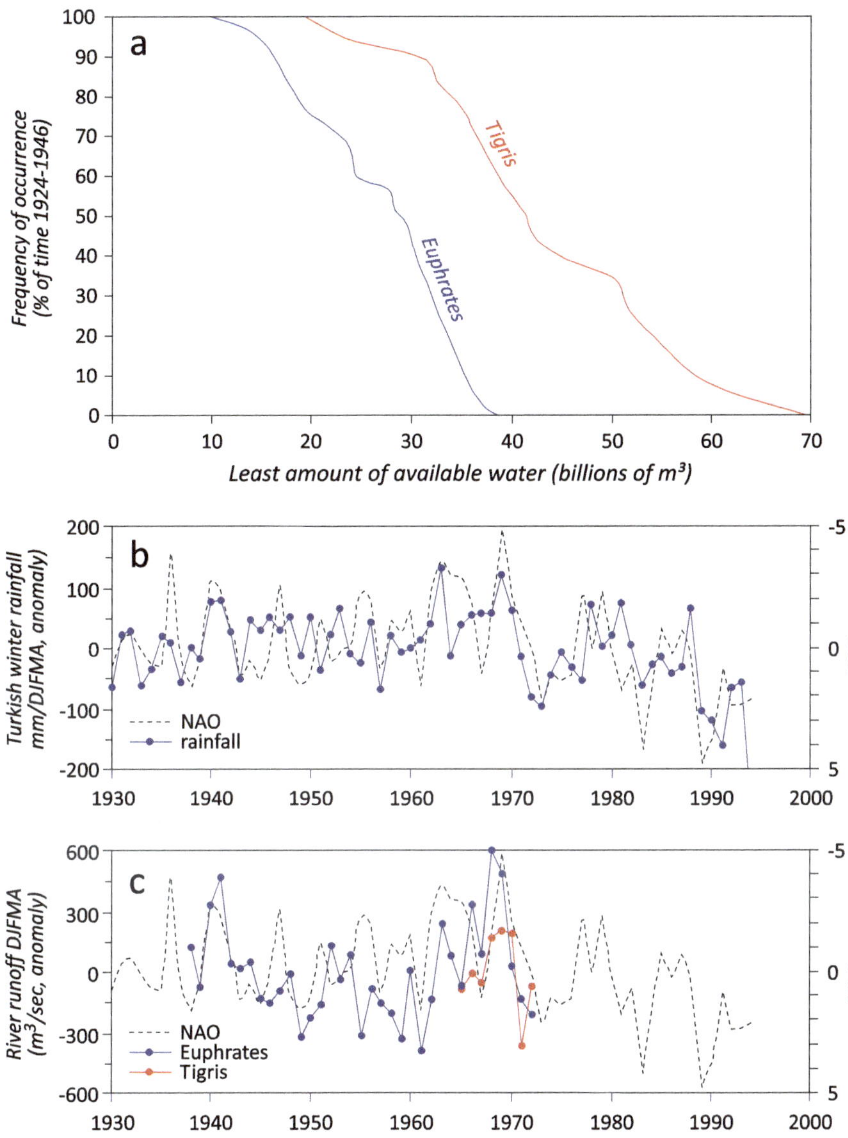

FIGURE 3.1 a) Frequency of occurrence of annual water supply by the Ephrates and Tigris rivers between 1924 and 1946 showing a broad range and high interannual variability;[a] b) correlation between the North Atlantic Oscillation (NAO) and winter rainfall over Turkey from December to April. Activity of the NAO is deduced from an index considering the difference between the normalized mean winter air pressure at sea level between Lisbon, Portugal (Azores High), and Stykkisholmur, Iceland (Icelandic Low) (NAO_{slp}). The NAO_{slp} is multiplied by -1 for better comparison; c) correlation between NAO_{slp} and average streamflow of the Euphrates and Tigris rivers during December–April.[b]

Notes:

a Al-Khashab, *Water Budget of the Tigris and Euphrates Basin*, 48–49.

b Cullen, and deMenocal, "Influence on Tigris—Euphrates Streamflow," 860.

increased during the late 1st millennium BCE to the early 1st millennium CE, in particular in Lower Mesopotamia, where water management practices provided the foundation for higher settlement densities and a moderate re-establishment of more centralised settlement structures. Demographic stability persisted until the turn of the millennium, when population across Mesopotamia decreased (fig. 3.2) in favour of newly expanding centres such as Baghdad, Samarra (both Iraq), and Raqqa (Syria).[25]

1.3 *Aims*

In this contribution, we follow the geoarchaeological approach, i.e. to investigate how the evolution of landscapes and ecosystems inferred from analyses of geo-bio-archives may have influenced settlement patterns, cultural developments, and land use.[26] We review state-of-the-art data on Holocene climatic changes in Mesopotamia and adjacent regions, summarise their impact on vegetation, hydrology, sedimentation, and landforms, and relate these natural dynamics to human responses as deduced from historical and archaeological sources. A local focus is put on Uruk in Lower Mesopotamia. In the sense of Verhoeven et al.,[27] we hypothesise that the "floodplain processes must have determined [...] the possibilities and limits of the Hydraulic Civilizations of the Mesopotamian Plain."

2 Holocene Climates of Mesopotamia and Adjacent Regions

Mesopotamia is situated within the subtropical high-pressure belt, associated with an arid to semiarid climate (map 3.2). Cyclones migrating from the Mediterranean into the Levant during winter provide rainfall for Upper Mesopotamia and, to a lesser extent, to Lower Mesopotamia. These low-pressure systems may even penetrate as far as Afghanistan in the east and Oman in the southeast. In summer, the moist westerlies move northwards and are replaced by downwelling air masses which induce high atmospheric pressure and generally hot and dry conditions.[28] The arid climate follows a

25 Wilkinson, "Regional Approaches to Mesopotamian Archaeology," 243–247.

26 Max Engel and Helmut Brückner, "Late Quaternary Environments and Societies: Progress in Geoarchaeology," *Zeitschrift für Geomorphologie N.F* 58.2 (2014): 1. doi:10.1127/0372-8854/2014/S-00168.

27 Verhoeven, "Changing Watercourses in Babylonia," 162.

28 Tony J. Wilkinson, *Archaeological Landscapes of the Near East* (University of Arizona Press, 2003), 17.

long-term, gradual trend: Flohr et al. found that the magnitude of the drought events of 1998–2000 and 2007–2010 were unprecedented since 950 CE based on a speleothem record of climate covering the Common Era.[29]

Rainfall in Upper Mesopotamia amounts to more than 250 millimetres per year in most parts. From Mosul in the north, situated on the Tigris, down to Baghdad at the border between Upper and Lower Mesopotamia, the rainfall gradient ranges from more than 400 to 150 millimetres per year, while interannual variability is high. Seasonal variability in mean temperatures is high as well, ranging from 7°C in January to 33°C in July for Mosul and 10°C in January to 35°C in July for Baghdad (map 3.2). Evaporation is substantial, in particular during the summer months, adding up to more than 3,000 millimetres per year.[30]

Lower Mesopotamia has a much drier climate. Annual rainfall decreases to 140 millimetres per year at Basra in the Shatt al-Arab region, where the two great rivers merge less than one hundred kilometres away from the Gulf coastline. Mean temperatures measured at Basra range from 12.2°C in January to 34.5°C in July,[31] when the area is subject to dust storms[32] and some of the most severe heat waves worldwide.[33] Evaporation is significantly lower in Lower Mesopotamia, reaching around 2,000 millimetres per year at Basra, where relative air humidity is among the highest in Iraq throughout the year due to its proximity to the Arabian Gulf coast.[34]

2.1 *General Climate Trends during the Holocene*

As climate strongly influences surface hydrology and strategies in crop cultivaton and pastoralism, especially in prehistorical and historical times, reliable records of climatic changes during the last millennia are crucial to understand ancient land-use patterns. The most complete datasets stem from

29 Pascal Flohr et al., "Late Holocene Droughts in the Fertile Crescent Recorded in a Speleothem from Northern Iraq," *Geophysical Research Letters* 44 (2017): 1534. doi:10.1002/2016GL071786.

30 Thalen, *Desert Shrub Rangelands in Iraq*, 56, 60, 66.

31 Hamdan et al., "Mesopotamian Wetlands," 178.

32 Kennett, and Kennett, "Holocene Marine Transgression and Climate Change," 231.

33 Christoph Schär, "Climate Extremes: The Worst Heat Waves to Come," *Nature Climate Change* 6 (2016): 129. doi:10.1038/nclimate2864.

34 Thalen, *Desert Shrub Rangelands in Iraq*, 64–66.

geo-bio-archives such as from (palaeo-)lake basins[35] and caves[36] to the north and west of the Mesopotamian heartland, and marine records from the eastern Mediterranean Basin[37] or the northern Red Sea.[38] Nearby records from the Zagros Mountains, such as those from the lakes of Mirabad and Zeribar, however, do not seem to adequately reflect lowland Mesopotamian climate as these sites received significant amounts of spring rainfall from continental

35 Lucia Wick, Gerry Lemcke, and Michael Sturm, "Evidence of Lateglacial and Holocene Climatic Change and Human Impact in Eastern Anatolia: High-resolution Pollen, Charcoal, Isotopic and Geochemical Records from the Laminated Sediments of Lake Van, Turkey," *The Holocene* 13 (2003): 665–675. doi:10.1191/0959683603hl653rp; Max Engel et al., "The Early Holocene Humid Period in NW Saudi Arabia—Evidence from Sediments, Microfossils and Palaeo-hydrological Modelling," *Quaternary International* 266 (2012): 131–141. doi:10.1016/j.quaint.2011.04.028; Thomas Litt et al., "Holocene Climate Variability in the Levant from the Dead Sea Pollen Record," *Quaternary Science Reviews* 49 (2012): 95–105. doi:10.1016/j.quascirev.2012.06.012; Michèle Dinies et al., "When the Desert was Green: Grassland Expansion during the Early Holocene in Northwestern Arabia," *Quaternary International* 382 (2015): 293–302. doi:10.1016/j.quaint.2015.03.007; Michèle Dinies et al., "Holocene Vegetation in Northwestern Arabia—Changing Natural Resources," in *Actualités des recherches archéologiques en Arabie*, ed. Julie Goy et al., (Paris: Routes de l'Orient, 2016), 1–19.

36 Miryam Bar-Matthews and Avner Ayalon, "Mid-Holocene Climate Variations Revealed by High-resolution Speleothem Records from Soreq Cave, Israel and their Correlation with Cultural Changes," *The Holocene* 21 (2011): 163–171. doi:10.1177/0959683610384165; Miryam Bar-Matthews et al., "Sea-land Oxygen Isotopic Relationships from Planktonic Foraminifera and Speleothems in the Eastern Mediterranean Region and their Implication for Paleorainfall during Interglacial Intervals," *Geochimica et Cosmochimica Acta* 67 (2003): 3181–3199. doi:10.1016/S0016-7037(02)01031-1; Sophie Verheyden et al., "Paleoclimate Reconstruction in the Levant Region from the Geochemistry of a Holocene Stalagmite from the Jeita Cave, Lebanon," *Quaternary Research* 70 (2008): 368–381. doi:10.1016/j.yqres.2008.05.004.

37 Martine Rossignol-Strick, "The Holocene Climatic Optimum and Pollen Records of Sapropel 1 in the Eastern Mediterranean, 9000–6000 BP," *Quaternary Science Reviews* 18 (1999): 515–530. doi: 10.1016/S0277-3791(98)00093-6; Kay-Christian Emeis et al., "Temperature and Salinity Variations of Mediterranean Sea Surface Waters over the Last 16,000 Years from Records of Planktonic Stable Oxygen Isotopes and Alkenone Unsaturation Ratios," *Palaeogeography, Palaeoclimatology, Palaeoecology* 158 (2000): 259–280. doi: 10.1016/S0031-0182(00)00053-5.

38 Helge W. Arz et al., "Mediterranean Moisture Source for an Early-Holocene Humid Period in the Northern Red Sea," *Science* 300 (2003): 118–121. doi:10.1126/science.1080325; Helge W. Arz, Frank Lamy, and Jürgen Pätzold, "A Pronounced Dry Event Recorded around 4.2 ka in Brine Sediments from the Northern Red Sea," *Quaternary Research* 66 (2006): 432–441. doi:10.1016/j.yqres.2006.05.006.

sources before and after the expansion of the 'classical' Mediterranean climate with typical winter precipitation during the Mid-Holocene.[39]

One of the most common geochemical indicators used to infer past climatic changes is the ratio of oxygen isotopes ^{18}O and ^{16}O ($\delta^{18}O$), which is driven by fractionation processes during evaporation and, therefore, is an independent measure of relative humidity.[40] Terrestrial $\delta^{13}C$ mostly reflects changes in phytomass and the ratio between C3 (most plant taxa; lower values) and C4 (grasses, sedges; higher values) type plants in the vicinity of the climate archive.[41]

A key site of Quaternary climatic change in the eastern Mediterranean and Near East is Lake Van, located in the eastern Taurus Mountains of southeastern Turkey (map 3.1). It underlies a climatic regime comparable to Mesopotamia—apart from higher annual rainfall and a longer rain period from autumn to spring—but is very sensitive to climatic fluctuations controlled by shifts of the jet stream and the subtropical high-pressure belt.[42] High $\delta^{18}O$ values from Lake Van strongly correlating with the Mg/Ca ratio—a measure of lake salinity—indicate very arid conditions during the Younger Dryas and the earliest Holocene, followed by a sudden shift to increased winter precipitation at around 10500 cal. BP (fig. 3.2).[43] $\delta^{18}O$ records from speleothems of the karstic caves of Soreq (Israel)[44] and Jeita (Lebanon),[45] both exposed to a genuine Mediterranean climate, show a more gradual increase in rainfall at that time. The Early Holocene shows a gradual increase in moisture availability over the entire Near East, as inferred from a rising lake level[46] and internal lake productivity reflected by total organic carbon (TOC), as well as a rise in grass pollen at Tayma (after 9300 cal BP)[47] on the northern Arabian Peninsula. While the

39 Lora A. Stevens et al., "Timing of Atmospheric Precipitation in the Zagros Mountains Inferred from a Multi-proxy Record from Lake Mirabad, Iran," *Quaternary Research* 66 (2006): 498–499. doi:10.1016/j.yqres.2006.06.008.

40 Wick, Lemcke, and Sturm, "Sediments of Lake Van," 667.

41 Bar-Matthews et al., "Paleorainfall during Interglacial Intervals," 3181.

42 Wick, Lemcke, and Sturm, "Sediments of Lake Van," 666.

43 Ibid., 673.

44 Bar-Matthews et al., "Paleorainfall during Interglacial Intervals," 3190.

45 Verheyden et al., "Stalagmite from Jeita Cave," 378.

46 Engel et al., "Holocene humid period in NW Saudi Arabia," 136; Anna Pint et al., "How to Discriminate Athalassic and Marginal Marine Microfaunas? Foraminifera and other Fossils from an Early Holocene Continental Lake in Northern Saudi Arabia," *Journal of Foraminiferal Research* 47 (2017): 179. doi:10.2113/gsjfr.47.2.175. The age model used in both papers is biased by the hard water effect; calibrated ^{14}C data are ~1000–1500 years too old based on the new age model from Tayma in Dinies et al. "Holocene Vegetation in Northwestern Arabia," 8.

47 Dinies et al., "Grassland Expansion in Northwestern Arabia," 297; Dinies et al., "Holocene Vegetation in Northwestern Arabia," 6.

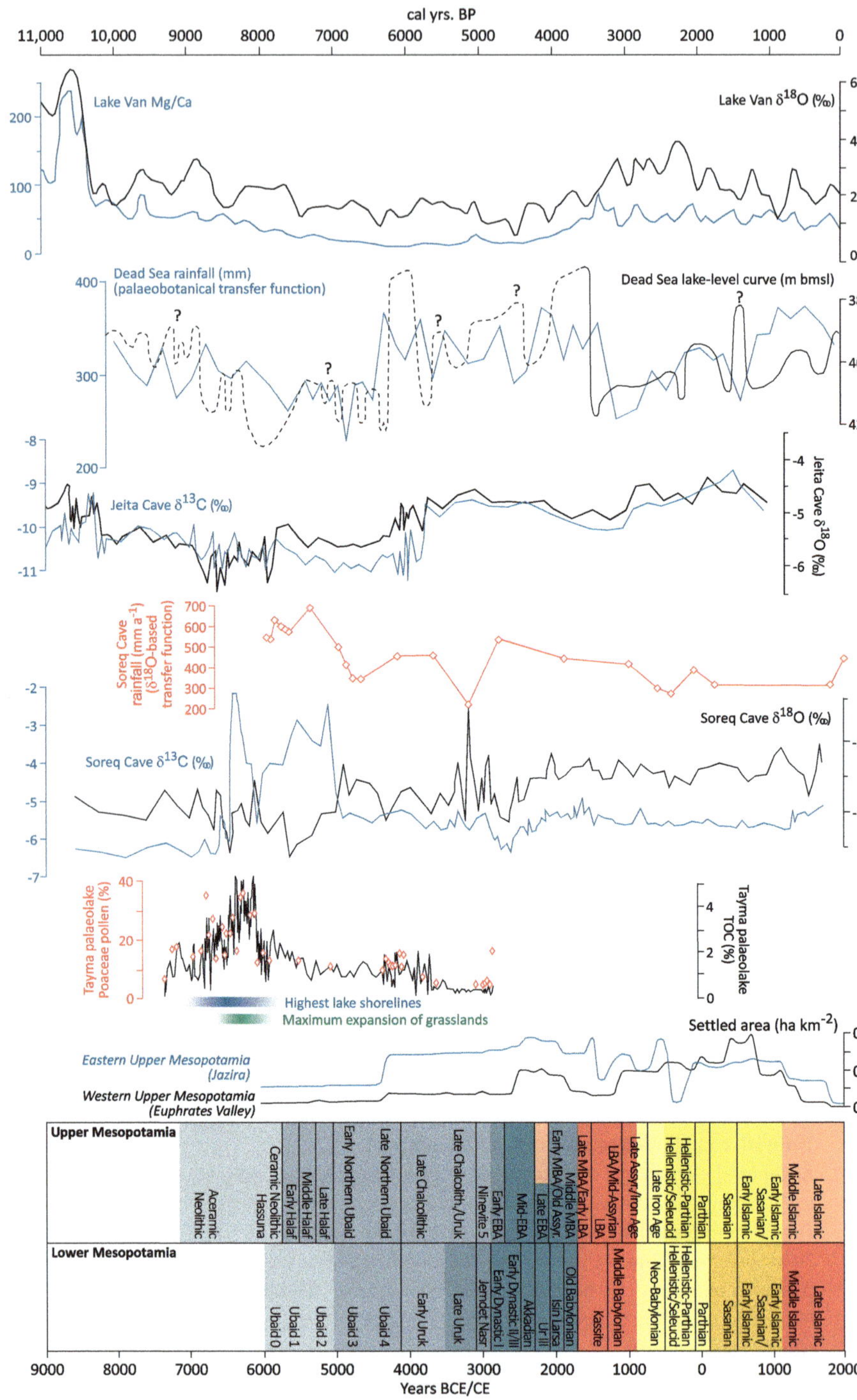
cal yrs. BP
Lake Van Mg/Ca
Lake Van δ18O (‰)
Dead Sea rainfall (mm) (palaeobotanical transfer function)
Dead Sea lake-level curve (m bmsl)
Jeita Cave δ13C (‰)
Jeita Cave δ18O (‰)
Soreq Cave rainfall (mm a-1) (δ18O-based transfer function)
Soreq Cave δ18O (‰)
Soreq Cave δ13C (‰)
Tayma palaeolake Poaceae pollen (%)
Tayma palaeolake TOC (%)
Highest lake shorelines
Maximum expansion of grasslands
Settled area (ha km-2)
Eastern Upper Mesopotamia (Jazira)
Western Upper Mesopotamia (Euphrates Valley)
Upper Mesopotamia
Aceramic Neolithic
Ceramic Neolithic Hassuna
Early Halaf
Middle Halaf
Late Halaf
Early Northern Ubaid
Late Northern Ubaid
Late Chalcolithic
Late Chalcolith./Uruk
Ninevite 5
Early EBA
Mid-EBA
Late EBA
Early MBA/Old Assyr.
Middle MBA
Late MBA/Early LBA
LBA
LBA/Mid-Assyrian
Late Assyr./Iron Age
Late Iron Age
Hellenistic/Seleucid
Hellenistic-Parthian
Parthian
Sasanian
Sasanian/ Early Islamic
Early Islamic
Middle Islamic
Late Islamic
Lower Mesopotamia
Ubaid 0
Ubaid 1
Ubaid 2
Ubaid 3
Ubaid 4
Early Uruk
Late Uruk
Jemdet Nasr
Early Dynastic I
Early Dynastic II/III
Akkadian
Ur III
Isin Larsa
Old Babylonian
Kassite
Middle Babylonian
Neo-Babylonian
Years BCE/CE

FIGURE 3.2 Holocene climate-related proxy records from sites around Mesopotamia (see map 3.1 for location of sites): A record of $\delta^{18}O$ and Mg/Ca ratios from Lake Van,[a] a lake-level curve and a rainfall transfer function based on pollen spectra from lacustrine deposits of the Dead Sea (Ein Gedi),[b] $\delta^{18}O$ and $\delta^{13}C$ records of a stalagmite from Jeita Cave,[c] $\delta^{18}O$ and $\delta^{13}C$ records and a $\delta^{18}O$-based transfer function for annual rainfall from Soreq Cave,[d] and a TOC (total organic carbon) record and relative abundance of Poaceae pollen (grasses)[e] combined with phases of highest lake shorelines[f] and maximum grassland expansion[g] at Tayma. The records are synoptically displayed with estimates of settlement density[h] and urbanization trends during prehistorical and historical periods of Upper and Lower Mesopotamia.[i] A key to colours indicating settlement trends is shown in Figure 3.3.

Notes:

a Wick, Lemcke, and Sturm, "Sediments of Lake Van," 670.
b Litt et al., "Dead Sea Pollen Record," 101.
c Verheyden et al., "Stalagmite from Jeita Cave," 377.
d Bar-Matthews et al., "Paleorainfall during Interglacial Intervals," 3195.
e Dinies et al., "Holocene Vegetation in Northwestern Arabia," 6.
f Engel et al., "Holocene Humid Period in NW Saudi Arabia," 135.
g Dinies et al., "Grassland Expansion in Northwestern Arabia," 298.
h Dan Lawrence et al., "Long Term Population, City Size and Climate Trends in the Fertile Crescent: A First Approximation," *PLOS ONE* 11 (2016): 8. doi:10.1371/journal.pone.0157863.
i Wilkinson, "Regional Approaches to Mesopotamian Archaeology," 243.

moisture pattern over time is generally similar to the classic Mediterranean climate sites, it is not yet entirely clear whether enhanced rainfall in northern Arabia is related to the African Summer Monsoon penetrating further eastward,[48] winter-spring tropical plumes crossing the Red Sea,[49] and/or Meditteranean winter rains.[50] Likewise, salinity decreased with the onset of the Holocene in the eastern Mediterranean Basin.[51] Conflicting evidence, however, arises from the Dead Sea record, where Litt et al. established a rainfall

48 Maria Guagnin, et al., "Rock Art Imagery as a Proxy for Holocene Environmental Change: A View from Shuwaymis, NW Saudi Arabia," *The Holocene* 26 (2016): 1829. doi:10.1177/0959683616645949; Max Engel, et al., "Lakes or Wetlands? A Comment on 'The Middle Holocene Climatic Records from Arabia: Reassessing Lacustrine Environments, Shift of ITCZ in Arabian Sea, and Impacts of the Southwest Indian and African Monsoons' by Enzel et al.," *Global and Planetary Change* 148 (2017): 265. doi:10.1016/j.gloplacha.2016.11.001.

49 Yehouda Enzel, Yochanan Kushnir, and Jay Quade, "The Middle Holocene Climatic Records from Arabia: Reassessing Lacustrine Environments, Shift of ITCZ in Arabian Sea, and Impacts of the Southwest Indian and African Monsoons," *Global and Planetary Change* 129 (2015): 71. doi:10.1016/j.gloplacha.2015.03.004.

50 Arz et al., "Early-Holocene Humid Period in the Northern Red Sea," 121.

51 Emeis et al., "Temperature and Salinity Variations," 259–280.

transfer function based on the pollen record correlating with reconstructed Dead Sea lake levels. The authors infer warm and dry conditions from 10000 to 6500 cal. BP and explain disparities by processes other than climate-driven isotope fractionation that may influence isotopic source-water composition in the eastern Mediterranen Sea.[52]

The Early to Mid-Holocene transition is associated with a period of more humid conditions in nearly any palaeoclimate record from the Near East, though its beginning and end vary. North of Upper Mesopotamia, Lake Van experienced relatively humid conditions until 4000 cal. BP with a climatic optimum around 6200–4000 cal. BP. At Soreq Cave, high precipitation occurred 8500–7000 years ago,[53] while the Jeita Cave stalagmite shows the lowest $\delta^{18}O$ and $\delta^{13}C$ values, the highest growth rates, and therefore most humid conditions between 9200–6500 years ago. Northern Arabia experienced the peak of the humid phase at 8700–8000 cal. BP, followed by an abrupt shift to drier conditions indicated by both a decline in TOC related to lower lake-internal bioproductivity[54] and a fall of the lake level at Tayma. A short window of slightly enhanced moisture availability at Tayma appears to have occurred 6300–5900 cal. BP.[55] Overlapping with this latter phase, the Dead Sea pollen record suggests highest Holocene rainfall amounts lasting for more than 3000 years from 6500 to 3300 cal. BP, before more arid conditions set in again. Both the humid phase and a long-term increase in rainfall during the subsequent drier phase strongly correlate with Dead Sea lake levels.[56]

The onset of significantly drier conditions during the Mid- and Late Holocene occurred around 4200–4000 cal. BP at Lake Van, when the Mediterranean winter rains decreased and a more continental climate gained influence.[57] At Jeita, rainfall decreased markedly from 6500 to 5800 BP and then more gradually until the end of the record (1100 years ago), with a temporary change to a somewhat more humid climate in the 4th millennium BP.[58] The long-term trend inferred from Soreq Cave shows increasing aridity during the last 7000 years,[59] corroborated by the high-resolution record for the last 2400 years from Geijkar Cave.[60] More recent insights into speleothem records at higher

52 Litt et al., "Dead Sea Pollen Record," 102–103.
53 Bar-Matthews et al., "Paleorainfall During Interglacial Intervals," 3195.
54 Dinies et al., "Holocene Vegetation in Northwestern Arabia," 11.
55 Engel et al., "Holocene Humid Period in NW Saudi Arabia," 135.
56 Litt et al., "Dead Sea Pollen Record," 102–103.
57 Wick, Lemcke, and Sturm, "Sediments of Lake Van," 673.
58 Verheyden et al., "Stalagmite from Jeita Cave," 378–380.
59 Bar-Matthews et al., "Paleorainfall during Interglacial Intervals," 3195.
60 Flohr et al., "Late Holocene Droughts in the Fertile Crescent," 1532.

resolution from Soreq Cave, however, reveal considerable millennial- as well as superimposed centennial to decadal-scale fluctuations within these long-term trends.[61] This reflects a highly sensitive and unstable climate regime.[62] While the cyclicity of 1500 years is related to Bond cycles driven by North Atlantic Oscillations (NAO),[63] decadal climatic changes, so-called Rapid Climate Change events (RCC) are associated with high-latitude cooling, low-latitude aridity, and major atmospheric circulation changes.[64]

2.2 *Rapid Climate Change Events* (RCC)

Compared to gradual climatic trends, short-term changes or climatic events, respectively, have often triggered strong responses in ancient societies. The 8200 BP global RCC, a climatic deterioration with clear imprints in Greenland ice core records,[65] seems to have coincided with an abrupt increase of moisture availability in the growing season based on the Lake Van record.[66] Bar-Matthews et al. inferred opposed conditions, i.e. sudden cooling and drying 8200–8000 years ago at Soreq Cave based on a slight increase in $\delta^{18}O$ and sharp decrease in $\delta^{13}C$ (fig. 3.2).[67] This discrepancy, however, was attributed to local soil denudation and considered not representative for regional climate variability by Verheyden et al., who observed no such drying and cooling signal at Jeita Cave.[68] A later higher-resolution speleothem analysis at Soreq Cave by Bar-Matthews and Ayalon[69] correlates the first stepdown in rainfall inferred at Jeita (6500–5800 BP)[70] with distinct decadal-scale arid events 6650–6600 and 6250–6180 years ago (fig. 3.3).

An abrupt drought event around 5200–5100 years ago was identified in the Late Pleistocene-Holocene record of Soreq Cave[71] and is corroborated by

61 Bar-Matthews, and Ayalon, "Mid-Holocene Climate from Soreq Cave," 168.
62 Bar-Matthews et al., "Regional Events at Soreq Cave," 91.
63 Gerald Bond, et al., "A Pervasive Millennial-scale Cycle in North Atlantic Holocene and Glacial Climates," *Science* 278 (1997): 1263. doi:10.1126/science.278.5341.1257.
64 Paul A. Mayewski et al., "Holocene Climate Variability," *Quaternary Research* 62 (2004): 251. doi: 10.1016/j.yqres.2004.07.001.
65 Elizabeth R. Thomas et al., "The 8.2 ka Event from Greenland Ice Cores," *Quaternary Science Reviews* 26 (2007): 72. doi:10.1016/j.quascirev.2006.07.017.
66 Wick, Lemcke, and Sturm, "Sediments of Lake Van," 673.
67 Bar-Matthews et al., "Regional Events at Soreq Cave," 91.
68 Verheyden et al., "Stalagmite from Jeita Cave," 379.
69 Bar-Matthews, and Ayalon, "Mid-Holocene Climate from Soreq Cave," 168.
70 Verheyden et al., "Stalagmite from Jeita Cave," 379.
71 Bar-Matthews et al., "Regional Events at Soreq Cave," 89; Bar-Matthews et al., "Paleorainfall during Interglacial Intervals," 3195.

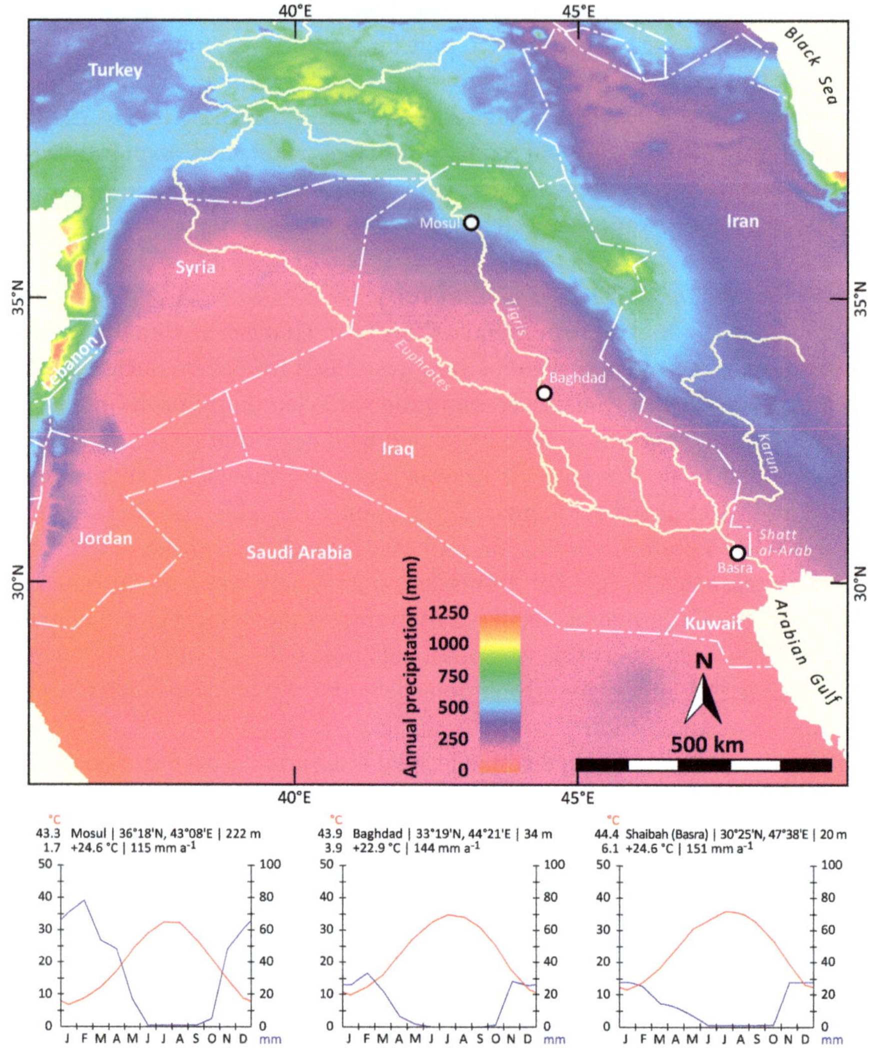

MAP 3.2 Distribution of annual precipitation in the Near East based on high-resolution, spatially interpolated monthly rainfall data of the WorldClim database (2.5 min. resolution).[a] Climate diagrams of Mosul, Baghdad and Basra were adapted from the Worldwide Bioclimatic Classification System.[b]

Notes:

a Stephen E. Fick and Robert J. Hijmans, "WorldClim 2: New 1-km Spatial Resolution Climate Surfaces for Global Land Areas," *International Journal of Climatology* (2017). doi:10.1002/joc.5086.

b Salvador Rivas-Martinez and Salvador Rivas-Saenz, "Worldwide Bioclimatic Classification System, 1996–2017. Phytosociological Research Center, University Complutense of Madrid, Spain," last modified January, 2017, accessed July 14, 2017. http://www.pendientemigracion.ucm.es/info/cif/plot/diagram.htm.

Bar-Matthews and Ayalon (slightly shifted: 5250–5170 BP)[72] as well as the $\delta^{13}C$ record of barley grains from 33 ancient Mesopotamian sites (5200 cal. BP).[73] Bar-Matthews and Ayalon[74] argue that this RCC is part of the widespread dry period recognised all over southeastern Europe and the Mediterranean around 6000–5000 years ago[75] and that the peak drought during this phase coincides with the broader peak in $\delta^{13}C$ and $\delta^{18}O$ dated to 5700–5600 years ago.

The RCC at around 4200 BP, with cross-continental recognition in Africa, West Asia and the North Atlantic area, was identified in sediments from the northern Red Sea as a distinct dry phase of short duration.[76] Its imprint at Lake Van[77] and Soreq Cave,[78] however, marks a long-term shift to drier conditions rather than a punctuated event.[79]

Another supraregional, Bond cycle-related arid event around 3200–2700 BP was previously linked with impacts on the Upper Mesopotamian civilisation[80] and may correlate with the abrupt decrease in rainfall and lake-level fall observed at Ein Gedi, Dead Sea.[81] Other regional records, however, fail to reflect this event.

Significant events of increased humidity have been inferred from the high-resolution speleothem analysis at Soreq Cave 6550–6450 and 4800–4700 years ago. Shorter, less intense humid events were identified at 6700–6680 BP, 6170–6100 BP, 5760–5740 BP, and 5500–5450 BP.[82]

---

72 Bar-Matthews, and Ayalon, "Mid-Holocene Climate from Soreq Cave," 168.

73 Riehl et al., "Drought Stress Evidenced by $\delta^{13}C$ in Barley Grain," 12349.

74 Bar-Matthews, and Ayalon, "Mid-Holocene Climate from Soreq Cave," 168.

75 Mayewski et al., "Holocene Climate Variability," 250; Bernhard Weninger et al., "The Impact of Rapid Climate Change on Prehistoric Societies during the Holocene in the Eastern Mediterranean," *Documenta Praehistorica* XXXVI (2009): 34–44. doi:10.4312/dp.36.2.

76 Arz, Lamy, and Pätzold, "A Dry Event Recorded Around 4.2 ka," 436.

77 Wick, Lemcke, and Sturm, "Sediments of Lake Van," 673.

78 Bar-Matthews et al., "Paleorainfall during Interglacial Intervals," 3196; Bar-Matthews, and Ayalon, "Mid-Holocene Climate from Soreq Cave," 169.

79 Simone Riehl et al., "Mid-to-Late Holocene Agricultural System Transformations in the Northern Fertile Crescent: A Review of the Archaeobotanical, Geoarchaeological, and Philological Evidence," in *Climates, Landscapes, and Civilizations*, ed. Liviu Giosan, et al., 119. Washington: American Geophysical Union, 2012. doi:10.1029/2012GM001221.

80 Riehl et al., "Holocene Agricultural System Transformations," 119–120.

81 Litt et al., "Dead Sea Pollen Record," 102–103.

82 Bar-Matthews, and Ayalon, "Mid-Holocene Climate from Soreq Cave," 169.

3 Holocene Landscape Changes

3.1 *Vegetation Changes*

Changes in Holocene vegetation, a pivotal resource for ancient societies, are driven by climate variability and, in particular since the Mid-Holocene, by anthropogenic impact. The state of knowledge for the Near East is mainly based on pollen analyses from lake[83] or marine[84] records and archaeobotanical studies.[85]

In the Early to Mid-Holocene, it has been observed that vegetation response to climatic changes lags behind up to three millennia.[86] As there are no robust and well-dated Holocene pollen spectra from the Mesopotamian heartlands, again Lake Van with its wide pollen catchment at the crossroads between the Indo-Turanian steppe and desert in the north and the Kurdo-Zagrosian oak and pistachio-almond steppe forest in the south provides a relevant reference for the northern basin.[87] Based on the Van record, the Early Holocene witnessed a dominance of non-arboreal pollen (NAP), such as of *Gramineae* and other steppe to desert-steppe herbs in combination with *Pistacia*, typically indicating summer drought and mild winters without frost. The shift towards deciduous oak forest seems to have gradually reached a maximum in arboreal pollen (AP) only around 6500 cal. BP lagging behind the much earlier onset of a more humid climate.[88] This lag could partially be explained by the underrepresentation of AP in the pollen rain, cooler summer months,[89] and changes in the seasonality of precipitation, which increased during winter but not in the growing season.[90] At around the same time, human clearing of deciduous oak in the Near East gave space to heliophilous taxa and led to increased

83 Wick, Lemcke, and Sturm, "Sediments of Lake Van," 670–672; Markus J. Schwab et al., "Holocene Palaeoecology of the Golan Heights (Near East): Investigation of Lacustrine Sediments from Birkat Ram Crater Lake," *Quaternary Science Reviews* 23 (2004): 1726–1727. doi:10.1016/j.quascirev.2004.05.001; Neil Roberts et al., "Climatic, Vegetation and Cultural Change in the Eastern Mediterranean during the Mid-Holocene Environmental Transition," *The Holocene* 21 (2011): 154–158. doi:10.1177/0959683610386819; Litt et al., "Dead Sea Pollen Record," 99–102.

84 Rossignol-Strick, "Pollen Records of Sapropel 1," 517–521.

85 Naomi F. Miller, "The Macrobotanical Evidence for Vegetation in the Near East, c. 18000/16000 BC to 4000 BC," *Paléorient* 23 (1997): 197–207. doi:10.3406/paleo.1997.4661; Riehl et al., "Holocene Agricultural System Transformations," 122–123.

86 Miller, "Macrobotanical Evidence for Vegetation in the Near East," 198; Wick, Lemcke, and Sturm, "Sediments of Lake Van," 671.

87 Roberts et al., "Climatic, Vegetation and Cultural Change," 155.

88 Wick, Lemcke, and Sturm, "Sediments of Lake Van," 670–672.

89 Roberts et al., "Climatic, Vegetation and Cultural Change," 156–157.

90 Wick, Lemcke, and Sturm, "Sediments of Lake Van," 673.

soil erosion at Birkat Ram in the Golan Heights.[91] The maximum expansion of oak-dominated forest-steppe vegetation lasted until the onset of the 4th millennium BP, when deciduous oak gradually declined and synanthropic taxa such as *Plantago lanceolata* set in.[92]

The archaeobotanical record of sites in the Zagros foothills, Upper Mesopotamia, and the Levant, mostly consisting of seeds and charcoal, also points to *Pistacia* being the only widespread tree taxon during the Younger Dryas, when steppe herbs dominated. Eastward expansion of woodland including oak, almond, and pistachio during the Early Holocene is reflected by seeds and charcoal from various sites. Significant human impact due to pastoralism is inferred only after 6000 BP,[93] while potential signs of overgrazing inferred from a meta-study of seed/grain, charcoal, and isotope data occurred during the Early Bronze Age.[94]

3.2 *Hydrological Variability of the Tigris and Euphrates Rivers and the Geomorphic Response*

The streamflow of the Tigris and Euphrates rivers is strongly controlled by the activity of the North Atlantic Oscillation (NAO),[95] the air-mass transfer between the Arctic and subtropical Atlantic which drives climate fluctuations in the Northern Hemisphere, including the eastern Mediterranean region, to a large extent[96] (fig. 3.2). Accordingly, Kay and Johnson present a Mid- to Late Holocene curve of Euphrates-Tigris streamflow strongly correlating with the oxygen isotope curve of Lake Van that is similarly tied to NAO-driven Mediterranean rainfall.[97]

Streamflow in combination with relative sea-level fluctuations of the Arabian Gulf are the main drivers of fluvial landscape formation on the

91 Schwab et al., "Holocene Palaeoecology of the Golan Heights,"1730.

92 Wick, Lemcke, and Sturm, "Sediments of Lake Van," 673.

93 Miller, "Macrobotanical Evidence for Vegetation in the Near East," 198.

94 Michelle de Gruchy, Katleen Deckers, and Simone Riehl, "A diachronic reconstruction of the Northern Mesopotamian landscape (4th to 2nd millennia BCE) from three separate sources of evidence," *Journal of Archaeological Science: Reports* 8 (2016): 265. doi:10.1016/j.jasrep.2016.05.047.

95 Cullen, and deMenocal, "Influence on Tigris—Euphrates Streamflow," 862.

96 Luigia Brandimarte, et al., "Relation between the North-Atlantic Oscillation and Hydroclimatic Conditions in Mediterranean Areas," *Water Resources Management* 25 (2011): 1270. doi: 10.1007/s11269-010-9742-5.

97 Paul A. Kay and Douglas L. Johnson, "Estimation of Tigris-Euphrates Streamflow from Regional Paleoenvironmental Proxy Data," *Climatic Change* 3 (1981): 258. doi: 10.1007/BF02423218.

Mesopotamian Plain.[98] High streamflow is generally associated with the formation of long, high-amplitude meanders, initial channel braiding and periods of aggradation or incision, whereas low streamflow is linked to a shift from braiding to meandering and dominant incision. While the *Ǧazīra* region of Upper Mesopotamia, dominated by broad terraces, gullies, large-scale gypsum crusts, endhoreic depressions, alluvial fans, and gravel-covered floodplains, is the result of relict depositional and erosional activity during the Pleistocene, Lower Mesopotamia comprising the meandering and anastomosing floodplains of the Euphrates, the meandering floodplain of the Tigris, the abandoned, inactive floodplains of the central plain, as well as the fluvio-lacustrine deltaic complex of the Euphrates-Tigris-Karun system is characterised by high morphodynamic activity over most of the Holocene until today. Dominating processes comprise vertical floodplain accretion, levee migration, and alluvial fan development along the margins of the plain directed perpendicular to its longitudinal axis. On the largest spatial scale, the most important evolution is the change from a Late Quaternary braided-river system to an anastomosing one and, related to the Mid-Holocene sea-level and base-level highstand, respectively, predominantly single-channel meandering system. The main active channels benefited from gradual siltation of the parallel, anastomosing watercourses.[99]

Riehl et al. compile sites within and around Mesopotamia where significant changes in streamflow and morphodynamic response associated with a pan-Mesopotamian aridisation trend after 4200 years ago were identified. A regime of predictable, steady streamflow seems to have changed to a lower total flow and more erratic pattern resulting in a shift from valley aggradation and floodplain stability to incision and the disappearance of wetland environments,[100] as, e.g., in the upper Khabur basin, a tributary of the Euphrates in northeastern Syria.[101] Further to the west, in the upper Middle Euphrates Valley of the Turkey-Syria border region, where the river enters the Syrian Plateau, the Early Holocene braided-river pattern led to the deposition of a predominantly silty terrace during high floods. After 6000 BCE, streamflow concentrated in very few main channels incising these deposits, which then formed terraces where soils developed. A shift to more unstable

98 Verhoeven, "Changing Watercourses in Babylonia," 181, 192; Morozova, "Holocene Avulsions," 412.

99 Verhoeven, "Changing Watercourses in Babylonia," 181.

100 Riehl et al., "Holocene Agricultural System Transformations," 124.

101 Katleen Deckers and Simone Riehl, "Fluvial Environmental Contexts for Archaeological Sites in the Upper Khabur Basin (Northeastern Syria)," *Quaternary Research* 67 (2007): 347. doi:10.1016/j.yqres.2006.11.005.

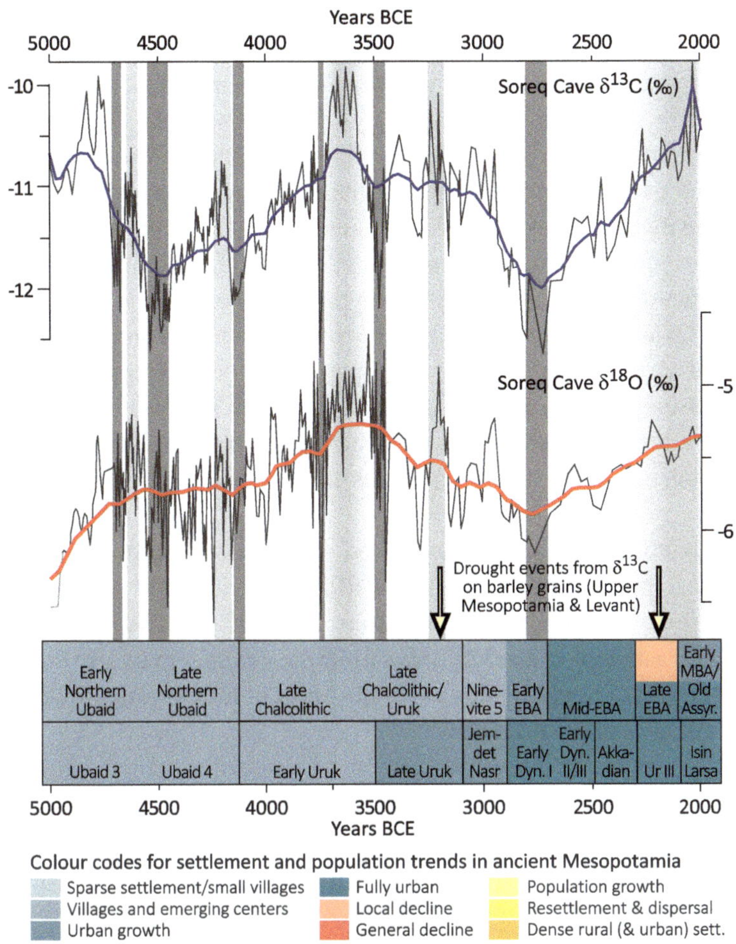

FIGURE 3.3 High-resolution $\delta^{18}O$ and $\delta^{13}C$ data from Soreq Cave, Israel (map 3.1), shown in combination with smoothed curves, inferred short-term dry (light grey sections) and humid (dark grey sections) periods,[a] cultural periods, and urbanisation trends.[b] Droughts reflected by $\delta^{13}C$ of barley grains from 33 archaeological sites of Upper Mesopotamia and the Levant are indicated by vertical arrows.[c] Colour codes for settlement and population trends also apply to fig. 3.2.

Notes:

a Bar-Matthews, and Ayalon, "Mid-Holocene climate from Soreq Cave," 167–168.

b Wilkinson, "Regional Approaches to Mesopotamian Archaeology," 243.

c Simone Riehl et al., "Drought Stress Variability in Ancient Near Eastern Agricultural Systems Evidenced by $\delta^{13}C$ in Barley Grain," *Proceedings of the National Academy of Sciences of the USA* 111 (2014): 12350. doi: 10.1073/pnas.1409516111.

conditions and erratic floods over the Middle Euphrates Valley occurred asynchronously depending on site characteristics (e.g. catchment order of tributaries, historical land use, etc.) and the general lag between climate shifts and fluvial response.[102]

Besides these gradual, longer-term changes driven by external factors, there are numerous short-term, intrinsic morphodynamic changes active on a more local scale. These include meander cut-offs where cut banks at the neck converge due to continuous erosion, downstream migration of meanders, or sudden avulsion processes.[103] Avulsion describes the diversion of a major river channel framed by levees to the lower-lying floodplain and initiation of a new channel, and can occur through abrupt reoccupation of a pre-existing channel or long-term channelling of a flooded plain, mostly in an anastomosing pattern, after a levee break.[104] Large-scale flooding associated with levee breaks represented a constant hazard to settlements in the floodplain throughout history.[105]

The oldest abandoned channels associated with avulsion processes are dated to the 4th millennium BCE (Uruk period), when Euphrates and Tigris coalesced near Sippar. After the 4th millennium BCE, the rivers separated as the Tigris shifted eastward, still exhibiting a multiple channel network. From the middle of the 2nd millennium BCE, with the onset of the two-channel system, these networks were abandoned.[106]

3.3 *Coastal Changes and Noah's Flood Hypothesis*

Until the end of the Pleistocene, the Arabian Gulf was dry and formed an extension of the Euphrates-Tigris drainage system. Initial marine incursions into the Gulf date to ca. 12500 BP whereas the current sea level was first reached ca. 7000 to 6500 years ago, followed by the Mid-Holocene relative sea-level highstand of at least two metres higher than today at around 6000–4500 years ago.[107]

---

102 Catherine Kuzucuoğlu, Michel Fontugne, and Damase Mouralis, "Holocene Terraces in the Middle Euphrates Valley, between Halfeti and Karkemish (Gaziantep, Turkey)," *Quaternaire* 15 (2004): 195–206. doi:10.3406/quate.2004.1767.

103 Verhoeven, "Changing Watercourses in Babylonia," 192–195.

104 Morozova, "Holocene Avulsions," 407–410; Jotheri et al., "Avulsions of the Euphrates River," 188.

105 Helmut Brückner, "Wasserstraßen im Wüstensand," *Antike Welt* 44.3 (2013): 11.

106 Morozova, "Holocene Avulsions," 407–410.

107 Gary A. Cooke, "Reconstruction of the Holocene coastline of Mesopotamia," *Geoarchaeology* 2 (1987): 19–20. doi:10.1002/gea.3340020102; Lambeck, "Shoreline Reconstructions for the Persian Gulf," 47; Engel, and Brückner, "SQSP—Holocene Coastal Changes," 299.

The rapid flooding of the ca. 1,000 kilometre-long and up to 350 kilometre-wide Gulf is explained by high rates of the global eustatic sea-level rise until 7000–6000 BP and its shallow and for most parts very gentle bathymetry. Extremely rapid lateral shifts of the Gulf coastline northward of up to one kilometre per year[108] led to flooding of the Shatt al-Arab region at ca. 9000–8000 BP, reflected by a thick sequence of brackish-marine deposits.[109] Marginal-marine sedimentary environments were reconstructed around 9000 BP along the current coastline near Fao, at around 8000 BP near Basra and at around 6000 BP during the Mid-Holocene highstand at least as far as the ancient city of Ur in the west and modern Amara in the east (fig. 3.1). In this area they override pre-transgressive fluvial to wetland deposits and possibly dunes in the east and evaporate-rich *sabkha* (salt flat) deposits towards the west.[110] After 5000 BP, during the highstand plateau and subsequent relative sea-level fall, hypersaline coastal *sabkhas* developed from salt marshes along the inland coastline under an increasingly arid climate. The subsequent initiation of delta formation can be inferred from the capping of the transgressive parasequence by coastal marsh facies, evaporates dominated by gypsum, dolomite and palygorskite, as well as fluvial silts and clays.[111] The latter terrestrial facies is associated with rapid post-highstand delta progradation controlled by river sediment loads, eustasy, and isostasy, while the role of vertical tectonic displacement seems to be of minor importance.[112]

The most detailed palaeogeographic scenarios for the local extent of the Mid-Holocene transgression were created by Pournelle[113] based on a synthesis of published borehole data of the Geological Survey of Iraq[114] and those

108 Jim T. Teller et al., "Calcareous Dunes of the United Arab Emirates and Noah's Flood: The Postglacial Reflooding of the Persian (Arabian) Gulf," *Quaternary International* 68–71 (2000): 303. doi:10.1016/S1040-6182(00)00052-5.

109 Adnan A.M. Aqrawi, "Stratigraphic Signatures of Climatic Change during the Holocene Evolution of the Tigris-Euphrates Delta, Lower Mesopotamia," *Global and Planetary Change* 28 (2001): 273–275. doi:10.1016/S0921-8181(00)00078-3.

110 Paul Sanlaville, "Considérations sur l'évolution de la Basse Mésopotamie au cours des derniers millénaires," *Paléorient* 15 (1989): 21. doi:10.3406/paleo.1989.4506; Aqrawi, "Holocene Evolution of the Tigris-Euphrates Delta," 275–276; Heyvaert, Baeteman, "Palaeocoastlines of the Lower Khuzestan Plain," 101–105.

111 Cooke, "Holocene coastline of Mesopotamia," 21; Aqrawi, "Holocene Evolution of the Tigris-Euphrates Delta," 275–276.

112 Sanlaville, "L'évolution de la Basse Mésopotamie," 11–14; Aqrawi, "Holocene Evolution of the Tigris-Euphrates Delta," 275–276; Pournelle, "Deltaic Landscapes," 112; Kennett, and Kennett, "Holocene Marine Transgression and Climate Change," 240–241.

113 Pournelle, "Deltaic Landscapes," 114–124.

114 Aqrawi, "Holocene Evolution of the Tigris-Euphrates Delta," 269.

compiled by Larsen and Sanlaville,[115] archaeological findings from the ancient cities of Lower Mesopotamia, and multispectral satellite-image analysis. In these scenarios, the maximum transgression reached even further up the two rivers mainly covering an area which is now occupied by wetlands (*ahwar*)[116] (map 3.1). The massive extent of the Mid-Holocene transgression is corroborated by cuneiform texts indicating that the inland sites of Lagash and Ur were port cities in the 3rd millennium BCE[117] and by Classical texts, which inspired J. DeMorgan to draw a 1st millennium BCE shoreline just south of Amara and Ahwaz as early as in 1900.[118] Excavations at the Ubaid sites of Eridu and Tell Oueili suggest an economy strongly adapted to coastal wetland environments, mostly based on fishing and cattle breeding,[119] while the archaeobotanical record of Oueili reflects marshy conditions around this tell as well.[120]

The high rates of lateral transgression until the Early to Mid-Holocene inferred from the sedimentary record in combination with a likely increase in rainfall at that time (see Part 2.1 and fig. 2.3) triggered associations with the great deluge of the Epic of Atrahasis, the Akkadian Epic of Gilgamesh, and the Genesis flood narrative,[121] all of which are believed to share a common oral tradition.[122] Teller et al. draw a scenario of a Neolithic community settling on a gentle rise of the pre-transgressive exposed floor of the Gulf which may have become isolated by the rapidly encroaching sea, a situation exacerbated by cataclysmic rainfall and runoff both by the Euphrates-Tigris system and from the nearby Zagros Mountains.[123] However, despite of the high rate of lateral

---

115 Curtis E. Larsen, "The Mesopotamian Delta Region: A Reconsideration of Lees and Falcon," *Journal of the American Oriental Society* 95 (1975): 54. doi:10.2307/599157; Sanlaville, "L'évolution de la Basse Mésopotamie," 15.

116 Pournelle, "Deltaic Landscapes," 123; Pournelle, "Physical Geography," 19.

117 Thorkild Jacobsen, "The Waters of Ur," *Iraq* 22 (1960): 184–185. doi:10.2307/4199683; Susan Pollock. *Ancient Mesopotamia* (Cambridge: Cambridge University Press, 1999), 34.

118 Larsen, "The Mesopotamian Delta Region," 44.

119 Jean-Louis Huot, "'Ubaidian Villages of Lower Mesopotamia," in *Upon these Foundations: The 'Ubaid Reconsidered, Proceedings from the 'Ubaid Symposium, Elisnore*, ed. Elizabeth F. Henrickson and Ingolf Thuesen (Copenhagen: Museum Tusculanum Press, 1989), 39; Pollock, *Ancient Mesopotamia*, 81–83.

120 Reinder Neef, "Plant Remains from Archaeological Sites in Lowland Iraq: Tell El 'Oueili," in *'Oueili—Travaux de 1985*, ed. Jean-Louis Huot (Paris: Éditions Recherche sur les Civilisations, 1991), 324.

121 Teller et al., "Postglacial Reflooding of the Persian (Arabian) Gulf," 304; Brückner, "Uruk—A Geographic and Palaeo-Ecologic Perspective," 246.

122 Andrew R. George, *The Babylonian Gilgamesh Epic: Introduction, Critical Edition and Cuneiform Texts, vol.* 1 (Oxford: Oxford University Press, 2003), 70.

123 Teller et al., "Postglacial Reflooding of the Persian (Arabian) Gulf," 304; Kennett, and Kennett, "Holocene Marine Transgression and Climate Change," 243.

transgression it is highly unlikely that communities were surprised in a way not permitting to shift their dwelling places accordingly. The origin of the flood narrative probably lies in a phase of extreme rainfall and flooding events related to avulsions in the alluvium of Lower Mesopotamia rather than a single, unique event,[124] which is corroborated by thick and sterile alluvial muds in several tells burying archaeological layers of pre-Ubaid civilisations.[125]

4 Social Response to Holocene Landscape Dynamics

The relevance of environmental changes for the ancient social systems of Mesopotamia is rooted in a strong dependence on locally available resources. While the range of farming activities and animals exploited were similar along the two rivers, subsistence techniques and specialisations strongly diverged from the headwaters to the Gulf shores.[126] We sum up how larger-scale environmental conditions and their changes over millennial time scales have influenced socio-economic, cultural, and political changes in different parts of Mesopotamia and how they help to explain the emergence of the first hierarchical, complex urban-type societies. Thereby, it becomes obvious that—apart from the Uruk case study—emphasis has to be placed on larger-scale dynamics and changes, as local developments may have varied significantly and may even have opposed broader trends, adding enormous complexity as we try to analyse the human-environment relationships in Mesopotamia over time in detail.[127]

4.1 *Upper Mesopotamia*

From the Neolithic until the Ubaid period, agricultural and agro-pastoral subsistence in Upper Mesopotamia benefited from a relatively stable climate with humid winters and reliable water availability best reflected by the Lake Van record (fig. 3.2).[128] These conditions facilitated the emergence of agricultural

124 Helmut Brückner and Max Engel, "Noah's Flood—Probing an Ancient Narrative Using Geoscience," in *Palaeohydrology—Traces, Tracks and Trails of Extreme Events*, ed. Jürgen Herget and Alessandro Fontana (Cham: SpringerNature, 2020), 14–15. doi:10.1007/978-3-030-23315-0_7.

125 C. Leonard Woolley, "Excavations at Ur of the Chaldees," *The Antiquaries Journal* 3 (1923): 315. doi:10.1017/S0003581500014980; R.L. Raikes, "The Physical Evidence for Noah's Flood," *Iraq* 28 (1966): 54. doi:10.2307/4199795; Morozova, "Holocene Avulsions," 416.

126 Pollock, *Ancient Mesopotamia*, 78–81.

127 Riehl et al., "Holocene Agricultural System Transformations," 129.

128 Roberts et al., "Climatic, Vegetation and Cultural Change," 151; Riehl et al., "Drought Stress Evidenced by $\delta^{13}C$ in Barley Grain," 12349.

practices and technologies since the Early Neolithic which relied mostly on different wheat, barley, and pulse varieties.[129] Apart from reduced functionality and even partial abandonment of Upper Mesopotamian sites as a response to the 8.2 ka RCC,[130] a stable climate and long-term development of rain-fed agriculture facilitated cultural continuity until the Late Chalcolithic.[131] Later RCCs as reflected by the high-resolution speleothem record of Bar-Matthews and Ayalon[132] in fig. 3.3, in particular the shift to a more continental climate with higher rainfall variability after the humid event around 4500 BCE, can also tentatively be linked with socio-political and agricultural adaptations. While during the 5th millennium BCE, only the disappearance of the Ubaid culture from the north seems to be worth mentioning in this context,[133] the first signs of urbanisation appearing during the late 5th to 4th millennium BCE parallel to the development of a dense network of rural settlements,[134] seem to be related to intensified land use.[135] Ongoing population increase in Upper Mesopotamia between 4400 and 3400 BCE may be linked to production surpluses during phases of enhanced rainfall,[136] and as yet stable perennial streamflows and fluvial wetlands, which can be inferred from the geomorphic record in the Upper Khabur basin of easternmost Syria.[137] A first larger-scale cultural break in Upper Mesopotamia linked to a temporary population decline due to southward migration was probably triggered by precipitation levels too low to sustain rain-fed cereal cultivation. This change is related to the 5200 BP drying and cooling event[138] which left its stable isotopic signature

---

129 Miller, "Macrobotanical Evidence for Vegetation in the Near East," 202; Simone Riehl, Mohsen Zeidi, and Nicholas J. Conard, "Emergence of Agriculture in the Foothills of the Zagros Mountains of Iran," *Science* 341 (2013): 65. doi:10.1126/science.1236743.

130 Michael Staubwasser, and Harvey Weiss, "Holocene Climate and Cultural Evolution in late Prehistoric-Early Historic West Asia," *Quaternary Research* 66 (2006): 378. doi:10.1016/j.yqres.2006.09.001.

131 Joanne Clarke, et al., "Climatic Changes and Social Transformations in the Near East and North Africa during the 'Long' 4th Millennium BC: A Comparative Study of Environmental and Archaeological Evidence," *Quaternary Science Reviews* 136 (2016): 114. doi:10.1016/j.quascirev.2015.10.003.

132 Bar-Matthews, and Ayalon, "Mid-Holocene Climate from Soreq Cave," 168.

133 Clarke et al., "Climatic Changes and Social Transformations," 114.

134 Lawrence et al., "Long-term Population, City Size and Climate Trends," 11.

135 Clarke et al., "Climatic Changes and Social Transformations," 114.

136 Lawrence et al., "Long-term Population, City Size and Climate Trends," 11.

137 Deckers, and Riehl, "Fluvial Environmental Contexts," 347.

138 Staubwasser and Weiss, "Holocene Climate and Cultural Evolution," 379; Clarke et al., "Climatic Changes and Social Transformations," 97.

in barley grains[139] and the Soreq speleothem record.[140] Significantly reduced and more variable streamflows after 2500 BCE, when both the Soreq and Lake Van records indicate the onset of a long-term drying trend (fig. 3.2 and 3.5), in combination with population growth and intensification of land use, resulted in increased erosion and correlating aggradation in subcatchments of the Euphrates Valley.[141]

The transition from the relatively humid Early Bronze Age to the Middle Bronze Age coincided with the 4200 BP aridification event of unprecedented magnitude expressed through reduced seasonal temperature, dust, and a decline in rainfall of circa twenty to thirty per cent,[142] which had a substantial impact on agricultural production and the selection of crops. A general decline in the presence and abundance of water-demanding crops and those sensitive towards soil salinity was observed in the archaeobotanical record of Upper Mesopotamia after the Early Bronze Age. While free-threshing wheat, garden pea, linseed and grape vine decrease, the cultivation of the much more stress-tolerant barley significantly increases all over Upper Mesopotamia at least in some areas combined with the resistant bitter vetch. Together, this represents unequivocal evidence of higher drought stress.[143] Furthermore, the significantly lower number of late Early Bronze Age/early Middle Bronze Age archaeological sites may relate to these challenges of extrinsic, environmental nature as well.[144] In the area of Tell Leilan in easternmost Syria, 73 per cent of all sites experienced abandonment while the total occupied area fell by 93 per cent.[145] Correlation between climate deterioration and the collapse of the pan-Mesopotamian Akkad Empire around 2200 BCE has been strenuously debated in light of strongly diverging resilience from site to site.[146] The high degree of political and economical centralisation

---

139 Riehl et al., "Drought Stress Evidenced by δ^{13}C in Barley Grain," 12349.

140 Bar-Matthews et al., "Regional Events at Soreq Cave," 89; Bar-Matthews, and Ayalon, "Mid-Holocene Climate from Soreq Cave," 168.

141 Kuzucuoğlu, Fontugne, and Mouralis, "Holocene Terraces," 204; Deckers, and Riehl, "Fluvial Environmental Contexts," 346.

142 Harvey Weiss et al., "The Genesis and Collapse of 3rd Millennium North Mesopotamian Civilization," *Science* 261 (1993): 1002. doi:10.1126/science.261.5124.995; Heidi M. Cullen et al., "Climate Change and the Collapse of the Akkadian Empire: Evidence from the Deep Sea," *Geology* 28 (2000): 382. doi:10.1130/0091-7613(2000)28<379:CCATCO>2.0.CO;2; Staubwasser and Weiss, "Holocene Climate and Cultural Evolution," 381.

143 Riehl et al., "Holocene Agricultural System Transformations," 129–130.

144 Riehl et al., "Drought Stress Evidenced by δ^{13}C in Barley Grain," 12350; Lawrence et al., "Long-term Population, City Size and Climate Trends," 10.

145 Staubwasser and Weiss, "Holocene Climate and Cultural Evolution," 382.

146 Weiss et al., "Genesis and Collapse," 999; Arlene M. Rosen, *Civilizing Climate. Social Responses to Climate Change in the Ancient Near East* (Lanham: Altamira Press, 2007), 143.

and agricultural specialisation may have prevented appropriate adaptation to the changing environments in the major cities leading to crisis, while more independent, peripheral regions with less complex socio-political hierarchies and more flexible production processes were able to adapt and kept thriving.[147]

Resettlement after the 4200 BP RCC from ca. 1900 BCE onwards occurred in the context of substantial socio-political reorganisation, where individual settlements had distinct functions within the region. The downfall of the Old Babylonian kingdom and onset of a Dark Age period with decreasing population densities in Upper Mesopotamia happened against the background of the aridisation trend continuing after the Middle Bronze Age, which fostered further emphasis on water stress-tolerant crop species, in particular barley. However, deurbanisation results from a complex interplay of different factors including social, political, economic, and ecological parameters, and the exact role of climatic changes is difficult to evaluate.[148] Lawrence et al., however, identify a decoupling of climatic changes and settlement patterns since 2000 BCE.[149]

The recovery phase of population and settlements during the Iron Age witnessed the emergence of new crop species such as cotton, pomegranate, or cucumber, whereas water demanding free-threshing wheat and grape increase again. In combination with ongoing gradual drying and salinisation indicators in the palaeobotanical record, this pattern indicates agricultural intensification through the introduction of irrigation techniques.[150] From a demographic point of view, cities grew to such a level in terms of size and power that they could not have been sustained by rain-fed agriculture only,[151] but may have been sustained by pastoralism and the establishment of an agro-pastoralist buffering strategy.[152]

4.2 *Lower Mesopotamia*

In Lower Mesopotamia, ambiguities also remain about the exact contribution of changes in climate, streamflow, and landscape dynamics to the success and downfall of complex societies. Similar to today's conditions, differences from Upper to Lower Mesopotamia in, for example, the rainfall gradient must also be anticipated for ancient times. Moreover, the Saharo-Arabian climate with strengthened African and Indian monsoonal activities driven by orbital

147 Roberts et al., "Climatic, Vegetation and Cultural Change," 152.
148 Riehl et al., "Holocene Agricultural System Transformations," 119.
149 Lawrence et al., "Long-term Population, City Size and Climate Trends," 11.
150 Riehl et al., "Holocene Agricultural System Transformations," 130.
151 Lawrence et al., "Long-term Population, City Size and Climate Trends," 11.
152 Wilkinson, "Contextualizing Early Urbanization," 96.

forcing during the Early Holocene,[153] as possibly reflected by the record from Tayma[154] (fig. 3.2), played a more important role and probably provided additional summer rainfall during the Neolithic. During this favourable pre-Ubaid and pre-transgressive phase of enhanced moisture availability, smaller, dispersed communities may have practised a semi-nomadic subsistence along the Ur-Shatt River in Lower Mesopotamia. Close to the coast, they were forced to relocate or at least adjust their roaming area due to the rapid lateral marine transgression. However, all of these scenarios are based on assumptions, as archaeological traces of this period are buried under thick alluvium.[155]

The gradual decline of available water resources during summer triggered competition for natural resources at the beginning of the Ubaid Period and led to population clustering, more intense agriculture, economic differentiation, and the establishment of new ideologies and socio-political structures. Growing communities emerged at sites such as Ur and Eridu,[156] which were influenced and shaped by transgressive wetlands and estuaries providing an important source of food as well as pathways for transportation.[157] Early Ubaid settlements show strong continuity in terms of their location, mostly covering river levees close to swamps and marshes. During Ubaid 3 and 4, new settlements were founded at elevated sites of remnant Pleistocene river terraces (turtlebacks) close to palaeo-shores,[158] which were high enough to protect from the ongoing transgression and recurring river flooding. The coastal and wetland resources provided resilience capacities countering a decrease in rainfall. Competition for resources increased during the Ubaid, magnified by migrating communities which, after merging with local groups, formed the core of what were to become urban centres.[159] The environmentally driven demographic pressure in combination with the rise in seafaring may have stimulated the spread of the Ubaid culture along the western margin of the Gulf.[160]

153 Martin Claussen et al., "Simulation of an Abrupt Change in Saharan Vegetation in the Mid-Holocene," *Geophysical Research Letters* 26 (1999): 2037. doi:10.1029/1999GL900494.

154 Engel et al., "Holocene Humid Period in NW Saudi Arabia," 135; Dinies et al., "Grassland Expansion in Northwestern Arabia," 298; Dinies et al., "Holocene Vegetation in Northwestern Arabia," 6.

155 Kennett, and Kennett, "Holocene Marine Transgression and Climate Change," 248.

156 Ibid., 250.

157 Pournelle, "Deltaic Landscapes," 252–253.

158 Pournelle, "Physical Geography," 22.

159 Kennett, and Kennett, "Holocene Marine Transgression and Climate Change," 250–251.

160 Robert A. Carter, "The Social and Environmental Context of Neolithic Seafaring in the Persian Gulf," in *The Global Origins (and Development) of Seafaring*, ed. Atholl Anderson, James H. Barrett, and Katherine V. Boyle (Cambridge: McDonald Institute of Archaeological Research, 2010), 198; Philipp Drechsler, "The Arabian Peninsula," in

Post-Ubaid community aggregation at the most favourable sites culminated in the first complex, urban-type societies of the Late Uruk period.[161] The importance of Lower Mesopotamia's wetlands and the marginal marine environment as a food resource and access gateway to transport and maritime trade is strongly reflected in the archaeological record[162] and in administrative texts reaching back to the earliest protoliterate lexical lists,[163] while longer-distance travel and exotic trading goods may have facilitated the exchange of ideas and the establishment of a clear social and political stratification. Natural watercourses were used and directed to make effective use of the amphibian environments in terms of large-scale irrigation (see also the example of Uruk in Chapter 4.3). Expanding floodplains after the onset of coastal regression provided fertile terrain for intensified irrigation agriculture which, in combination with fishing, represented an important base for population growth.[164] The nuclei of early irrigation practices may be found on proximal crevasse splays resulting from levee breaks with their coarser texture being less susceptible to salinisation and flooding. Morozova regards the complex avulsion belts created by successions of levee breaks as "oases" of easy cultivation, high agricultural productivity, and therefore targets of migration, all representing critical factors for the emergence of the complex, urban-type societies.[165] Reduced streamflows of the two rivers related to the 5200 BP RCC had severely affected irrigation agriculture in Lower Mesopotamia and may be regarded as a root cause for societal collapse.[166]

During the Early Dynastic Period, settlements decreased in numbers but increased in size as wetlands slowly contracted with coastal regression and lower groundwater tables, and advancing fluvial fans from the valley margins potentially started to bury older sites. While the urban centres grew, smaller settlements followed the prograding coastline, still connected to the inland centres by waterways. Wetland resources such as reed and fish remained important throughout the Ur III period and beyond, while irrigation agriculture on elevated grounds protected from seasonal flooding became even more significant in a continuously drying climate.[167] The 4200 BP RCC seems to have impacted

A Companion to the Archaeology of the Ancient Near East, vol. I, ed. Daniel T. Potts (Hoboken: Wiley-Blackwell, 2012), 491. doi:10.1002/9781444360790.ch25.

161 Kennett, and Kennett, "Holocene Marine Transgression and Climate Change," 252.

162 Huot, "'Ubaidian Villages of Lower Mesopotamia," 39.

163 Pournelle, "Physical Geography," 23.

164 Kennett, and Kennett, "Holocene Marine Transgression and Climate Change," 252.

165 Morozova, "Holocene Avulsions," 414–416.

166 Staubwasser and Weiss, "Holocene Climate and Cultural Evolution," 379.

167 Pournelle, "Physical Geography, 25–27.

Lower Mesopotamia and coincides with reduced crop yields, political deterioration and instability as well as migration from the Upper Mesopotamian plains as reflected in cuneiform inscriptions.[168] However, the population decrease since 2100 BCE is considered to be gradual over a time period of ca. 1500 years.[169]

4.3 *Case Study: Uruk*

Uruk (modern Warka) is located ca. 15 kilometres east of today's main course of the Euphrates River and rises ca. 16 metres above the alluvial plain, which has an elevation of ca. ten metres a.s.l. The city was founded at the end of the 5th millennium BCE and grew rapidly. Uruk played a pivotal role in the evolution of the world's first urban-type societies and is seen as a major origin of writing and bureaucracy during the 4th millennium BCE.[170] By the beginning of the 3rd millennium BCE, the city already covered ca. 5.2 square kilometres,[171] protected by a nine-kilometres-long city wall (map 3.3), and kept its status as the largest city in Lower Mesopotamia until the 6th century BCE. At some point in the early 3rd millennium BCE, Uruk lost its exceptional political position as other cities emerged in the region. Afterwards, the city persisted with fluctuating regional importance,[172] before the area became deserted around the 7th century CE.[173]

Considering present geomorphic and archaeological evidence, Uruk seems to have been founded on the proximal part of an interior delta where the Euphrates River debouched into extensive wetlands during the Late Ubaid.[174] Watercourses were systematically used since the mid-4th millennium BCE[175] and networks of intramural canals for transportation and irrigation purposes

168 Weiss et al., "Genesis and Collapse," 1002; Staubwasser, and Weiss, "Holocene Climate and Cultural Evolution," 382.

169 Brinkman, "Mesopotamian Demography," 173.

170 Hans J. Nissen, "*The Early History of the Ancient Near East, 9000–2000 B.C.*," Chicago/London: Chicago University Press, 1988, 85–90; van Ess, "Uruk—Verortung in Raum und Zeit," 40.

171 Margarete van Ess, "Die topographische Entwicklung der Stadt Uruk im 4. und 3. Jt. v. Chr.," in: *Uruk. Altorientalische Metropole und Kulturzentrum* (=*Colloquien der Deutschen Orient-Gesellschaft Band 8*), ed. Margarete van Ess (Wiesbaden: Harrassowitz, in press).

172 van Ess, "Die topographische Entwicklung der Stadt Uruk."

173 Robert McC. Adams and Hans J. Nissen, *The Uruk Countryside* (Chicago, London: The University of Chicago Press, 1972), 59.

174 Ibid., 243.

175 van Ess, "Uruk—Verortung in Raum und Zeit," 42.

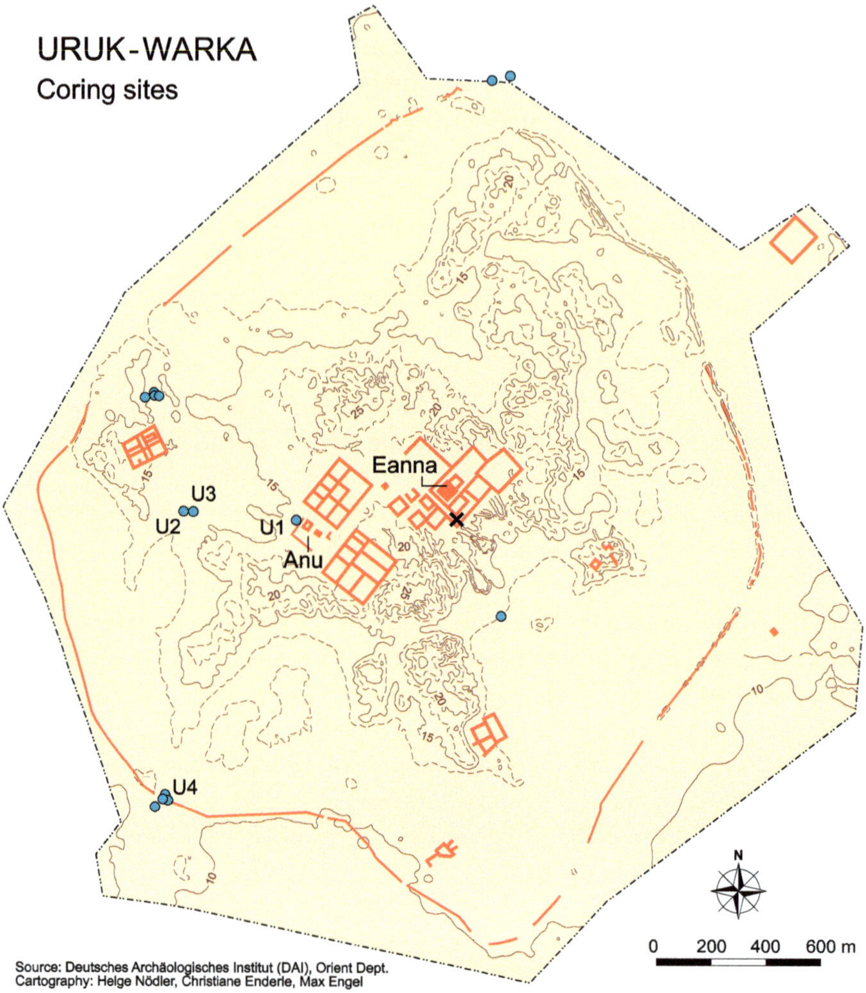

MAP 3.3 Topographic map of the archaeological zone of Uruk (modern Warka), with sediment cores (blue circles), city wall and main buildings (red structures). Altitudes in metres a.s.l.
Note: Modified after Brückner, "Uruk—A Geographic and Palaeo-Ecologic Perspective," 243.

identified by geomagnetic surveys[176] as well as sediment cores were established during the 3rd millennium BCE.[177]

The role of environmental changes as a driver of cultural change inside the city and in the region has been strenuously debated. It is anticipated that the presence of the Euphrates controlled the growth of the city[178] and that watercourses shifting away from the city contributed to its decline, in particular in the middle of the 2nd millennium BCE.[179] Here, we present original geoarchaeological data from inside the city limits of Uruk providing clues to location factors for the city's foundation, i.e. evidence for changing morphodynamics and environments such as the 4th millennium BCE canal networks or catastrophic floods.[180]

Percussion cores were taken inside the city, e.g. right at the settlement core next to the "*Steingebäude*" (U1 in map 3.3 and fig. 3.4), in the area of the canal networks (U2, U3), and inside the main canal outlet at the southern city gate (U4).[181] Well-sorted, sterile fine to medium sand is present at the base of all cores representing the undulating surface of Late Pleistocene to Early Holocene aeolian dunes, most likely connected to those identified at the bottom of the Arabian Gulf before it became flooded.[182] The rising sea level drowned lower river courses and triggered the deposition of alluvial mud, intercalated by episodic sandy flood deposits. The transgression of the Gulf shifted the shoreline and a belt of wetlands northwestwards. While these wetlands reached Uruk during the peak of the transgression coinciding with the relative sea-level highstand at around 4000 BCE as indicated by cores U1 and U4 (upper boundary ca. two metres a.s.l.), the coastline never did.[183] The inland extent of the marine incursion passing Uruk along the main course of

176 Helmut Becker, Margarete van Ess, and Jörg Fassbinder, "Uruk: Urbane Strukturen im Magnet- und Satellitenbild," in *Uruk—5000 Jahre Megacity. Begleitband zur Ausstellung „Uruk—5000 Jahre Megacity" im Pergamonmuseum. Hrsg. von der Curt-Engelhorn-Stiftung für die Reiss-Engelhorn-Museen Mannheim, dem Deutschen Archäologischen Institut—Orient-Abteilung, der Deutschen Orient-Gesellschaft e. V. und dem Vorderasiatischen Museum—Staatliche Museen zu Berlin*, ed. Nicole Crüsemann, et al., 357–358. Petersberg: Imhof Verlag, 2013.

177 Brückner, "Wasserstraßen im Wüstensand," 14.

178 Tony J. Wilkinson, "Early Channels and Landscape Development around Abu Salabikh, a Preliminary Report," *Iraq* 52 (1990): 75–83. doi:10.2307/4200320.

179 Adams, and Nissen, *The Uruk Countryside*, 59.

180 Brückner, "Uruk—A Geographic and Palaeo-Ecologic Perspective," 230.

181 Ibid., 234–242.

182 Michael Sarnthein, "Sediments and History of the Postglacial Transgression in the Persian Gulf and Northwest Gulf of Oman," *Marine Geology* 12 (1972): 259. doi:10.1016/0025-3227(72)90002-3; Uchupi, "Stratigraphy, Paleoclimate and Neotectonism," 9.

183 Brückner, "Uruk—A Geographic and Palaeo-Ecologic Perspective," 242.

FIGURE 3.4 a) The tell of Uruk with the preserved core of the Eanna-Ziggurat (21st century BCE), the main sanctuary of the goddess Inanna (Ishtar), in the far background. The foreground shows excavations of the Seleucid sanctuary of Ishtar (Irigal); b) View from Uruk towards two other former mounds. The vegetated, marshy floodplain surrounding Uruk during the late 5th millennium BCE is today replaced by a desert landscape with encroaching sand dunes in the background. The foreground shows colluvium which accumulated at the foot of the tell due to the decay of the air-dried bricks; c) Coring site U1 (map 3.3) in front of the limestone-built "*Steingebäude*". Mud-brick layers of Anu Ziqqurrat in the middleground; d) Former irrigation canals of Mesopotamian agriculture, today preserved as 'inverted channels'. Black ball pen for scale.

the Euphrates at 4000–3300 BCE as reconstructed by Pournelle[184] (map 3.1) definitely never reached the city proper. Slightly before the time of the maximum transgression, Uruk was founded, initiating the rich archaeological stratigraphy embedded into an alluvial matrix as detected in cores U1 and U3, the former potentially covering even the earliest settlement period. At the fringes of the city (U4), increasing precipitation of gypsum crystals reflects the arid trend since the Late Ubaid[185] and possibly also increasing soil salinisation due to irrigation agriculture.[186] Preliminary examination of artifacts in the cores indicates that the Late Ubaid until at least the Early Dynastic I are covered in the

184 Pournelle, "Deltaic Landscapes," 124.
185 Brückner, "Uruk—A Geographic and Palaeo-Ecologic Perspective," 242–245.
186 Brückner, "Wasserstraßen im Wüstensand," 11.

central part of the city (U1), the presumed core of the initial settlement. At U2, a thick layer of sterile alluvium overlies the basal archaeological layer. Its origin is unclear, even though it shares similarities with the massive flooding deposits already identified at Ur and other tells which were associated with Noah's Flood hypothesis.[187] Sediment cores U2 and U4 verify the Uruk-age canal networks inside the city as they comprise facies units of low-energy aquatic deposition, laminated and slightly coarser than the sterile alluvium, and numerous (sub-)rounded ceramic fragments. From the western-central part of the city (6.38 metres a.s.l.) to the southwestern outflow gate (5.57 metres a.s.l.), a clear gradient of the base of the canal can be inferred confirming flow direction and interconnection of the canal system.[188] The fauna, detected in U4 at the base of the former canal (*Unio tigridis, Melanopsis nodosa, Theodoxus* [*Neritaea*] *doriae*), indicates clear, running freshwater.[189]

The Sumerian irrigation canals are still well visible in today's landscape as 'inverted channels' (see fig. 3.4 d). During their use, the canals were cleaned regularly. After they were given up (probably due to the salinization of the fields), they were gradually filled with sediments which were then cemented as a consequence of the high evaporation and the reprecipitation of calcium carbonate. Thus, the canals became more resistant to wind erosion than the degraded soils of the former fields. This created the obvious relief inversion.

5 Conclusion

A number of links between changes in climate, landscape, and socio-technical adaptation have been established based on a large body of interdisciplinary research. Many of the relationships presented above seem to be straightforward, especially where they are confirmed by cuneiform inscriptions, Classical texts, or archaeological evidence. However, a significant gap in chronological resolution between very precise information on socio-historical development and constraints on climatic changes with a much higher uncertainty exists.[190] Moreover, RCC impacts appear to have been complex and site-specific. Resilience to RCCs such as the 4200 BP event strongly depended both on physical sensitivity towards climatic changes and political governance. It appears

187 Woolley, C. Leonard, "Excavations at Ur of the Chaldees," 315; R.L Raikes, "The Physical Evidence for Noah's Flood," 54.
188 Brückner, "Uruk—A Geographic and Palaeo-Ecologic Perspective," 242–245.
189 Ibid.," 242.
190 Riehl et al., "Holocene Agricultural System Transformations," 129.

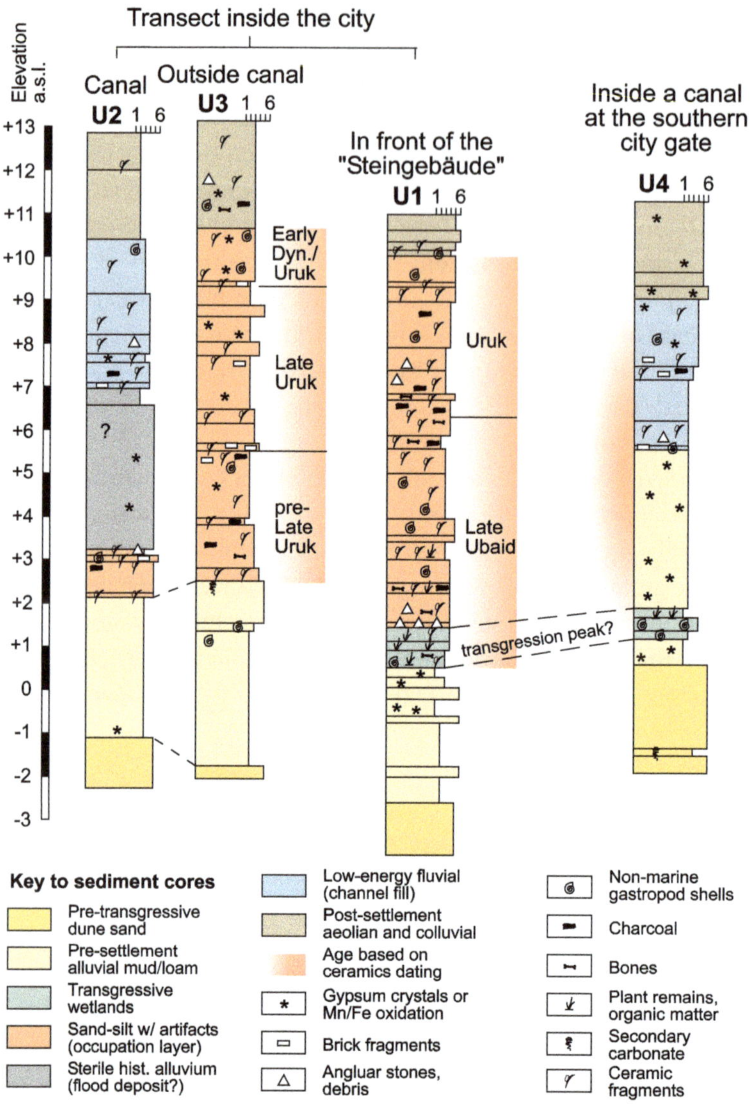

FIGURE 3.5 The stratigraphy from intramural Uruk based on sediment cores U1 to U4 (map 3.3) revealing local environmental changes from Early Holocene dune formation to the establishment of freshwater marshes, and the spread of the alluvial plain due to the Early to Mid-Holocene sea-level rise, followed by almost five thousand years of urban history. Width categories of the core log: 1=clay to silty clay; 2=clayey silt; 3=silt to clay-rich loam; 4=sandy silt to loam; 5=silty fine sand to silty sand; 6=fine to medium sand.
Note: Brückner, "Wasserstraßen im Wüstensand," 12; with modifications from van Ess, "Die topographische Entwicklung der Stadt Uruk".

that in semi-arid regions and in cities where political leadership failed to implement adaptation strategies in the light of changing environments, negative responses occurred more rapidly and were more severe.[191] The role of environmental change in the emergence of complex urban-type societies of Mesopotamia—emphasised by some, downplayed by others—cannot be separated from other strongly interrelated cultural factors such as population growth, environmental and societal circumscription, increasing conflicts, development of irrigation agriculture, and the spread of trade in a clear and distinctive manner.[192]

However, it was shown that the rainfall gradient between Upper and Lower Mesopotamia first benefited rain-fed agriculture and settlement in Upper Mesopotamia while the emergence of irrigation practices in combination with the Mid-Holocene arid events and, later, long-term aridisation, represented a pull factor and motor of the development of complex societal structures. Shifting coastal and alluvial environments such as river channels and wetlands provided crucial resources for irrigation and fishing/hunting, respectively. First-hand sedimentary data from Uruk provides direct evidence for the changing landscapes of Lower Mesopotamia and strategies for efficient use of resources and landforms to sustain large urban populations. These data also show how salinisation of the fields due to high evaporation and the lack of a proper drainage system may have forced the abandonment or shrinking of the southern settlements and their shift towards the north.

191 Roberts et al., "Climatic, Vegetation and Cultural Change," 152.

192 Kennett, and Kennett, "Holocene Marine Transgression and Climate Change," 248.

CHAPTER 4

Destructive Waters—Refreshing Waters: Ancient Near Eastern Aquatic Symbolism at the Dawn of History

Rosel Pientka-Hinz

Abstract

In search of the most ancient references to the legendary Mesopotamian Flood and thus the mother of all climatic catastrophes, a cylinder seal from the Uruk period (ca. 3300–2900 BCE) depicting apparently dangerous hybrid animals that are arranged in a remarkably harmonic equilibrium caught my eye. Interpreting them as symbols of natural forces in cosmic interaction, this essay aims to highlight the crucial role of the symbolic snake twist in combination with powerful and aggressive animals like eagles and lions. Hypothesising that the beneficial curling of the snake not only symbolises fertility and a secure water supply, but also develops its harmonising potential by generating the notion of a balanced hybridity, places special emphasis on the twin snakes' integrative effect. As a veritable 'intercultural symbol' that emerged in a cultural contact situation, the mediating role of the snake twist becomes obvious in different fields of human creativity, namely in the invention of writing and the simultaneous development of an iconographic program. Conflict resolution strategies were developed in various media not only in order to avert climatic disasters. In the course of this, the lines between technical, symbolic or magical spheres blur.

•••

> "[…] Like water I do not know where I am going, like a boat I do not know at which quay I put in […]" (Babylonian incantation against evil)[*,1]

* This study has markedly benefited from a one-year fellowship at the International Consortium for Research in the Humanities (Käthe-Hamburger-Kolleg) "Dynamics in the History of Religions", under the direction of Prof. Dr. Volkhard Krech, Ruhr-University Bochum. Previous reflections on snake symbolism were put up for discussion in the course of presentations given in Bochum, Ghent, Rome, Berlin, Würzburg and Marburg. I am most grateful to my friend Dr. Heather D. Baker (University of Toronto) for correcting my English in large parts of this article. Clearly any errors are entirely my responsibility.

1 Wilfred G. Lambert, "Dingir.šà.dib.ba Incantations," *Journal of Near Eastern Studies* 33 (1974): 290, ll. 17–18.

 | DOI:10.1163/9789004444973_006

1 Introduction

From when humans first entered the Middle Eastern region of the so-called Fertile Crescent, traces of material as well as symbolic production indicate both a strong segregation of natural objects and cultural achievements, and a close interpenetration of these fields. In this context, both the practical approach to plants and animals and related ideological implications played an important role. In practice, people were engaged in hunting and gathering, in breeding and cultivating, in nourishing and cherishing. The earliest depictions of animals show threatening birds of prey and lions as well as snakes and scorpions contrasting with peaceful sheep and cattle, donkeys and goats. The most ancient Near Eastern texts written in proto-cuneiform archaic script which does not specify its underlying language already register wild animals as separate from domestic animals, not least boasting about breeding successes. In later periods when people produced texts written in unambiguously Sumerian or Akkadian, we read of kings as strong as a lion, heroes as impetuous as an aurochs, or women as fecund as a cow, surely to be understood as metaphorical descriptions for the purpose of highlighting 'superhuman' features.

The demarcation of a sheltered human world from the dangerous outside world undoubtedly played an important role through the ages and was a topic of interest in all media. It shaped social interaction as much as communication with the gods. In the following contribution, the problem will be approached from its 'weak points'. With a critical eye to the frequently invoked nature-culture-dichotomy, we intend to look at the boundaries between the civilised world and the wilderness, for the transitions between security and danger. We will look for a 'third space' where the spheres are blended and impressive animal symbolisms help to express this very specific realm of experience. Here, a world of monsters opens up and dragons are born. Here, allegedly contrasting forces mingle and the impetus for the development of rituals originates.

2 The Cultural Setting

At the dawn of history, around the end of the 4th to the beginning of the 3rd millennium BCE, when the first urban cultures of Sumer, Elam, and Egypt developed in the Ancient Near East, dynamic centralisation processes gave rise to an extensive redistribution of resources. Raw materials, manufactured goods, and people with or without professional expertise traveled from the hinterland as well as from remote areas into the centres where they were more or less successfully integrated into the prevailing order, often bringing along new information and ideas that influenced existing standards. As a matter of particular interest for the alluvial plain of Southern Mesopotamia, specific living conditions dictated practical procedures aimed at extensive trading activities in one of the most important urban centres of this early period, commonly known as archaic Uruk:[2] although productive land and rich hunting as well as fishing grounds were available, natural resources apart from bitumen, salt, clay, and water, especially the much sought-after durable and resistant materials like timber, stone, and mineral ores, were almost entirely lacking. This stood in marked contrast to the adjacent highlands of Iran that were rich in such resources. In the course of time and in tandem with the developing urbanisation processes, Sumerian demand for raw materials increased continuously and as a consequence, apart from the role of looting in warfare, the need for satisfactory trade relations became a major concern for local authorities.[3] It is important to realise that the urbanisation of southern Mesopotamia on the one hand and of the Iranian highlands on the other must not be considered in separation since both cultures developed in a mutual cultural exchange at all times.

The region in the westernmost part of Iranian Elam called Susiana (modern-day Ḫūsistān) occupied an intermediate position at the time since it also lay west of the Zagros mountain range, geographically an extension of the Tigris and the Euphrates river valleys, not far from ancient Sumer.[4] It becomes apparent that specific processes of acculturation and demarcation can be observed exactly in this contact zone at the boundary between the Sumerian and Elamite cultures. Susa, the capital of this Elamite region, always was in close contact with the inhabitants of Mesopotamia on the one hand and with the nomadic peoples who descended from the Northern mountain valleys on the

2 For a historical overview see, Hans J. Nissen, *Geschichte Altvorderasiens* (Munich: R. Oldenbourg Verlag, 1999), esp. 38–59 and 140–163; John N. Postgate, *Early Mesopotamia: Society and Economy at the Dawn of History* (London: Routledge, 1992/1994); Robert K. Englund, "Texts from the Late Uruk Period," in *Mesopotamien: Späturuk-Zeit und Frühdynastische Zeit*, ed. Pascal Attinger and Markus Wäfler, 13–233 (Freiburg,Göttingen: Universitätsverlag,Vandenhoeck and Ruprecht, 1998).

3 Cf. Peter R.S. Moorey, "Iran: A Sumerian El-Dorado?," in *Early Mesopotamia and Iran: Contact and Conflict 3500–1600 BC*, ed. John Curtis, 31–43 (London: British Museum Press, 1993); Gebhard J. Selz, "'Elam' und 'Sumer'—Skizze einer Nachbarschaft nach inschriftlichen Quellen der vorsargonischen Zeit," in *Mesopotamie et Elam*, ed. Léon. de Meyer and Hermann. Gasche, 23–43 (Ghent: University of Ghent, 1991).

4 Less than one-hundred kilometres; see Selz, "'Elam' und 'Sumer,'" 29.

other hand. The rich region of Susiana thus alternated between the primacy of two centres of control and influence: Southern Mesopotamia with intense trading ambitions to the west, and the great tribal heartlands of Iran, meaning the central Zagros mountain region and the highland plateau to the north and east.[5] This was the region that provided the stimulus for the creation of the Mesopotamian dragon.

3 The Main Structural Orders in Archaic Uruk

In archaic Uruk, the central figure of all essential negotiations was the so-called Priest-King (Sum. en), represented in cultic scenes as the celebrated 'man in the net kilt' (fig. 4.1) whose attentive and protective abilities were focused on the city's inhabitants and the main fields of conflict resolution. Just as in later literary texts, where the ruler was referred to as "the good shepherd of the people"[6] and "the keeper of righteousness and justice," the *en* was responsible for both the internal stability of his realm (some kind of temple community) and the adequate defense of its borders. In addition, he functioned as chief mediator between the human and the divine sphere. Thus the office of the archaic *en* of Uruk combined religious and secular, that is to say politico-military powers in equal measure.[7]

5 Pierre Amiet et al., "Susa in the Ancient Near East," in *The Royal City of Susa: Ancient Near Eastern Treasures in the Louvre*, ed. Prudence O. Harper, Joan Aruz, and Françoise Tallon (New York: Harry N. Abrams, 1992), 2.

6 See most recently Eva Cancik-Kirschbaum, "'Menschen ohne König …': Zur Wahrnehmung des Königtums in sumerischen und akkadischen Texten," in *Das geistige Erfassen der Welt im Alten Orient: Sprache, Religion, Kultur und Gesellschaft*, ed. Claus Wilcke, 167–190 (Wiesbaden: Harrassowitz Verlag, 2007), esp. 172–173.

7 See in detail Piotr Steinkeller, "On Rulers, Priests and Sacred Marriage: Tracing the Evolution of Early Sumerian Kingship," in *Priests and Officials in the Ancient Near East*, ed. Kazuko Watanabe, 103–137 (Heidelberg: Universitätsverlag C. Winter, 1999); Gebhard J. Selz, "Über Mesopotamische Herrschaftskonzepte: Zu den Ursprüngen mesopotamischer Herrscherideologie im 3. Jahrtausend," in *dubsar anta-men: Studien zur Altorientalistik. Festschrift für Willem H.Ph. Römer zur Vollendung seines 70. Lebensjahres mit Beiträgen von Freunden, Schülern und Kollegen*, ed. Manfried Dietrich and Oswald Loretz, 281–344 (Münster: Ugarit-Verlag, 1998), compares the Sumerian concept of rulership ("ein bürokratisch-sakrales") with the Semitic opposite ("ein dynastisch-charismatisches", p. 283). He emphasises the major importance of the archaic economic system of stockpiling ('Vorratswirtschaft') with the *en* being responsible for correct management: "[…] daß die Funktion des 'En' in seiner sakralen und ökonomisch-organisatorischen Verantwortung und Kontrolle der (Tempel-)güter bestand. Rein wirtschaftlich gesprochen besaß der 'En' die Verantwortung für Verwaltung und Verteilung der Nahrungsmittel" (see ibid., 292).

4 The Cosmic Equilibrium

We may see a notion of this cosmic totality depicted on a cylinder seal of green stone from the Uruk period (ca. 3300–2900 BCE), one of the most ancient cylinder seals at all, which can be considered a real masterpiece of this early age. It shows two different hybrid creatures, lion-headed eagles in a prey-catching position and pairs of snake-necked lions whose necks are affectionately intertwined and whose elegantly elevated tails nearly embrace the flying creatures (fig. 4.2).

Starting from this early masterwork of ancient Near Eastern imagery, the following look at the early historical world of symbols aims to highlight an extremely creative phase of human history. The period of archaic Uruk was a time when the development of symbolic systems was essential, especially in regard to the invention of proto-cuneiform writing and the elaboration of an iconographic program. Both fields represented highly dynamic systems whose evolution took place in comprehensive interaction.[9] At the cusp of the emergence of anthropomorphic divinities, quite a number of major agents were animals.

FIGURE 4.2 'Balanced hybridity.' Illustration of a Late Uruk cylinder-seal impression, ca. 3300–2900 BCE.
DRAWING BY ROSEL PIENTKA-HINZ
Note: Cf. Dominique Collon, *First Impressions. Cylinder Seals in the Ancient Near East* (London: British Museum Publications, 1987), 185 fig. 885.

9 Cf. Jean-Jacques Glassner, *Écrire à Sumer: L'invention du cunéiforme* (Paris: Édition du Seuil, 2000), and the review article by Gebhard J. Selz,"Schrifterfindung als Ausformung eines reflexiven Zeichensystems," *Wiener Zeitschrift für die Kunde des Morgenlandes* 90 (2000): 169–200.

The eagle, the lion, and the snake—all three of them can be considered mighty animals and therefore rulers of their particular habitats. As the most powerful animal of the air, the eagle was the "perfect animal" to represent the realm above the Earth's surface.[10] Similarly, the lion controlled the open country, especially the inhospitable steppe while the chthonic snake was at home in the realm beneath the surface, be it earthy or aqueous or muddy. But there is more to this than meets the eye. These ancient Near Eastern animals were also promoted to privileged symbolic vehicles because of their physical abilities and remarkable patterns of behaviour.

Considering the archaic pictographs already found in the most ancient texts from Uruk, the eagle, or perhaps rather the large bird of prey (Sum. ti_8), was depicted by a sign that represented a bird's wing (fig. 4.3, left). Interestingly enough, the Sumerian word á that also was written by this sign not only meant 'a bird's wing' but also 'power'. Thus the prominent feature that became manifest in the bird's graphic character obviously was its strong wing.

By contrast, the archaic pictograph representing a lion (Sum. piriĝ or ug) was a wide-open lion's mouth (fig. 4.3, middle). This sign was later used to write the Sumerian word ḫuš which means 'fearsome' or 'furious', thus emphasising the animal's characteristic feature.

Finally, the archaic pictograph showing a slithering snake with a forked tongue (fig. 4.3, right) was already used to illustrate 'length' and 'distance' as well as 'snake' (Sum. muš) in the most ancient texts from South Mesopotamian Uruk.[11] Although the semantic identification of a great number of ideograms forming part of this proto-cuneiform writing system still proves difficult,[12] the pictograph's unambiguous shape[13] as well as the great frequency of the snake

10 Elena Rova, "The Eagle and the Snake. Remarks on the Iconography of Some Archaic Seals," *KASKAL* 3 (2006): 5–6.

11 Whereas in later periods the very similar cuneiform signs expressing 'snake' (MUŠ), 'length' (BU) and 'distance' (SUD) are properly distinguished, at an early stage of writing graphic variants still alternate, all of them originating from one single prototype depicting a 'snake'; cf. Rosel Pientka-Hinz, "Schlange. A. In Mesopotamien," in *Reallexikon der Assyriologie und Vorderasiatischen Archäologie*, ed. Michael P. Streck (Berlin: Walter de Gruyter, 2009), vol. 12, 204, and now Robert K. Englund, "BU!," in *Why Should Someone Who Knows Something Conceal It? Cuneiform Studies in Honor of David I. Owen on His 70th Birthday*, ed. Alexandra Kleinerman and Jack M. Sasson (Bethesda: CDL Press, 2010), 112.

12 Englund, "Texts from the Late Uruk Period," esp. 65–72.

13 As opposed to the preferred representation of only distinctive parts of the body (head, ear, leg, wing etc.), in this case the total view of the animal was obviously crucial and intentional; cf. Yushu Gong, *Studien zur Bildung und Entwicklung der Keilschriftzeichen* (Hamburg: Verlag Dr. Kovač, 1993), esp. 13, 102–105, 126–127.

FIGURE 4.3 Left: a bird's wing, archaic sign Á/DA.[a] Middle: a lion's mouth, archaic sign PIRIĜ.[b] Right: a slithering snake, archaic sign MUŠ/BU.[c]

Notes:

a Margret W. Green and Hans J. Nissen, *Zeichenliste der archaischen Texte aus Uruk* (Berlin: Gebr. Mann Verlag, 1987), 170: 6 (A_2), 183: 65 (DA).

b Green and Nissen, *Zeichenliste der archaischen Texte*, 263: 428 (PIRIG, $ALIM_x$); cf. Catherine Mittermayer, *Die Entwicklung der Tierkopfzeichen. Eine Studie zur syro-mesopotamischen Keilschriftpaläographie des 3. und frühen 2. Jahrtausends v. Chr.* (Münster: Ugarit-Verlag, 2005), 23.

c Green and Nissen, *Zeichenliste der archaischen Texte*, 248: 373 (MUŠ), 181: 56 (BU, GID_2, SU_{13}).

sign in the archaic lexical and administrative text corpus[14] indicates its prominent position in the ancient scribe's mental mapping of the world around 3000 BCE. Both salient features, the characteristic animal behaviour (wriggling or meandering) and its body shape, inspired the snake's selection for an ideograph as part of one of the most ancient writing systems.

If the archaic eagle and lion distinguished themselves by their unbridled power and terrifying ferocity, what then exactly was the qualifying factor that allowed the snake to rise to a leading position in our seal's composition?

5 The 'Aqueous' Snake

From the late Chalcolithic Period (5th millennium BCE) on, snakes were a prominent element in the imagery of stamp seals and painted pottery all over the Fertile Crescent, most notably in snake-loving Iran.[15]

14 Green and Nissen, *Zeichenliste der archaischen Texte*, 56; Englund, "Texts from the Late Uruk Period," 70.

15 Pierre Amiet, *La glyptique mésopotamienne archaïque* (Paris: Centre Nationale de la Recherche Scientifique, 1980); Alwo von Wickede, *Prähistorische Stempelglyptik in Vorderasien* (München: Profil Verlag, 1990); Mahmoud Rahsad, *Die Entwicklung der vor- und frühgeschichtlichen Stempelsiegel in Iran: im Vergleich mit Mesopotamien, Syrien und Kleinasien—von ihren Anfängen bis zum Beginn des 3. Jahrtausends v.Chr.* (Berlin: Dietrich

We have an idea concerning one of the central meanings of snake images in these early days. At a symbolic level, snakes used to represent water courses, according to their specific locomotion often stylised as wavy lines, clearly reveal their aquatic or chthonic character.[16] As creatures of the water and the humid underground, snakes were considered a powerful symbol of fertility since time immemorial[17]—most notably in ancient Iran and Northern Mesopotamia, where people were much more dependent on periodic rainfall and flowing bodies of water emerging from the mountains than in adjacent Southern Mesopotamia, where the alluvial plain enabled a controlled exploitation of the rivers with an elaborate irrigation system developed for the purpose.[18]

In addition, another ophidian symbol, again of Iranian origin, depicting two entwined snakes held great attraction. In the archaic period, this motif was incorporated in the iconography of Elamite Susa in combination with one of the most important Sumerian symbols at all, the 'rosette' (fig. 4.4, left). Though we are far from sure of the precise meaning conveyed by the rosette, an association with fertility symbolism in the broadest sense, possibly combined with

---

Reimer Verlag, 1990); Elisabeth von der Osten-Sacken, "Schlange. C. In der Bildkunst," in *Reallexikon der Assyriologie und Vorderasiatischen Archäologie*, ed. Michael P. Streck (Berlin: Walter de Gruyter, 2009), vol. 12, 219–220. For the conspicuously extensive Iranian evidence cf. Elizabeth Carter et al., "Prehistoric Susa," in *The Royal City of Susa: Ancient Near Eastern Treasures in the Louvre*, ed. Prudence O. Harper, Joan Aruz, and Françoise Tallon, 43–46 (New York: Harry N. Abrams, 1992); Margaret C. Root, "Animals in the Art of Ancient Iran," in *A History of the Animal World in the Ancient Near East*, ed. Billie J. Collins, 169–209 (Leiden: Brill, 2002), esp. 173–180; François Bridey, *L'iconographie du décor peint de la céramique de Suse I: Les coupes des collections du musée du Louvre et du musée d'Archéologie nationale de Saint-Germain-en-Laye* (Paris: RMN, 2011), 42–44, 91–92; Heidemarie Koch, *Frauen und Schlangen: Die geheimnisvolle Kultur der Elamer in Alt-Iran* (Mainz: Philipp von Zabern, 2007).

16 Elizabeth van Buren, D., "Entwined Serpents," *Archiv für Orientforschung* 10 (1935–36): 54; Rova, "The Eagle," *KASKAL* 3 (2006): 9, fig. 10 with n.8 (with further literature); Bridey, *L'iconographie du décor peint*, 91: "très tôt le serpent a été associé à l'élément liquide, à « l'eau des sources qui serpentent sur le sol »." Cf. Elisabeth von der Osten-Sacken, *Der Ziegen-'Dämon': 'Obed- und Uruk-zeitliche Götterdarstellungen* (Kevelaer: Butzon and Bercker, 1992), 75–80.

17 Diana K. McDonald, *Serpent Imagery on Ancient Near Eastern Pottery* (Ann Arbor: Columbia University, 1989), 121: "[…] its appearance in connection with underground water or springs, as well as its seasonal appearance in spring, might make it seem connected to the chthonic forces which renew vegetation and fertility. Additionally, the snake's sinuous curves and undulating movement are reminiscent of water and its motion. The gleaming surface of the reptile also reinforces this association."

18 Cf. Selz, "Über Mesopotamische Herrschaftskonzepte," 286 "Als entscheidende Quelle für die Ausformung einer Herrschaftsorganisation gelten die naturräumlichen Gegebenheiten und die sich daraus entwickelnden sozio-ökonomischen Verhältnisse".

FIGURE 4.4 Left: The snake twist with 'rosettes.' Illustration of a Late Uruk cylinder-seal impression.[a] Right: Two male adders involved in a ritualised struggle.[b]
DRAWING BY ROSEL PIENTKA-HINZ.
Notes:
a Amiet, *La glyptique mésopotamienne archaïque*, Pl. 14 bis E; Reinhard Dittmann, "Seals, Sealings and Tablets: Thoughts on the Changing Pattern of Administrative Control from the Late-Uruk to the Proto-Elamite Period at Susa," in *Ǧamdat Naṣr, Period or Regional Style?: Papers given at a symposium held in Tübingen*, ed. Uwe Finkbeiner and Wolfgang Röllig (Wiesbaden: Reichert Verlag, 1986), 359: fig. 5: 3 and 4; Rova, "The Eagle," 4, fig. 3 a, 6, fig. 5; Boehmer, *Uruk*, 115.
b Roland Bauchot, ed., *Schlangen: Evolution, Anatomie, Physiologie, Ökologie und Verbreitung, Verhalten, Bedrohung und Gefährdung, Haltung und Pflege* (Augsburg: Bechtermünz Verlag, 1994), 96.

the notion of divine authority,[19] may be accepted. Thus the inventive Elamites living in Susa with close trade links to both the nomadic people of the hinterland and to the urban culture of Sumerian Uruk developed a new symbol composed of two elements: an old one known from the mountainous regions (two entwined serpents) and a new one whose circulation might be related to the emerging Sumerian culture (the rosette).[20]

19 In this regard seal impressions of the third millennium BCE showing a gracious face or just two eyes keeping a watching brief over the whole scene (e.g. a temple) in combination with large rosettes seem to be of special interest; see Amiet, *La glyptique mésopotamienne archaïque*, fig. 681; Collon, *First Impressions*, 22: fig. 53. Some scholars have connected the rosette with Inana, assuming some kind of exchangeability with the goddess's star-symbol. Objections were raised by Jeremy Black and Anthony Green, *Gods, Demons and Symbols of Ancient Mesopotamia: An Illustrated Dictionary* (London: British Museum Press, 1992) 156–157; Ursula Seidl and Manfred Krebernik, "Rosette," in *Reallexikon der Assyriologie und Vorderasiatischen Archäologie*, ed. Michael P. Streck, 442–447 (Berlin: Walter de Gruyter, 2007), esp. 443: "[…] scheinen die Mesopotamier mit der formalen Ähnlichkeit gelegentlich gespielt zu haben", and 446: "Die R[osette] konnte anscheinend an allem, was Gott, Mensch und Tier umgibt, angebracht sein; sie scheint ein allgemein positives, vielleicht schützendes Zeichen zu sein".

20 For an evaluation of the outstanding creative potential of the Elamites, who might have been the first to invent an early script as well as the cylinder seal, see Selz, "Schrifterfindung," esp. 170–171; Rainer N. Boehmer, *Uruk: Früheste Siegelabrollungen* (Mainz: Philipp von Zabern, 1999), 115.

Terrestrial snakes are usually solitary unless they are willing to mate: In that case they hug and squeeze each other in a coiling embrace that can last for several days, more or less resembling a twist-laid rope. A similar spectacle is presented by two love-crazed competing males who rear up and twist in the same posture in a ritualised struggle trying to force down the opponent's head, sometimes assisted by fierce bites (fig. 4.4, right). Presumably both versions of snake twists found their way into ancient Near Eastern pictorial representation, combined into one single form called the 'snake-guilloche motif' (*Flechtband*).[21] Since both situations are closely connected to procreation, the impossibility of distinguishing between them was likely incidental or even intentional.

In Southern Mesopotamia, the entwined serpents' motif that had been adopted from Elamite Susa through intensive trade through which passed authorising and thus identity-establishing seal impressions along with the various goods exchanged became an integral part of the archaic symbolic system. The two snakes clearly showed their rich fertility, emphasising the motif of life as regular embodiments of the male and female principle.[22] For instance, two entwined snakes in close proximity to a frog-like creature in an unambiguously sexual or birthing posture certainly evoke connotations of animal fertility.[23] More specifically yet, snake twists emerging from storage vessels (fig. 4.5, left)

21 Van Buren, "Entwined Serpents"; Karel G. Stevens, "Eine ikonographische Untersuchung der Schlange im vorgeschichtlichen Mesoptamien," in *Archaeologia Iranica et orientalis: miscellanea in honorem Louis Vanden Berghe*, ed. Léon de Meyer and Ernie Haerinck (Gent: Peters, 1989), 5; for its popularity in northern Mesopotamia and Syrian glyptic see Adelheid Otto, *Die Entstehung und Entwicklung der Klassisch-Syrischen Glyptik* (Berlin: Walter de Gruyter, 2000), 274–275, and "Wasser. B. Archäologisch," in *Reallexikon der Assyriologie und Vorderasiatischen Archäologie*, ed. Michael P. Streck (Berlin: Walter de Gruyter, 2016), vol. 15, 10–14; for its Iranian evidence see Ulrike Löw, *Figürlich verzierte Metallgefäße aus Nord- und Nordwestiran* (Münster: Ugarit-Verlag, 1998), 493–496.

22 Van Buren, "Entwined Serpents," 54. Another animal couple that from earliest times onwards represented sexuality or rather sexual instincts were the female and the male scorpion with their exceptional mating behavior, an elaborate 'wedding dance'; see Antoine Cavigneaux, "La Pariade du Scorpion dans les Formules Magiques Sumériennes (Textes de Tell Haddad V)," *Acta Sumerologica* 17 (1995): 75–99, esp. 88 et seq.; Rosel Pientka-Hinz, "Skorpion," in *Reallexikon der Assyriologie und Vorderasiatischen Archäologie*, ed. Michael P. Streck (Berlin: Walter de Gruyter, 2011), vol. 12, 577 et seq. The literary juxtaposition of a 'white' and a 'black bull', most probably describing the female and the male scorpion while mating (Cavigneaux, "La Pariade," 89), bears analogy to Iranian depictions of snake couples with both sexes differentiated by colour; cf. Löw, *Figürlich verzierte Metallgefäße*, 495 et seq., who speculates on the notion of dualism.

23 Stevens, "Eine ikonographische Untersuchung," 6 (C7); Boehmer, *Uruk*, 54 with Abb. 66 ('wohl tierschützende Dämonen'); Collon, *First Impressions*, 182: 863.

already symbolised one of the most important Sumerian concepts of economic stability, the idea of 'plenty and abundance', literally "what hopefully exists" or "what should be" (Sum. ḫé-ĝál)[24] in this early period. Here, the concurrent development of writing and iconographic symbols becomes obvious as we realise that the composite sign for Sumerian ḫé-ĝál and every other item that "hopefully should be" begins with the picture of a storage vessel (fig. 4.5, middle). In historical periods, this early stage of development led to the notion of overflowing vases, the so-called aryballoi, held by favourable water spirits named *'Laḫmu'* (fig. 4.5, right).[25]

Further explicit pairings in the Uruk glyptic imagery depict eagles or vultures catching prey next to entwined snakes (fig. 4.6). At first glance, these seem to illustrate the hostile relationship between predator and prey,[26] but on closer examination they are more likely to depict an allegorical view of the opposition between two different principles.[27] They affect the viewer emotionally due to the "strong decorative effect of the composition, in which symmetry and repetition of the same elements undoubtedly play a large role."[28] A harmonising or neutralising effect with the snakes 'flowing round' the birds may also

---

24 Abraham H. Jagersma, *A Descriptive Grammar of Sumerian* (Leiden: Internet Publication, 2010), 129.

25 Eva A. Braun-Holzinger, "Apotropaic Figures at Mesopotamian Temples in the Third and Second Millennia," in *Mesopotamian Magic: Textual, Historical, and Interpretative Perspectives*, ed. Tzvi Abusch and Karel van der Toorn (Groningen: Styx Publications, 1999), 159 with n.82; *cf.* Otto, "Wasser", *Reallexikon der Assyriologie und Vorderasiatischen Archäologie* vol. 15, 10–14 § 6. On the interconnection of streaming water with Mesopotamian theology of creation and royal mediation in later times see now Stéphanie Anthonioz, "Water(s) of Abundance in the Ancient Near East and in Hebrew Bible Texts: A Sign of Kingship," in *Thinking of Water in Judah*, ed. Ehud Ben Zvi and Christoph Levin (Berlin: de Gruyter, 2014), 41–68; and more detailed Stéphanie Anthonioz, *L'eau, enjeux politiques et théologiques, de Sumer à la Bible* (Leiden: Brill, 2009), 399 et seq.

26 The fight between eagle (or vulture) and snake was definitely one of the most characteristic iconographic themes of the Iranian glyptic material since the early fourth millennium BCE; cf. Rova, "The Eagle," 14.

27 Stevens, "Eine ikonographische Untersuchung," 16 n.1; Mark A. Brandes, *Siegelabrollungen aus den archaischen Bauschichten in Uruk-Warka* (Wiesbaden: Steiner Verlag, 1979), 211: "[…] ebenso gut der ursprünglich zwischen ihnen herrschende Zustand des Friedens"; Rova, "The Eagle," 23: "[…] the association of the two animals may thus be interpreted in terms of polarity rather than of hostility, i.e. in terms of constant tension between two opposite poles which are both necessary for the balance of the universe, and finally as coincidentia oppositorum".

28 Rova, "The Eagle," 2 with fig. 1; cf. Amiet, *La glyptique mésopotamienne archaïque*, 70: "relèvent du goût des anciens orientaux pour l'antithèse, l'équilibre et la symétrie, notions abstraites susceptibles d'évoquer de façon très large l'ordre cosmique".

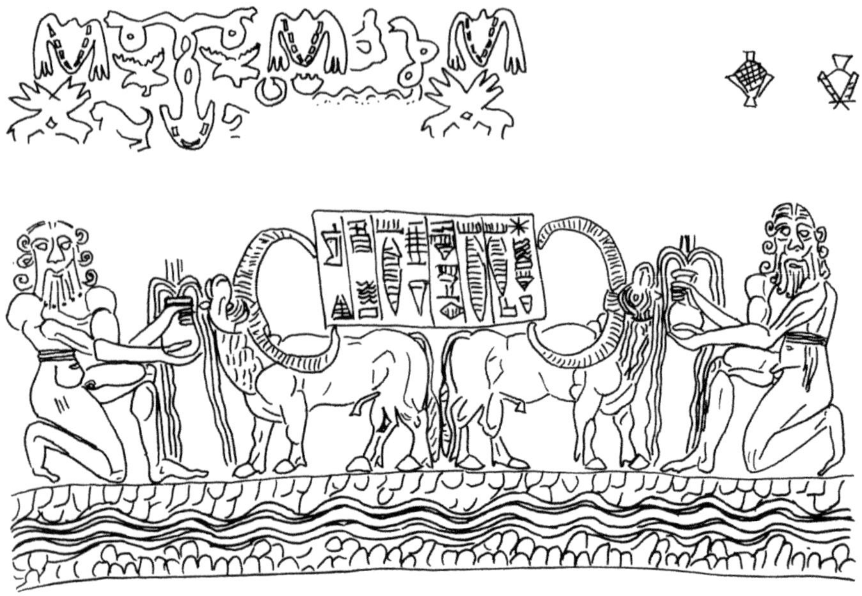

FIGURE 4.5 Left: 'Overflowing vessels'. Illustration of a Late Uruk cylinder-seal impression.[a] Above right: Archaic sign GAN (Sum. ḫé).[b] At the bottom: Water spirit with overflowing vase and water buffaloes. Illustration of an Akkadian cylinder-seal impression, ca. 2200 BCE.[c]
DRAWING BY ROSEL PIENTKA-HINZ.
Notes:
a E.g. Amiet, *La glyptique mésopotamienne archaïque*, fig. 1631; Boehmer, *Uruk*, 85, 119; Rova, "The Eagle," 4 fig. 3 a.
b Green and Nissen, *Zeichenliste der archaischen Texte*, 208: 190 (GAN).
c Cf. Donald P. Hansen, "Art of the Akkadian Dynasty," in *Art of the first Cities*, ed. Joan Aruz and Ronald Wallenfels (New Haven: Yale University Press, 2003), 209.

have been intended. With the snakes swallowing their own tails and thus, additionally, defusing the situation, the animals finally lose their potential threat. Simultaneously this initiates a stylistic development to entirely headless snake coils merely symbolising endless currents of water.[29]

29 Compare the almost endless entwined snake from Uruk, which was only brought into being by rolling a remarkable seal at least three times. The conception of this complex motif indeed required a high faculty of abstraction; cf. Christoph Gerber, "Die Schlange auf der Tonkugel," in *Beiträge zur Kulturgeschichte Vorderasiens: Festschrift für Rainer Michael Boehmer*, ed. by Uwe Finkbeiner, Reinhard Dittmann, and Harald Hauptmann, 175–176 (Mainz: Philipp von Zabern, 1995).

FIGURE 4.6 Different variants of the eagle-snake theme. Illustration of Late Uruk cylinder-seal impressions.
DRAWING BY ROSEL PIENTKA-HINZ.
Note: Cf. Brandes, *Siegelabrollungen*, Tafel 27–29; Rova, "The Eagle", 2.

Another composition showing bull heads instead of eagles (fig. 4.7) might be evaluated in the same way, only replacing the most potent animal of the air by its equivalent on the ground, namely the bull. These animals certainly played an important part in the religious ceremonies of archaic Uruk[30] and therefore were also appropriate choices to be combined with the previously alien snake couple.

On a different level, in one of the earliest rebus-like instances of writing,[31] both constellations can also be read as a graphic rendition of the two most important notions associated with abundance in this early period: beneficent rain (eagle + snakes = wind + water) and prosperity through cattle breeding (bull head + snakes = [breeding] bull + semen).[32] In this sense, the eagles and the snakes are understood as symbols for wind and water in a harmonic encounter.[33]

---

30 Rosel Pientka-Hinz, "Bunte Kühe? Zu den frühesten Farbbezeichnungen im Alten Orient," in *The Empirical Dimension of Ancient Near Eastern Studies: Die empirische Dimension altorientalischer Forschungen*, ed. Gebhard J. Selz and Klaus Wagensonner (Wien: LIT Verlag, 2011), 349–356.

31 For further examples see Ursula Moortgat-Correns, "Die Rosette—ein Schriftzeichen?: Die Geburt des Sterns aus dem Geiste der Rosette," *Altorientalische Forschungen* 21 (1994): 359–371; Pientka-Hinz, "Bunte Kühe?," 355; for a comparable in-depth analysis of the development of Ancient Egyptian pictographs and symbolic signs see Ludwig D. Morenz, *Bild-Buchstaben und symbolische Zeichen: Die Herausbildung der Schrift in der hohen Kultur Altägyptens* (Freiburg/Göttingen: Universitätsverlag/Vandenhoeck and Ruprecht, 2004).

32 In Sumerian, the word for 'water' and 'semen' is the same (a or e_4). For the extreme concentration of effort with regard to most ancient cattle-breeding that can be observed especially in archaic Uruk, see Pientka-Hinz, "Bunte Kühe?."

33 The difference in character of the four Mesopotamian winds (the rainy east wind, the cool north wind, the hot and humid south wind, and the virtually ignored west wind) was recently pointed out by Frans A.M. Wiggermann, "The Four Winds and the Origins of Pazuzu," in *Das geistige Erfassen der Welt im Alten Orient*, ed. Claus Wilcke (Wiesbaden:

FIGURE 4.7 Bull—snake theme. Illustration of a Late Uruk cylinder-seal impression.
DRAWING BY ROSEL PIENTKA-HINZ.
Note: Boehmer, *Uruk*, Abb. 68, 6.

Accordingly, the Sumerian sign combination for 'rain' (Sum. /šeĝ/) later came to be a combination of the signs for 'water' (A), 'heaven' (AN) and 'wind' (IM). Without the windy element, it is read as 'rain' (šèĝ = A.AN), and with the windy element it can also be read as 'rain' ($šeĝ_x$ = IM.A, IM.A.AN) though the writer may have had a stormy weather in mind.[34] We have seen that, as the most powerful animal of the air, the eagle was the 'perfect animal' to represent the realm above the Earth's surface, including atmospheric phenomena like wind and rain. Significantly we do not find the lion-headed eagle Anzu that was to

Harrassowitz Verlag, 2007), 125–165. It is striking that from the group of the partly anthropomorphic figures representing the four winds in the second millennium BCE, only the south wind shows ophidian features, namely intertwined legs sometimes ending in snake heads.

34 Cf. the telling Sumerian designations for the rainy east wind, tu_{15}-im-šèĝ-ĝá 'rain bringing wind' and tu_{15}-ḫé-ĝál-la 'abundance bringing wind' on the one hand, and for the south wind, tu_{15}-mar-uru5 'flood wind', on the other hand; Konrad Volk, *Inanna und Šukaletuda: Zur historisch-politischen Deutung eines sumerischen Literaturwerkes* (Wiesbaden: Harrassowitz Verlag, 1995), 46–47 n.306; 182. Kamran V. Zand has recently detailed the complexity of the later logographic writing for the monstrous Anzu bird (AN.IM.MI^{mušen}), embodiment of mighty storms; Kamran V. Zand, "Zu den Schreibungen des Anzud-Vogels in der Fāra-Zeit," in *Von Göttern und Menschen. Beiträge zu Literatur und Geschichte des Alten Orients. Festschrift für Brigitte Groneberg*, ed. Dahlia Shehata, Frauke Weiershäuser and Kamran V. Zand, 415–442 (Leiden: Brill, 2010). As is well known that constructive playing with ideograms continued to be a constitutive element of the cuneiform writing system.

become so famous later (see below) depicted here, but a more harmless creature without the aggressive lion head.

In summary it can be said that these Urukean animals are rather pictured in a distinct state or condition than in explicit action.[35] We could also speak of 'natural states' that can be brought about by 'culturally formed' reality-constituting measures. A similar conception of universal relations can also be traced in the depiction of those remarkable hybrid creatures from the beginning (fig. 4.2).

6 The Cosmic Equilibrium (Resumed)

Here we can observe real processes of de-escalation since the terrestrial as well as the atmospheric aggressors are neutralised by the concept of the harmonic snake twist and further symmetry-generating features. In this context it must be emphasised that the lion's aggressive power appears on earth and in the sky. The lion-headed eagle must be interpreted as an early incarnation of the monstrous bird Anzu,[36] the symbol of disastrous thunderstorms and sandstorms in historical times. In analogy, the creatures on earth, fusions of lions and snakes, should be seen as representing aggressive or chaotic environmental forces precisely on the land. Critical for the agricultural economy of Southern Mesopotamia were not only devastating storms, but above all floodwater and soil salinisation. Accordingly, relevant environmental forces, possibly symbolised by the lion, could have been aggressive currents of water, i.e. changes of riverbeds, flooding disasters or dike breaches, but also soil degradation as a result of salinisation and finally extreme heat combined with drought damages. The invasion of pests like locusts or rodents and last but not least hostile attacks by hordes of people should not be forgotten either.[37] In analogy to the lion-headed Anzu bird, further concepts of leonine monsters were developed over the centuries, initially representing turbulent weather or other

35 Brandes, *Siegelabrollungen*, 211: "Die Bannung der beiden Tiere in einen Zustand, nicht in eine Handlung". Comparable efforts concerning the conception of universal relations might be seen in the development of the earliest terminology of colors that also reveals the tendency to describe changeable states ('bright' vs. 'dark', 'flourishing' vs. 'dried up', etc.); Pientka-Hinz, "Bunte Kühe?,"

36 Eva A. Braun-Holzinger, "Löwenadler," in *Reallexikon der Assyriologie und Vorderasiatischen Archäologie*, ed. Dietz O. Edzard (Berlin: Walter de Gruyter, 1987–1990), vol. 7, 94–97. And see above note 46.

37 For an overview of the agricultural conditions in early Sumer see Tony J. Wilkinson, "Hydraulic Landscapes and Irrigation Systems of Sumer," in *The Sumerian World*, ed. Harriet Crawford, 33–54 (London/New York: Routledge, 2013); Magnus Widell, "Sumerian Agriculture and Land Management," ibid. 55–67.

awe-inspiring environmental phenomena.[38] Since these composite beings representing water-spitting monsters (the so-called 'lion-griffins' or 'lion-dragons')[39] as well as veritable 'flood monsters'[40] were predominantly composed of leonine and aquiline body parts, it seems not unreasonable to assume we are seeing the depiction of a previous concept here, with its individual characters (a lion accompanied by an eagle) even clearer identifiable.

Concerning the snake twist, it is certainly no coincidence that the Sumerian language refers to a watercourse's 'mouth' (its beginning or branch-off point), its 'tail' (the ending or outfall) and most notably its 'neck', the banks for both rivers and canals.[41] Keeping in mind the 'aqueous' nature of the snake , there is even further philological evidence with regard to its beneficial 'curling' in association with the 'tying up' of the evildoer, since both actions are designated by the same term 'curling, twisting' (Sum. gú-gilim, lit. 'to intertwine the neck or the rope').[42] Two intertwined snakes, which in a certain way symbolise a balanced and stable interplay of the water forces and at the same time guarantee 'plenty and abundance'— at first glance they act like a stabilising rope[43] that keeps the powerful forces under control. To a certain extent, they even tranquillise the dangerous environmental phenomena in the air and on earth.

To sum up, since Uruk's agriculture was highly dependent on well-functioning canal management and good weather conditions, the people lived in constant fear of devastating storms as well as catastrophic floods, i.e. having their harvest drowned by spring floods bursting the dikes. And last but not least, people feared crop losses due to soil salinisation. Aside from agrarian measures maintained through hard community labour, further conflict resolution strategies were developed in various media. We have seen at the iconographic level the ancient artists succeeded in finding a stylistic tool to neutralise chaotic or

38 Frans A.M. Wiggermann, "Mischwesen. A. Philologisch. Mesopotamien," in *Reallexikon der Assyriologie und Vorderasiatischen Archäologie*, ed. Dietz O. Edzard (Berlin: Walter de Gruyter, 1983–1997), vol. 8, 222–246, esp. § 2.1.

39 Cf. F.A.M. Wiggermann, "Mischwesen," § 7.25; and Eva A. Braun-Holzinger, "Löwendrache," in *Reallexikon der Assyriologie und Vorderasiatischen Archäologie*, ed. Dietz O. Edzard (Berlin: Walter de Gruyter, 1987–1990), vol. 7, 97–99.

40 Ursula Seidl, "Das Flut-Ungeheuer abūbu," *Zeitschrift für Assyriologie und Vorderasiatische Archäologie* 88 (1998): 100–113.

41 Marten Stol, "Kanal(isation). A. Philologisch," in *Reallexikon der Assyriologie und Vorderasiatischen Archäologie*, ed. Dietz O. Edzard (Berlin: Walter de Gruyter, 1976–1980), vol 5, 355–365.

42 Cf. Pascal Attinger, *Eléments de linguistique sumérienne: la construction de du$_{11}$/e/di „dire"* (Fribourg, Suisse: Ed. Universitaires, 1993), 519.

43 In historical times, ropes were still likened to the ophidian god Niraḫ in association with water and boat equipment; cf. Frans A.M. Wiggermann, "Niraḫ, Irḫan," in *Reallexikon der Assyriologie und Vorderasiatischen Archäologie*, ed. Dietz O. Edzard (Berlin: Walter de Gruyter), vol. 9, 1998–2001, 572 § 2.2.

FIGURE 4.8 The feeding of 'the dragon' depicted on a Late Uruk cylinder-seal impression.
DRAWING BY ROSEL-PIENTKA-HINZ.
Note: Amiet, *La glyptique mésopotamienne archaïque*, Pl. 13 fig. L.

FIGURE 4.9
'Royal hybridity.' Illustration of the lion-hunt stele from Uruk, ca. 3300–3000 BCE.
DRAWING BY ROSEL PIENTKA-HINZ
Note: Watanabe, *Animal Symbolism*, Fig. 1.

uncontrollable forces in order to create a balanced harmony, a restructured and peaceful world.[44] In order to maintain this balance, the creatures tamed in this way—early embodiments of the Mesopotamian clearly apotropaic 'snake-dragon' (Sum. muš-ḫuš, lit. 'fearsome serpent')[45]—were even fed by skilled personnel sometimes (fig. 4.8).

Such feedings surely took place in particular localities where ritualised sacrifices were offered and the boundaries between nature and culture dissolved in order to activate all forces to the benefit of its participants. We do not know which authority was represented by the seal. And although there is only little information about ritual acts in connection with the feeding of snakes or snake-like animals in Uruk,[46] the seal of the so-called 'snake-charmer' is testimony to the presence of such a specialist.[47]

It is in this context, too, that the predominant role of the ruler as an opponent of a lion in a real duel to the death, (fig. 4.9) must be pointed out. The royal hunt continued to be the most famous Mesopotamian ritual for three millennia.[48] It was the particular task of the ruler to neutralise the evil energy of potential opponents embodied by the animal—be it a foreign ruler or other dangerous environmental phenomena—and simultaneously make the power and vitality of the beast available to the kingdom. In a way, the ruler became effectively 'intertwined' with the lion. In his function as conqueror of the enemy,

---

44 Cf. Christoph Uehlinger, "Drachen und Drachenkämpfe im alten Vorderen Orient," in *Auf Drachenspuren: Ein Buch zum Drachenprojekt des Hamburgischen Museums für Völkerkunde*, ed. Bernd Schmelz und Rüdiger Vossen (Bonn: Holos Verlag, 1995), 57: "Die Verschlingung der Hälse zeigt an, daß ihre Macht in einem dynamischen Gleichgewicht gehalten ist."; Manfred Hutter, *Religionen in der Umwelt des Alten Testaments I: Babylonier, Syrer, Perser* (Stuttgart: Verlag W. Kohlhammer, 1996), 52: "Miteinander kämpfende Dämonen oder Mischwesen neutralisieren sich dabei im Zusammenprall ihrer positiven und negativen Wirkungen, so dass das Böse eliminiert wird." For further iconographic attempts to control chaotic states by inventing mixed creatures cf. for instance Rova, "The Eagle," 4 fig. 3 b (fighting griffins in company with a huge coiling snake).

45 A composite being with snake's body and neck, lion's forlegs and bird's hindlegs; Frans A.M. Wiggermann, "mušḫuššu," in *Reallexikon der Assyriologie und Vorderasiatischen Archäologie*, ed. Dietz O. Edzard (Berlin: Walter de Gruyter, 1993), vol 8, 455–462.

46 In this regard the Old Sumerian goddess of incantations Nin-girima is of considerable interest, since apart from her ophidian character one of her main cult centers was Uruk with the famous 'Steinstifttempel' that enclosed a suspicious water basin. This basin possibly functioned as a dwelling for snakes and fish, representing the snake nourishing toponym muš-bùluĝ-bùluĝ "snake feeding"; see Manfred Krebernik, "Nin-girima. I. Beschwörungsgöttin," in *Reallexikon der Assyriologie und Vorderasiatischen Archäologie*, ed. Dietz O. Edzard (Berlin: Walter de Gruyter, 1998–2001), vol 9, 363–367 § 3.

47 See Pientka-Hinz, forthcoming.

48 Cf. Chikako E. Watanabe, *Animal Symbolism in Mesopotamia. A Contextual Approach* (Vienna: Institut für Orientalistik der Universität Wien, 2002).

who ritually merged with the lion, the king himself became a kind of dragon representing the concept of 'royal hybridity'.[49]

7 Conclusion

Through the following millennia of ancient Near Eastern history, the ambivalence of water symbolism remained omnipresent. A rich textual tradition demonstrates that the motifs presented here were preserved for a long time. Though they were not immune to further development and reinterpretation, the principal dichotomy of life-giving water and destructive water was maintained throughout the entire period.[50]

On the one hand, the 'waters of abundance' trope found its way into various mythological tales. In the course of the elaboration of anthropomorphic divine concepts, they were associated with different water deities, especially with the Sumerian god of wisdom Enki (the Akkadian Ea) and his entourage. Furthermore, various snake gods whose relation to the Eastern mountains often remained recognisable can subsequently be assigned to the variously connoted chthonic realms.[51] The rulers praised themselves in their inscriptions as the suppliers of water to their country. The maintenance of the canal systems as well as the provision of the people with 'plenty and abundance', along with the worship of the gods, was a top priority in their reign. According to the famous prologue of the Laws of Ḫammurabi, even in the second millennium BCE, the programmatic title of the ruler as 'water supplier' remained inseparably linked with the care of Uruk above all.[52]

---

49 And although it has not become a standard designation for the ruler, it remains to be examined to what extent dragon epithets became part of the royal self-image. In later periods especially the Sumerian king Šulgi boasted about his lion's courage and his dragon's blood: "I am a fierce-looking lion, begotten by a dragon" (piriĝ igi ḫuš ušumgal-e tud-da-me-en; Shulgi A l. 3); The Electronic Text Corpus of Sumerian Literature (ETCSL) 2.4.2.01 (http://etcsl.orinst.ox.ac.uk).

50 See most recently Anthonioz, *L'eau, enjeux politiques et théologiques*, 2009; and id., *The Water(s) of Abundance*, 2014.

51 They were related to vegetation and agriculture, growth and decay, see Frans A.M. Wiggermann, "Transtigridian Snake Gods," in *Sumerian Gods and their Representations*, ed. Irvine L. Finkel and Markham J. Geller (Groningen: Styx Publications, 1997), 33–55. In Elamite culture, the connection of beneficial water and snakes remains particularly obvious; see Pierre de Miroschedji, "Le dieu élamite au serpent et aux eaux jaillissantes," *Iranica Antiqua* 16 (1981): 1–25.

52 "The lord who revitalizes the city of Uruk, who provides abundant waters for its people;" Martha T. Roth, *Law Collections from Mesopotamia and Asia Minor* (Atlanta: Scholars Press, 1997), 77.

On the other hand, environmental catastrophes of all kinds were integrated into a mythological context of combat and defeat, where dragons are slain by heroic gods in order to resolve the tension between chaos and order. Helpful hybrid beings can be seen performing cultural work in their proper spheres of action and agency, i.e. as attendants of their gods or in an apotropaic function as doorkeepers, with the snake-dragon representing the most powerful of its kind. It was this creature that rose to become the companion of the Babylonian god Marduk and henceforth was considered a guarantee of stability and peace.

And the Mesopotamian Flood—the very starting point of my reflections? Cataclysmic waters, according to literary texts from the second millennium BCE on, were always accompanied by bad weather and torrential rain. They should therefore be read more as a 'flood storm' than a huge inundation (Sum. /amaru/, Akk. *abūbu*). This became a metaphor and mythologeme especially associated with warlike devastation. As the personification of divine wrath and aggression, it transformed into the monstrous weapon of the gods. In addition, the motif was integrated in myth as a cosmic re-creative act that was part of an infinite cycle of creation and destruction. As such it was instrumentalised as a boundary between an antediluvian primordial era and the following historical time—the beginning of civilisation.[53]

The Mesopotamian way of life always remained under threat from environmental disasters and warlike attacks. The cuneiform texts tell of devastating storms, dike breıaches, harvest losses, famine, plagues, drought, armed conflicts, etc. Whether any of these are references to a real climate change is, however, subject to debate.[54]

The constructive new beginning, the re-establishment of the all-embracing balance, was embraced by the people as a call not to be idle: "It is a deluge!"[55]

53 Cf. Claus Wilcke, "Weltuntergang als Anfang," in *Weltende*, ed. Adam Jones (Wiesbaden: Harrassowitz-Verlag, 1999), 63–112; Stefan M. Maul, "Ringen um göttliches und menschliches Mass. Die Sintflut und ihre Bedeutung im Alten Orient," in *Schönheit und Mass. Beitäge der Eranos Tagungen 2005 und 2006*, ed. Erik Hornung und Andreas Schweizer (Basel: Schwabe, 2007), 161–183; Anthonioz, *L'eau, enjeux politiques et théologiques*, 2009; Beate Pongratz-Leisten, "Sintflut (flood). A. In schriftlichen Quellen, in *Reallexikon der Assyriologie und Vorderasiatischen Archäologie*, ed. Michael P. Streck (Berlin: Walter de Gruyter, 2011), vol 12, 525–526; Yi S. Chen, *The Primeval Flood Catastrophe* (Oxford: Oxford University Press, 2013); Yi S. Chen, "Major Literary Traditions Involved in the Making of Mesopotamian Flood Traditions," in *Opening Heaven's Floodgates*, ed. Jason M. Silverman (New York: Gorgias Press, 2013), 141–190.

54 Cf. James Kinnier Wilson, "On the UD-ŠU-BALA at Ur Towards the End of the Third Millennium BC," *Iraq* 67 (2005): 47–60.

55 Walther Sallaberger, *'Wenn Du mein Bruder bist …' Interaktion und Textgestaltung in altbabylonischen Alltagsbriefen* (Groningen: STYX Publications, 1999), 159–163.

CHAPTER 5

Deserts—Wind—Water: Strategies of Human Adaptations and Renewable Resources in Extremely Arid Environments. A Historical Retrospective

Eckart Ehlers

Abstract

Qanats are a millenia-old traditional technology to tap and to exploit subterraneous groundwater tables by construction of underground water channels in the arid desert regions of the Middle East and North Africa (MENA). Historically first traceable in the beginning of the first millennium BCE in northwestern Iran, this technique has spread all over the MENA region in connection with the expansion of Achaemenid and Sassanian Iran and the spread of Islam into North Africa and the Mediterranean. Via colonial Spain qanat technology reached Latin America; its diffusion to the east and its historical and geographical pathways are still under discussion. Qanats have survived until today as a highly sustainable and unique irrigation system in traditional agricultural societies of the MENA region. Due to modern over-exploitation of the hardly renewable groundwater resources through pump irrigation, however, qanats are on the brink of extinction.

This paper analyses the traditional engineering aspects of qanat construction and its historical origins and spreads. Qanats were preconditions for isolated rural and urban habitats in the midst of otherwise hostile desert environments. In combination with climatically well adapted forms of vernacular architecture they have even enabled the development and perseverance of formidable desert cities. On the other hand, they have cemented unfavorable socio-economic relationships between land and water owning landlords and their dependent rural agricultural workers and peasants due to the mechanisms of rent-capitalism. Today, the Persian qanat technology and qanat irrigation are registered under UNESCO's World Heritage Sites as humanity's cultural property.

•••

 | DOI:10.1163/9789004444973_007

In memoriam
Michael E. Bonine (1942–2011)
Professor of Geography and Near Eastern Studies,
University of Arizona Tucson
A dear friend—a wonderful colleague—a great scholar

∵

1 Natural Environments and Human Development—Some Introductory Considerations

One of the most constant factors in the history and development of humankind is its permanent struggle with nature. All societies at all times have been faced with this challenge—and they still are. In this context, nature is more than those extreme events that are nowadays experienced as so-called 'catastrophes' with losses of life and destruction of material culture. In a more general as well as both historical and philosophical context "nature stands for that which is elemental, autonomous, spontaneous, organic, accessible, and non-manufactured".[1] Its antithesis "is artificiality, technology, order based on conventions and agreements, the made and coerced, the created and the cultivated: in short, culture".[2] In spite of these clear distinctions, even antagonisms, nature and culture can hardly be separated from each other. In a philosophical perspective and in Western thought and tradition, the evolution of human societies and civilisations has often been interpreted as the outcome of a struggle between nature and humankind, i.e. as a kind of interdependent entity. Nature and its elements are seen as the primordial environment, culture as the result of its conquest or transformation by people.[3]

From a geographical perspective and in regard to the specific topic of this volume, two paradigms and historical approaches may serve as examples of how the conflicts between nature and society/culture were theorised in the history of science and in different cultural contexts. The first example is based

1 Rolf P. Sieferle, "The Ends of Nature," in *Environment across Cultures. Wissenschaftsethik und Technikfolgenbeurteilung*, ed. Eckart Ehlers, and Carl F. Gethmann (Berlin: Springer, 2003), 14.

2 Sieferle, "The Ends of Nature," 14.

3 Clarence J. Glacken, *Traces on the Rhodian Shore. Nature and Culture in Western Thought from Ancient Times to the End of the Eighteenth Century* (Berkeley, Los Angeles, London: University of California Press, 1967), 763.

on the hypothesis that nature dominates all and everything, including human activities and their results. Even worse: human characters themselves, human attitudes and human behaviour are determined by climate and their natural environments. Ever since Gobineau's "Essai sur l'inégalité des races humaines" (1851–1853) the idea of nature's decisive influence on human societies has been accorded a great importance in the interpretations of the close (inter-)relationships between nature and culture. In the late 19th and early 20th century, environmental determinists argued that "man is a product of the earth's surface" and that earth "has entered into the bone and tissue, into his mind and soul".[4] This school of environmental determinists found one of its strongest advocates in Huntington's influential work on *Civilization and Climate* (1915), later incorporated into his much discussed *Mainsprings of Civilization* (1954/1959). Somewhat less deterministic approaches and almost simultaneously developed were concepts of people's "creative adjustment" to their specific natural environment: Cultural anthropologists and especially French geographers argued that people made choices in adjusting to nature and coping with their specific environments. Based on their social organisation, their traditional knowledge systems, their technological know-how, and/or their cultural values, they had different possibilities to develop their own "genres de vie".[5] Such possibilities, theorised as 'possibilism', offer choices to adjust to their environment, thus reconciling nature and culture within limits. It goes without any further discussion that especially extreme natural environments—the arctic, high altitudes or deserts for example—pose specific challenges to human societies and their endeavours to adjust, to cope and to survive until today.

Nowadays, determinism and/or possibilism as geographic explanations for the interactions between nature and society have become more or less obsolete. Scientific and technological developments seem to have conquered nature, and nature no longer poses obstacles to human interference in or occupation of natural environments. On the contrary: human inventiveness, the developments of science, and innovative forms of sophisticated knowledge systems seem to have overcome nature—and not always to the benefit of both nature and humanity.[6] Our case study, the creative adaption and use of nature's limited resources by people over centuries, on the one hand, and its

4 Ellen C. Semple, *Influences of Geographic Environment* (New York: 1911), 1.

5 P. Vidal de la Blache, "Les genres de vie dans la géographie humaine," *Annales de Géographie* 20 (1921): 193–212 and 289–304; P. Vidal de la Blache, *Principes de géographie humaine* (Paris, 1922).

6 For detailed analyses in different cultures see: Ehlers, and Gethmann, *Environment across Cultures*; for the Middle East see: William C. Brice, ed., *The Environmental History of the Near and Middle East* (London, New York, San Francisco: Academia Press, 1978) and;

destruction by modern technologies in a very short time, on the other, underpins the following deliberations.

2 Deserts and Desert Societies: Challenges and Responses

In spite of these fundamental changes in the relationship between nature and society today, extreme natural environments still pose major challenges for human occupation and exploitation. All the more we have to give due respect to the great achievements of earlier generations and older societies in coping with the adversities of nature and to develop sustainable forms of human survival. A striking example of this successful experience can be found in the millennia-old conquest of the deserts of North Africa and the Near and Middle East by the ingenious use of their very limited natural resources.

2.1 *Qanats: Hidden Waters and Desert Culture*

Whoever has had the possibility to cross North Africa or the Near and Middle East by air and to fly from the Maghreb to Tehran, Kabul, Islamabad or even further to Xinjiang (新疆) may have observed linear pock-marked structures in the deserts of northern Africa and western Asia. Focused especially on the foothills of mountain ranges that are snow-covered in winter and rain-fed, these obviously artificial structures terminate in settlements and/or small plots of irrigated gardens and fields in the middle of nowhere. Settlements, gardens and fields, however, cannot exist and flourish without water. Thus, questions arise: Where does the water come from? How is it 'produced'? And how has it influenced the development of a specific desert culture?

The answers to these questions are connected with the term qanat (Arab. *qanāt*), the common expression used in Iran for these conspicuous structures. Iran is very obviously the birthplace of this unique technology of 'mining the water'—a technology developed more than 2500 years ago and, in the course of history, distributed all over the dry belt of North Africa and the Near and Middle East. Qanats are—in the words of Ann K.S. Lambton—"mining installations or techniques" that work as follows:

> By means of a gently sloping tunnel, which cuts through alluvial soil and passes under the water table into the aquifer, water is brought by gravity

---

Malcolm T. Wagstaff, ed., *Landscape and Culture. Geographical Archaeological Perspectives* (Oxford, New York; Blackwell, 1987).

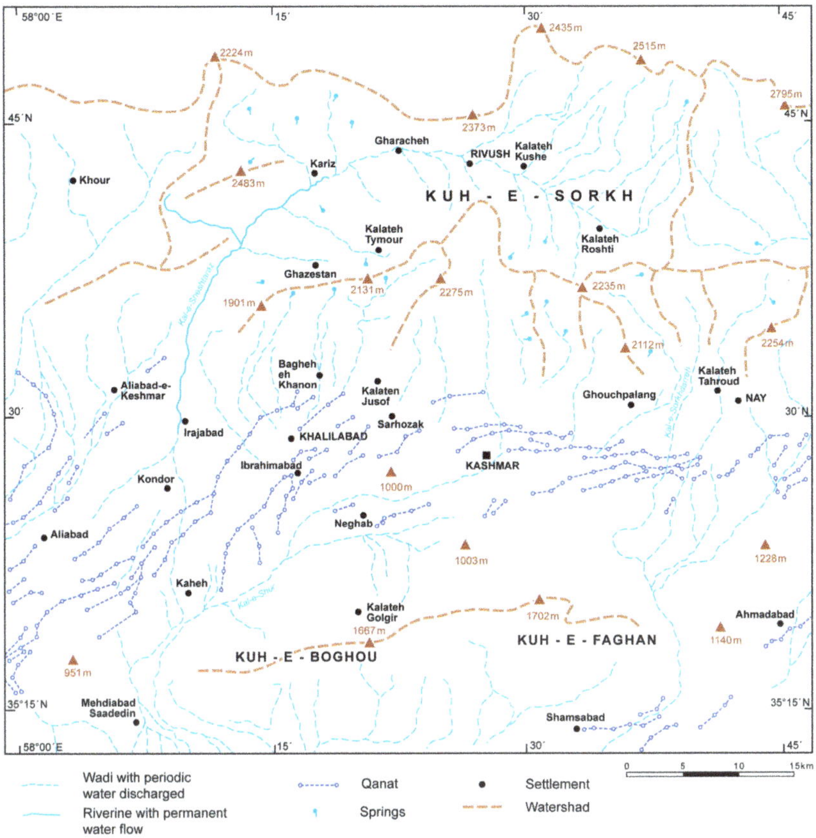

MAP 5.1 Qanat structures in the Basin of Kašmar, Central Iran.
SOURCE: SEE FOOTNOTE 68.

flow from the upper end, where it seeps into the gallery, to a ground service outlet and irrigation canal at its lower end.[7]

Mining and harvesting of subterraneous waters may be defined as a climate- and culture-specific phenomenon of North Africa and the Near and Middle East with a high degree of regional authenticity and uniqueness. The hypothesis of this paper is: the millennia-old techniques of mining and harvesting of hidden waters are a historically and culturally unique example of highly sustainable exploitation of limited natural resources. As such, qanat culture

7 Ann K.S. Lambton, "The Origins, Diffusion and Functioning of the Qanat," in *Qanat, Kariz & Khattara. Traditional Water Systems in the Middle East and North Africa*, ed. Peter Beaumont, Michael Bonine, and Keith McLachan (London: University of London, 1998), 5.

has survived for almost 3000 years until today, and it has enabled societies to turn barren deserts into spatially limited yet flourishing oases of settlement and agriculture.

It is the geological, topographical, and hydrological setting that constitutes the environmental preconditions for this very specific mode of water production and management. Its technique looks simple, yet it is very refined taking into account the traditional implements used for qanat construction.[8] The starting point of all 'engineering' and water exploration activities is digging a trial shaft (Pers. *gamāne*) into the gravelly and sandy alluvial fans covering the forelands of hills and mountains. This is purely manual work. Shovels, hoes, buckets, and windlasses are the main tools. Shafts are dug until they reach a depth at which the trial shaft hits a water table, ensuring a constant flow of water atop a stratigraphically and hydrologically impermeable stratum. In order to determine the location where the subterranean water flow is to surface, a meticulous calculation of the subterranean gradient of this flow has to be undertaken to ensure a smooth and steady flow of the tapped waters. Special surveyors calculate the gradient of the subterraneous channels in order to avoid too rapid a flow of the water, which would erode the conduits and potentially collapse them. Thus, the depth of the "mother well" (Pers. *mādar-e čāh*) and the elevation of the water outlet (Pers. *sar-e češmeh*), determine the length of any conduit (Pers. *kār/rāh-ye āb*). In other words: geology, topography, and hydrology are the determinants for the depth and length of qanats. Mother wells may be as deep as three hundred metres or more and underground channels may be as long as seventy kilometres.[9] These are extremes, however, the average size of a qanat varies between ten and 25 metres in depth and five to 15 kilometres in length, many being much shorter.

8 Henri Goblot, *Les Qanats. Une technique d'acquisition de l'eau* (Paris, La Haye, New York: Mouton Editeur, 1979), 239; Hans E. Wulff, *The Traditional Crafts of Persia. Their Development, Technology, and Influence on Eastern and Western Civilizations* (Cambridge: MIT Press, 1966), 405–406.

9 Mohammad H. Papoli-Yazdi, et al., *The Quassabeh Qanat in Gonabad* (Mashad: University of Mashad, 2000), 292–293; Ali A. Semsar-Yazdi and Majid Labaf-Khaneiki, ed., *Qanats in its Cradle. Situation of Qanat (Kariz, Karez, Falaj) in the World* (Yazd: International Center on Qanats & Hydraulic Structures, 2012), 91–92. Contradictions in data accuracy reflect the problems of exact registrations, esp. in view of the fact that many qanats are out of function or no longer recognisable. The qanat of Gonabad/Qasabe in Khorassan is often mentioned with a depth of three hundred metres for its mother well (Semsar-Yazdi and Labaf-Khaneiki, 2012). According to Semsar-Yazdi and Labaf-Khaneiki, the longest qanat with approximately eighty kilometres is that of Zareh/Yazd (Ibid.). The same authors, however, speak of "more than 150 m" depth of the Gonabad motherwell one year later: Ali A. Semsar-Yazdi and Majid Labaf-Khaneiki, ed., *Veins of Desert. A Review of the Technique of Qanat/Falaj/Karez.* (Iran Water Resources Management Company, 2013), 300—an indicator for the sometimes questionable reliability of published data.

Mother well and mouth of the qanat are connected by the afore-mentioned horizontal, traditionally hand-dug tunnels. The material to be excavated in order to establish a smooth flow of water in the qanat is lifted up through vertical shafts. These shafts have an average diameter of eighty centimetres; tunnels or water-bearing galleries have a flat bottom with a width of approximately one metre and more or less elliptical vaults with a height of barely over 1.2 metres. The distance between these shafts, which also ensure an appropriate ventilation of the whole structure, is easily discernible on the ground or from above by the small rings of excavated material in distances between twenty and one hundred metres apart. Fig. 5.1 summarises the main aspects of both geological and hydrological preconditions for qanat development as well as the details of their technological implementation.

2.2 *Qanats: Social and Economic Aspects and Cultural Implications*

Natural and technological aspects are one side of the coin. The other is economic and social considerations. Qanat ecology, economy, and sociology were (and are) bound up in a highly differentiated process, which creates a unique cultural cohesion of all parties involved in the construction of qanats and the use of their waters. In this context it must be noted that neither land nor cultivable soils are at the heart of any desert society and culture. It is water! Water is the most precious asset for any economic activity for indeed human existence in an extremely arid environment. Its outstanding economic and social

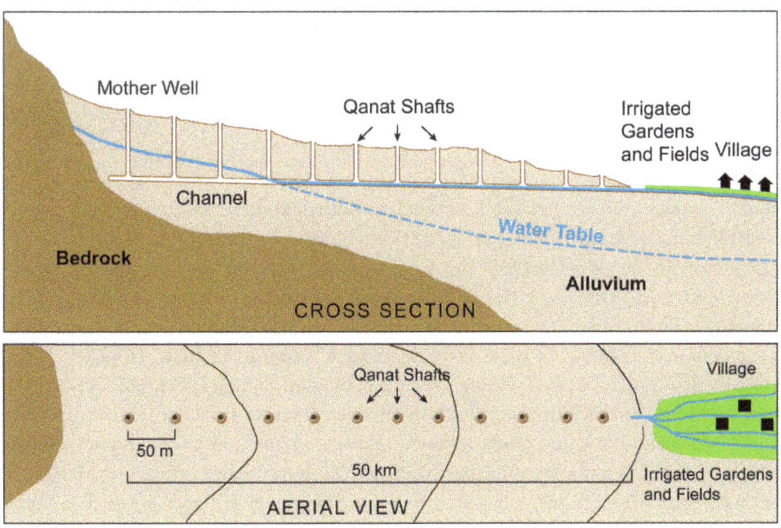

FIGURE 5.1 Qanat: cross section and aerial view of geological and hydrological structure (modified by Eckart Ehlers based on different sources).

importance is reflected in its traditional valuation not only in Iranian society. Ownership and control over water is central for any understanding of local or regional power structures, and it is equally essential for our understanding of social and economic hierarchies in agricultural production processes in desert regions.

Qanat irrigation of fields and gardens as well as qanat water supply for houses and settlements cover a wide range of activities, all of which are closely interconnected and are part of distinct socio-economic microcosms in rural/urban Iran.[10] The construction and maintenance of qanats requires a wide range of different activities, specialisations, and professions. They coincide with interdependent social stratifications and are part of a highly stratified rural hierarchy, based on a number of different preconditions and actors. First of all, the construction of a qanat is costly and needs financially potent investors. Secondly, it needs experts for its construction and maintenance. And thirdly, the ultimate management of water distribution and usage is a responsible task, demanding reliable and trustworthy local helpers.

Investors are the indispensible precondition for any qanat activity. The exploratory search for appropriate locations, the sinking of trial shafts (*gamana*) and of a suitable mother well, the digging of shafts and the construction of the subterraneous water-bearing galleries: all these activities are not only time-consuming, but also costly. Thus it is not surprising that from the very beginning of qanat construction privileged and/or rich investors have promoted and funded qanat-building and its technology. Historically, feudal rulers, the nobility, militaries, and bureaucrats as well as rich and large landowners (Pers. *mālek-e bozorg/arbāb*) have been the financiers of these hydraulic systems. More recently and especially since World War II, urban businessmen and entrepreneurial landlords became the main investors in constructing and

10 Michael E. Bonine, "From Qanat to Kort: Traditional Irrigation Terminology and Practices in Central Iran," *Iran* 20 (1982): 145–159; Eckart Ehlers, "Rent Capitalism and Unequal Development in the Middle East: The Case of Iran," in *Work, Income and Inequality. Payment Systems in the Third World*, ed. Frances Stewart (London: Macmillan, 1983), 32–61; Eckart Ehlers, "The Iranian Village: A Socio-Economic Microcosm," in *Agricultural Development in the Middle East*, ed. Peter Beaumont, Keith McLachlan (Hoboken: Wiley & Sons, 1985), 151–170; Eckart Ehlers, "Agrarreform und agrarsozialer Wandel im islamischen Orient. Entwicklungsprobleme ländlicher Räume zwischen Modernisierung und Re-Islamisierung," in *Geographische Forschung in Marburg.*, ed., Wolfgang Andres, Ekkehard Buchhofer, Günther Mertins (Marburg: Selbstverlag der Marburger Geographischen Gesellschaft e.V., 1986), 19–37; Eckart Ehlers and Javad Safi-Nejad, "Formen kollektiver Landwirtschaft in Iran: Boneh," in *Beiträge zur Kulturgeographie des islamischern Orients*, ed. Eckart Ehlers (Marburg: Selbstverlag des Geographischen Instituts der Universität Marburg, 1979), 55–82.

maintaining qanats. Often considering their financial engagement as a speculative investment in the hope of high financial returns (see below), investors in qanat construction hardly ever became agricultural producers themselves or residents in rural areas.

Constructors and qanat builders constitute the second necessary group. The key person in the construction process of a qanat is the *muqanni*, a skilled and experienced specialist (see fig. 5.2). Depending on the size and depth of a qanat a group of several *muqanni*—many of them trained members of traditional family businesses—carry out the 'engineering' part. They are assisted by a number of unskilled labourers, mostly landless peasants (Pers. *ḫošnešin*). These are engaged in operating windlasses and in disposing sand, gravel, and debris from the construction process in crater-like circles around the shafts. It is interesting to note that the technology and methodology of qanat construction have hardly changed over hundreds of years, as authors have noted.[11] They all refer to the treatise by Karaji, published in the 5th/11th century, in which he reflects earlier writings on qanat technology and praises the "exploitation of subterraneous waters as beneficial human achievements, because by its help the earth was made to flourish, men's lives achieved order, and abundant benefit accrued".[12] Karaji describes basically the same techniques still in use until the middle of the 20th century.[13]

Water management: The successful 'mining' activities and the provision of water mark the end of the technological part of qanat construction. The distribution and allocation of the available water begins the third phase of qanat water management. It is in many respects the most sophisticated one. Different parties and agents are involved. Conflicts over water shares are not uncommon. Depending on the client—rich landlords (Pers. *mālek-e bozorg/ ʿomdeh-ye mālek*) or a village community of smaller land-holdings (Pers. *ḫordeh-ye malek*)—water distribution and allocation used to be and still is a hierarchical process. As mentioned before, the 'harvested' waters flow by gravity. When the water comes to the surface at the outlet and mouth of the qanat (Pers. *sar-e češmeh*), it is directed through a more or less extensive network

11 Goblot, *Les Qanats*; Kuros, Gholam-Reza, *Irans Kampf um Wasser. Die Vergangenheit und ihre Lehren, die Zukunft und ihre Aufgaben in der iranischen Wasserwirtschaft*. Archiv für Wasserwirtschaft Nummer 70 (Berlin, 1943), 93; Lambton, "The Origins, Diffusion and Functioning of the Qanat."

12 Ibid., 6.

13 Muhammad al-Karaji, *Inbaṭ al-miyāh al-ḫafīya* [The Extraction of Hidden Waters] (Tehran: Bonyad Farhang Iran Publication, 1966); M. Nadji, "Karadjis 'Erschließung verborgener Gewässer.' Ein Lehrbuch der Geowissenschaften aus dem 11. Jahrhundert," *Zeitschrift Deutsche Geologische Gesellschaft* 123 (1972): 1–13.

FIGURE 5.2 *Muqanni* and Windlass for qanat construction (reconstruction in the Water Museum of Yazd).

of channels (Pers. *ǧub*) into the fields for irrigation or into urban quarters for household consumption. In line with the afore-mentioned distribution system via the main irrigation channel (Pers. *šahǧub*), the water then was and is directed into a number of parallel branch channels (Pers. *pošteh ǧub*), from which further and smaller irrigation channels are diverted. The final result of these highly sophisticated distribution systems are basically linear cultural landscapes following the gradients of the terrain. The capacity of the qanats and their discharge of water, the fertility of the irrigated land, the slopes and their gradients: these and other factors determine the shape, size and extent of these cultural islands amidst seas of sand and stones (see also fig. 5.1).

Due to the extraordinary value of water as the basis of all agricultural activity and human existence in extremely arid environments, qanat water is a highly coveted production factor. Land is abundant, water is scarce and therefore often causes conflicts. The central authority is an official in charge of water distribution for irrigation (Pers. *mirāb*), mostly appointed by the owners of a qanat. The *mirab*'s work is assisted and complemented by other persons who are responsible for water distribution at community level (Pers. *ābyār*). Both positions are crucial for the fair and exact allocation of water in an environment where water is more valuable than cultivable land. *Mirāb* and *ābyār* are not only accountable to the owners of a qanat, but must also be accepted and trusted by the users of the scarce water resources.

The scarcity and value of water are reflected in a sophisticated organisation of meticulous water division at field level. Based on property rights in land and water, expressed in shares of a total, and depending on the volume of available water, traditions have developed a great variety of locally very different allocation practices. It would go beyond the aim and scope of this paper to dig into the details of these practices—for more details see Bonine (1982).[14] It is, however, remarkable to see the refinement of their execution in terms of times, tools, and legally differentiated claims to irrigation water. These tasks are entrusted to yet another group of specialists, appointed by the *mirāb* or elected by the owners of the qanat water. In Yazd and central Iran, for instance, they are called *ṭāqdār*s/*ṭāštebān*s. Other subordinate functions include *ǧubāni* in charge of canal (Pers. *ǧub*) cleaning, *daštbāni* for field protection against wild animals or thieves[15] and similar tasks.

This rough list of people and professions engaged in the construction of qanats, their maintenance, the management of the waters they provide and their appropriate distribution reveals a great variety of individuals with different functions and of different social status. The situation described, however, remained more or less static in central Iran until the middle of the 20th century and can be considered the prototype of qanat-based irrigation systems all over arid Iran—and beyond.

2.3 *Rent-capitalism—A Theoretical Attempt as an Explanation?*

The technological and organisational details of qanat construction and water management—as shown above—can well be interpreted as a well- adapted example of an intricate socio-economic system typical for the Near and Middle East in general—and dating back to early pre-Islamic civilisations. This region, its societies, and their historical foundations have always been different from other cultures. Marx and Marxists coined the term "Asiatic mode of production" to emphasise their particular nature. Wittfogel stressed the specific role of water and irrigation as well as its management for his identification of "hydraulic civilizations" as driving forces of bureaucracies, social hierarchies and

14 Bonine, "From Qanat to Kort."

15 Ibid., 156.

the development of statehoods and power structures.[16] In any case: water is a decisive factor in both analyses.[17]

Whether certain components of Karl Marx's Asiatic mode of production or Wittfogel's hydraulic civilisations can be meaningfully applied to our case study may be questionable. It is certain, however, that a third and less well-known, yet highly interesting and applicable theoretical approach may be of interest. It is the concept of rent-capitalism that the Austrian geographer Hans Bobek presented in an influential essay on "Die Hauptstufen der Gesellschafts- und Wirtschaftsentfaltung in geographischer Sicht".[18] Derived from his own field research in Iran and his insights into its urban—rural relationships, Bobek argued that especially in the East (in German: *Orient*!) rent capitalism can be seen as an intermediate phase of historical and socio-economic development between feudally organised predecessor societies and the following industrial capitalism.[19]

According to Bobek, rent-capitalism can be defined as a genuine socio-economic stage of its own. As such it emerged in the early Middle East in connection with the development of the early city and temple-states both in Egypt and Mesopotamia: The theory of rent-capitalism is well suited to explain the seemingly complicated organisational ramifications of qanat water provision and management. At the same time it may be suited to substantiate the basis for the foundations of the term "desert culture". Rent-capitalism is interpreted as a culture specific "socio-economic form of its own". It is defined as a specific Near and Middle Eastern mode of production with a specific impact on society

16 Karl A. Wittfogel, "The Hydraulic Civilizations," in *Man's Role in Changing the Face of the Earth*, ed. William L. Thomas Jr. (Chicago: The University of Chicago Press, 1956), 152–164.

17 Neither Marx nor Wittfogel ever claimed the validity of their theories specifically for the case of Iran. Both are mentioned here because of their references to the importance of economically 'unproductive' bureaucracies in traditional agriculture (see both table 5.1 and 5.2 in this chapter).

18 Hans Bobek, "Die Hauptstufen der Gesellschafts- und Wirtschaftsentfaltung in geographischer Sicht," *Die Erde: Journal of the Geographical Society of Berlin* 90 (1959): 259–298.

19 For critical statements against Bobek see e.g. Günter Leng, "'Rentenkapitalismus' oder 'Feudalismmus'? Kritische Untersuchungen über einen (sozial)geographischen Begriff," *Geographische Zeitschrift* 62 (1974): 119–137; Eugen Wirth, "Die Beziehungen der orientalischen Stadt zum umgebenden Land," in *Erdkuntliches Wissen* (Stuttgart: Steiner Verlag, 1973), 323–332; Hans Bobek, "Zum Konzept des Rentenkapitalismus," *Tijdschrift voor Economische en Sociale Geografie* 65 (1974): 73–78; Hans Bobek, "Rentenkapitalismus und Entwicklung in Iran," in *Interdisziplinäre Iran-Forschung. Beiträge aus Kulturgeographie, Ethnologie, Soziologie und Neuerer Geschichte* (Wiesbaden: Dr. Ludwig Reichert, 1979), 113–124.

and social hierarchies.[20] And it is identified as a socio-economic system with a millennia-old history and living tradition:

> Rent-capitalism arose through commercialisation and the transformation, undertaken in a plain profit-seeking spirit, of the original lordly (or feudal) claims on income from the peasant and artisan under-strata. Its elaboration was definitely promoted by the keeping of accounts and other forms of rationalization of rent drawing that were developed in the large temple estates.[21]

Traditional forms of qanat construction and water management—as shown in the preceding chapters—and their importance for agricultural production in the desert areas of MENA fit well into the overall concept of rent-capitalism as a distinct socio-economic stage. Since ancient times and until quite recently, it was common for the ruling classes, i.e. notables, landowners, urban elites, and merchants, to invest in land and water and then to have the work done by others. In line with Bobek's concept of rent-capitalism and in application of its basic assumption, i.e. the split of agricultural production into four or five equally weighted production factors, qanat-based agriculture depends on: (1) land; (2) water; (3) seed; (4) human labour and; (5) equipment/animal labour.

This system of agricultural production, known in Arabic-speaking countries under the term *ḫāmessāt*, is based on the general assumption that each of these five factors is accountable for about twenty per cent of any harvest. Due to the fact that large landowners were the only ones who could supply land and undertake the costly provision of water (and sometimes also had to provide

20 Bobek's theory of rent-capitalism has found a positive reaction in both Biblical and Islamic Studies. See e.g. Bert G. Fragner, "Der Rentenkapitalismus als islamwissenschaftliche Forschungsaufgabe," *Zeitschrift Deutsche Morgenländische Gesellschaft* (ZDMG) 5 (1983): 168–175); Bert G. Fragner, *Rent-capitalism in the Light oft the Economic History of Early Iran. Proceedings of the Symposium held on the Occasion of the 50th Anniversary of the Wiener Institut für Völkerkunde* (Wien: Wiener Beiträge zur Ethnologie und Anthropologie 1 (1984): 205–227; Rainer Kessler, *Staat und Gesellschaft im vorexilischen Juda. Vom 8. Jahrhundert bis zum Exil* (Leiden: Brill, 1992); Bernhard Lang, "The Social Organization of Peasant Poverty in Biblical Israel," *Journal for the Study of the Old Testament* 7 (1982); Oswald Loretz, "Die prophetische Kritik des Rentenkapitalismus," *Ugarit-Forschungen* 7 (1975): 271–278.

21 Quoted after Hans Bobek, 1962, i.e. the English translation of his article, Hans Bobek, "The Main Stages in Socio-economic Evolution from a Geographical Point of View," in *Readings in Cultural Geography*, ed. P.L. Wagner, 218–247 (Chicago: University of Chicago, Dept. of Geography, 1962), note 18.

TABLE 5.1 Distribution of a winter grain harvest in the village of Goldasteh

Recipient	Percentage of harvest (%)
Total harvest	100
Blacksmith	0.16
Carpenter	0.16
Field-guard	1.0
Village-headman (*kadkhuda*)	1.0
Mullah	0.2
Ox-drover	0.08
Boneh/farm labourers	5.0
Total deductions	**7.6**
Rest	**92.4**
Landlord:	
two shares for soil and water: 2 × 18.48	36.96
Urban capitalist (*gavband*):	
two shares for seed, plough, and tools: 2 × 18.48	36.96
Share-croppers (*raiyat*) and farm labourers:	
one share for human labour: 1 × 18.48	18.48
share-cropped harvest	92.4

seeds) they were *a priori* entitled to forty per cent of any harvest without being engaged in the agricultural production process themselves. The remainder was traditionally distributed among the rural peasants (Pers. *ḫordeh-ye malek*), tenants (Pers. *raʿyat*) or landless labourers (Pers. *ḫošnišin*). Whenever or wherever irrigation was part of the agricultural production, *mirāb*, *ābǧār*, *ṭāqdār* or *dašt-bāni* also became beneficiaries of this rent-capitalist production and distribution system by receiving certain shares of the harvest in cash or in kind.

In order to make this theoretically and practically complicated system more understandable, two examples may serve for clarification. A first simplified example is the distribution of a winter grain harvest (Pers. *kešt-e shetfi*) in the village of Goldasteh near Tehran in 1964 (tab. 5.1).[22] Without taking into consideration the high investment costs of qanat construction and its maintenance (borne by the owners of the irrigation waters) and without considering deductions for water management, the survey reveals a remarkable participatory flexibility in the distribution of the harvest. After paying off different minor

22 Source: Ehlers and Safi-Nejad, 1979.

recipients for their specific contributions to the production process, the remaining 92.4 per cent were split among the five traditional production factors, each one accounting for 18.48 per cent of the harvest. These were allocated to the owners of the land and the water rights and to an urban capitalist providing seed and equipment. Less than twenty per cent of the harvest remained for those who were working on the land.

In more general terms, the following survey explains the rent-capitalist mechanisms in a comparative historical context (tab. 5.2).[23] It shows the great variability of its application within one region.[24] The case of the Varamin district near Tehran shows the great flexibility and diversity of rent-capitalist share-cropping arrangements between the urban owners of land and water and their rural dependents (Pers. *ḫordeh-ye malek, raʿyat, ḫošnešin*). In all cases, landlords and water-owners were remunerated with fifty per cent of the gross proceeds of the harvest or more while, the dependent peasants, i.e. the workers of the land and thus decisive for agricultural productivity, had to be content with 25 to 45 per cent of the harvest.

Altogether it is probably fair to say that qanat irrigation and rural desert society can well be understood within a framework of rent-capitalism as a specific socio-economic expression of human existence under extreme natural conditions, not only in the desert regions, but also in neighbouring countries. Even if we have to concede that the general acceptance and applicability of Bobek's theory is sometimes questioned, we must also acknowledge its explanatory potential for the specific situation of qanat irrigation.[25] At the same time, it is very well suited to explaining the close context of work, income and socio-economic inequality.[26] The conditions of qanat technology in the dry highlands of Iran have created a whole ecological, economic, and social microcosm, which is not comprehensible without the existence, 'mining' and harvesting of subterraneous waters.[27] Thus, it is understandable why some authors go so far as to describe the qanat-based agricultural system of the Central Iranian Plateau as an "irrigation culture".[28] In our understanding of the term

23 Cyrus Gharatchedaghi, *Distribution of Land in Varamin: An Opening Phase of the Agrarian Reform in Iran* (Oplanden: Leske-Verlag, 1967); Ehlers, "Agrarreform und agrarsozialer Wandel im islamischen Orient."

24 Gharatchedaghi, *Distribution of Land in Varamin*, 181.

25 Bobek, "Zum Konzept des Rentenkapitalismus," 73–73; Bobek, "Rentenkapitalismus und Entwicklung in Iran," 113–124.

26 Ehlers, "Rent Capitalism and Unequal Development in the Middle East," 32–61.

27 Ehlers, "The Iranian Village: A Socio-Economic Microcosm," 151–170.

28 Keith McLachlan, *The Neglected Garden. The Politics and Ecology of Agriculture in Iran* (London, J.B. Tauris and Co Ltd, 1988), 90.

TABLE 5.2 Organisation of agricultural production and harvest sharing in Varamin district

		Land	Water	Seed	Plough	Human Labour	Fertiliser	Pesticides	Transport	Harvest Share of Landlord vs. Peasants
Grain Cultivation (*kesht shatfi*)										
Ex. 1	Landlord	x	x	x	x		x	x	x	75
	Peasant					x	x	x	x	25
Ex. 2	Landlord	x	x	x	x		x	x	x	66
	Peasant					x	x		x	33
Ex. 3	Landlord	x	x				x	x	x	55
	Peasant			x	x	x		x	x	45
Cotton and Legumes (*kesht saifi*)										
Ex. 1	Landlord	x	x	x	x		x	x	x	75
	Peasant					x		x	x	25
Ex. 2	Landlord	x	x	x	x		x	x	x	66
	Peasant					x		x	x	33
Ex. 3	Landlord	x	x		x		x	x	x	55
	Peasant			x		x	x	x	x	45
Ex. 4	Landlord	x	x		x		x	x	x	50
	Peasant			x	x	x	x	x	x	50

"culture" we, however, propose to go one step further: Qanats did not only enable irrigation culture. Their waters have in the past sustained the livelihood of villagers and urbanites. And they have been crucial for the development of the scattered rural and urban landscapes in the midst of barren wastelands and deserts. Thus, we define qanat ecology, economy, and related social consequences as a complex entity that exerts its influence on rural and urban environments, their inhabitants and their mode of life as well as on parts of social structures and their hierarchies. Probably more than in many other ecological settings, climate and culture in extremely arid environments are inseparably interwoven. As such they represent ingenious human adaptations to the challenges of harsh and hostile conditions of life. Climate and culture have formed a symbiotic relationship which, as a consequence, has created a specific kind of desert society and economy.

2.4 *Qanats: Rise and Fall—Origins and Dispersals*

Readers may have noted that our analysis of the 'detection' of underground waters and their use for irrigation and human consumption has been expressed somewhat ambivalently by using past and present forms of grammar. As a matter of fact: qanat detection, development and use are very old techniques. Qanats are still in use in Iran and in many neighbouring countries as well as in North Africa. However, their overall number is decreasing rapidly as they are replaced by other, less sustainable forms of irrigation and water management.

It is a widespread belief that the origins of the qanat culture are located somewhere in northwestern Iran, more specifically in the former kingdom of Urartu. This kingdom, flourishing in the border region between Iran, Turkey, and present-day Armenia from about 1000 BCE onwards and reaching its peak of power in the 9th and 8th centuries BCE, was well known for its mining activities. Many archaeologists and historians claim that subterraneous veins of water were tapped and used for irrigation purposes in connection with primitive forms of ore mining. Other researchers argue that qanats may have been "discovered" as a by-product of early copper mining in the Zagros mountains "à la fin du deuxième millenaire".[29] Early uses of qanat irrigation are documented more precisely in clay tablets and inscriptions of the neighbouring conquerors of the Urartu civilisation, the Assyrians, the Medes, and the Persians. Further testimony to these facts is given by different hydraulic structures which are, however, not clearly identifiable as genuine qanats. Famous and archaeologically documented irrigation structures are those commissioned by the Assyrian king Sennacherib (705–680 BCE) near his capital Niniveh,[30] constructed to water the royal gardens. Genuine qanat structures, on the other hand, are found in Achaemenid Persia. Again and first of all, qanats were developed as devices for the irrigation of royal palaces and gardens. Famous examples are the parks and gardens of King Cyrus the Great (550–529 BCE) in Pasargadae in the midst of semi-arid terrain and those commissioned by Darius II (522–486 BCE) in and around Persepolis in the Iranian province of Fars. Both Xenophon and Polybios give vivid accounts of these early cultural adaptations to the harsh climatic environments of the first Persian Empire, calling them paradeisos.[31]

---

29 Goblot, *Les Qanats*, 61.

30 Ariel M. Bagg, *Assyrische Wasserbauten. Landwirtschaftliche Wasserbauten im Kernland Assyriens zwischen der 2. Hälfte des 2. und der 1. Hälfte des 1. Jahrtausends v. Chr.* (Mainz: Verlag Philipp von Zabern, 2000), 435.

31 Penelope Hobhouse, *Persische Gärten. Paradiese des Orients* (München: Knesebeck Verlag, 2005). For further details see: Stefano Bianca, *Hofhaus und Paradiesgarten: Architektur und Lebensformen in der islamischen Welt* (Munich: C.H. Beck, 1991); John Brookes, *Gardens of*

It would go far beyond the purpose and limits of this paper to go into further details of the historical development of qanat technology. However, we have an important document—a textbook—on qanat technology and qanat construction from medieval times. This treatise by Karaji, mathematician and engineer (953–1029 CE), on the extraction of groundwater refers to qanat practices under the Umayyad and Abbasid caliphates—which also indicates its distribution and wide application at that time. More importantly, however, the work contains meticulous instructions for the construction of qanats and the mathematical calculations to ensure the steady and unimpeded flow of their water. Karaji's book not only shows the refined techniques of qanat construction one thousand years ago, but also the unbroken continuation of this culture until quite recently.[32]

As mentioned earlier, qanats are very sensitive human interventions into natural ecosystems. Their geological setting in easily erodible material like alluvial sediments made and makes them extremely vulnerable to erosion by water or by severe earthquakes. On the other hand, regular costly cleaning and maintenance of the sophisticated system ensured the stability of the social hierarchies and the elevated roles of the investors and the specialists in qanat construction, e.g. the *muqanni*. Alongside the impact of climate, geology, and geomorphology, it was especially the devastating effects of warfare and wanton destruction of the galleries that caused far-spread collapses of qanat culture in the deserts and their fringes. The medieval invasions by Djinghiz Khan and Timur in the 13th and 14th century and their devastations of urban and rural cultures in the conquered regions are notorious examples.

Altogether it is hard to give exact numbers of qanats in Iran. Such an observation holds true not only for the present, but even more so for earlier periods. Referring to al-Karaji's treatise on the extraction of hidden waters, we can conclude with good reason that qanat technology and qanat culture must have flourished in Iran and beyond, interrupted again and again by natural disasters or military destruction. More precise data are available for the 20th century. Referring to Iranian sources, Kortum speaks of approximately 50,000 qanats in Iran in the early 20th century with about 22,000 remaining towards the end

Paradise. The History and Design of the Great Islamic Gardens (London: G. Weidenfeld and Nicholson, 1987); Elizabeth B. Moynihan, *Paradise as a Garden in Persia and Mughal India* (London: Scolar Press, 1979); Attilio Petruccioli, ed., *Der islamische Garten. Architektur—Natur—Landschaft* (Stuttgart: DVA. 1995).

32 al-Karaji, *Extraction of Hidden Waters*; Nadji, "Karadjis 'Erschließung verborgener Gewässer.'"

of the century.[33] Other sources speak of 42,000 qanats for the whole country in 1942, including an unspecified number of dry channels.[34] The same authors mention 28,054 qanats for 1992/1993 and 33,691 for 2002/2003.[35] Confusing and contradictory as some of these figures may be—not to speak of the probably great numbers of unproductive water-bearing galleries—it remains true that hidden waters still play a role. For the early 1940s, Kuros quotes a source according to which ninety per cent of Iran's irrigated fields and gardens were watered by qanats.[36]

Reconsidering our remarks concerning the geological and climatic preconditions for qanat construction (see Part 2.1), it is clear that qanats were and are distributed unevenly within the country. They cluster in the foothill regions of mountain ranges, especially the southern fringes of the Alborz and the eastern forelands of the Zagros systems as well as practically all the foothill regions of the Central Iranian mountain ranges. Kortum refers to the province of Tehran: out of a total of 4,121 village communities, 1,574 were irrigated by qanats.[37] After the proclamation of Tehran as capital city of Persia in 1786, many new qanats were built in order to ensure a water supply for the rapidly growing city. Braun refers to a total of more than four hundred functioning qanats in the immediate region of Tehran in 1968; in the 1950s, 23 were still providing water to the city itself.[38] In central Iran (Najafabad near Esfahan) 106 qanats were still in operation in the early 1970s.[39] The inner Iranian desert regions along a line from Qom via Kashan, Yazd, and Kerman to Bam have always been model regions of qanat culture where qanat irrigation plays a major role even today.[40] Among the more recent analyses and case studies, three deserve

---

33 Gerhard Kortum, "Die Qanat-Bewässerung auf dem Hochland von Iran und ihre Ausbreitung über die Erde," in *Historische und zeitgenössische Probleme und Perspektiven in asiatischen und afrikanischen Gesellschaften*, ed. Hübner Ulrich and Antje Richter (Schenefeld: EB, 2004), 181.

34 Semsar-Yazdi, and Labaf-Khaneiki, ed., *Qanats in its Cradle*, 85.

35 Ibid., 87.

36 Gholam-Reza Kuros, *Irans Kampf um Wasser. Die Vergangenheit und ihre Lehren, die Zukunft und ihre Aufgaben in der iranischen Wasserwirtschaft*, 77.

37 Kortum, "Die Qanat-Bewässerung auf dem Hochland von Iran und ihre Ausbreitung über die Erde," 175.

38 Cornel Braun, *Teheran, Marrakesch und Madrid. Ihre Wasserversorgung mit Hilfe von Qanaten. Eine stadtgeographische Konvergenz auf kulturhistorischer Grundlage* (Bonn: Ferd. Dümmlers, 1974), 30.

39 Martin Hartl, *Das Nadjafabadtal. Geographische Untersuchung einer Kanatlandschaft im Zagrosgebirge (Iran)* (Regensburg: Institut für Geographie an der Universität Regensburg, Selbstverlag, 1879), 172.

40 Michael E. Bonine, *Yazd and its Hinterland. A Central Place System of Dominance in the Central Iranian Plateau* (Marburg. Institut für Geographie der Universität Marburg,

special mention: those of Safi-Nejad on the village of Talebabad and on Boneh, and of Papoli-Yazdi on one of the country's largest qanat in Gonabad, not to speak of the general surveys by Semsar-Yazdi and Labbaf Khaneiki.[41]

Time and space may be considered as the decisive paradigms for the evaluation of qanats and qanat technology as an outstanding cultural and civilisational response to the challenges of climate and extreme arid natural environments. Our outline of the historical development of qanat culture has so far shown the persistence and perseverance of qanat culture in Iran for more than 2500 years until its final phase today. However, chronology is only one side of the coin. This unique irrigation technology has to prove its acceptance as an outstanding human invention and achievement in chorological terms. And in this case both chronology and chorology, time and space speak for human inventiveness and ingenuity. The very fact that qanats are concentrated on the arid to semi-arid Eurasian drybelt stretching over an almost 10,000 kilometres and comprising an area between approximately 17° West to 100° East is a clear indication of the close connection between climate

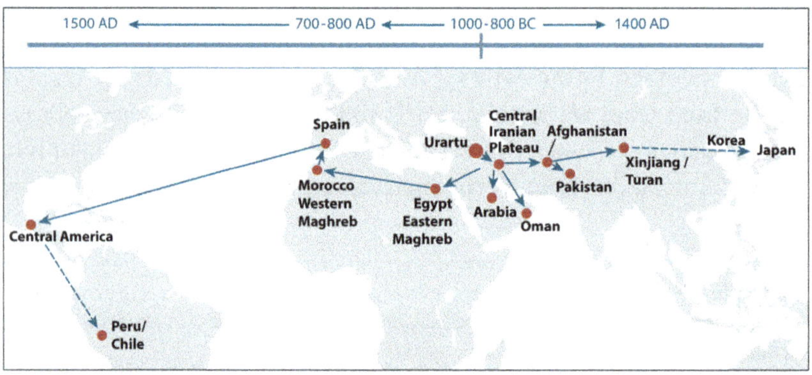

MAP 5.2 Spread of the qanat system in time and space.

Selbstverlag, 1980); Eckart Ehlers, "The Dying Oases of Central Iran. Causes and Consequences," in *Desertification in Extremly Arid Environment*, ed. Wolfgang Meckelein, 65–72 (Stuttgart, Geographische Studien, 1980); Paul W. English, *City and Village in Iran: Settlement and Economy in the Kirman Basin* (Madison, Milwaukee, London: The University of Wisconsin Press, 1966); Paul W. English, "The Origins and Spread of Qanats in the Old World," *Proceedings American Philosophical Society* 112 (1968): 73–90.

41 Javad Safi-Nejad, *Talebabad. Case Study of a Village. Tehran* (Teheran: University of Tehran: Institute of Social Research, 1966); Papoli-Yazdi, et al., *The Quassabeh Qanat in Gonabad*; Semsar-Yazdi, and Labaf-Khaneiki, ed., *Qanats in its Cradle. Situation of Qanat (Kariz, Karez, Falaj) in the World*; Semsar-Yazdi, and Labaf-Khaneiki, ed., *Veins of Desert.*

and culture—and it is in this area that the rise and fall, origin and diffusion of qanats and qanat technology took place.

As mentioned earlier, the first comparatively well documented period of qanat construction was under the Assyrian and early Persian Empire, the result of the discovery of the hidden waters in their realms and the consistent further development of this innovative technology. It was also in this period that qanat technology was obviously 'exported' to the fringes of the Achaemenid Empire to Egypt in the west and to their periphery in present-day Afghanistan and Pakistan in the east.[42] The spread of Islam and the expansion of Arab realms and empires, especially in North Africa and around the Mediterranean, mark the second distinct diffusion process of qanats. In spatial terms it covers North Africa and here, above all, the Maghreb, but also the Iberian Peninsula, Sicily, and even the Canaries. It speaks for the great cultural and civilisational importance of this elaborate irrigation system that from 1492 CE onwards, the Spanish not only maintained qanat technology in their homeland, but that they transferred it to the New World. Mexico as well as possibly some specific locations in the foothills of the Andes in Peru and northern Chile are locations where qanats have been found and described.[43] The very fact that there are many names for the same structure and technique of water mining and harvesting shows the wide distribution of this unique technology all over MENA to this day. While the term *qanat* is Persian and has become a *terminus technicus*, there are a great number of synonyms.[44] In North Africa and the Maghreb, terms like *ḫettara* (Morocco) or *foggara* (Algeria) are common, in Afghanistan and neighbouring regions the term *karez* is being used, and in parts of the Arab world (e.g. Oman) the word is *falaǧ*. All these regional terminologies—not to speak of many more at local levels—refer to one and the same method of water mining and harvesting.[45] The fact that these structures are not only relics of a great cultural history is documented by their rapidly diminishing

42 Braun, *Teheran, Marrakesch und Madrid*, 2–4.

43 Ibid.; Christopher S. Beekman, Phil S. Weigand, and John J. Pint, *Old World Irrigation Technology in a New World Context. Qanats in Spanish Colonial Western Mexico. Antiquity 73/279* (Cambridge: Cambridge University Press, 1999), 440–446; Johannes Humlum, "Underjordiske vandingskanater: Karese, Qanat, Foggara. Dens konstruktion, virkemase, geografiske udbedrelse og circumglobale spredning," *Kulturgeografi* 16 (1964): 81–123; Carl Troll, "Qanatbewässerung in der Alten und Neuen Welt," *Mitteilungen der Österreichischen Geographischen Gesellschaft*, 313–330.

44 Comprehensive surveys of the qanat terminology and their synonyms are to be found in Baumont, Bonine, and McLachlan, *Qanat, Kariz and Khattara*; Ali A. Semsar-Yazdi, and Majid Labaf-Khaneiki, *Qanats in its Cradle*; Ali A. Semsar-Yazdi, *Veins of Desert*.

45 George B. Cressey, "Qanats, Karez and Foggara," *Geographical Review* 48 (1958): 27–44.

presence in Afghanistan,[46] in Pakistani Baluchistan[47] or in the fringe areas of the Near Eastern Mediterranean (esp. Lightfoot 1997).

A very special case is presented by the extensive karez systems in the Xinjiang Uigur Autonomous Region in the west of the People's Republic of China (PRC), especially in the Turfan Basin. This extremely dry basin located approximately one hundred metres below mean sea level counts a total of 1784 karez with an overall length of 5,272 kilometres and more than 170,000 vertical shafts. As in other cases—and according to official tourist information—only one third of the karez are at present operational. However, beyond their commendable presentation in a museum-like garden, the Turfan karez may have played a central role in the eastward spread of Iranian qanat culture.

Such an assumption can be seriously substantiated by a very recent and detailed analysis of qanat structures in the Turfan basin by Bertil Mächtle et al.[48] Based on an in-depth study of eight karez systems in the centre of the Turfan Basin and C14-dates, the results deliver a highly interesting and well documented reconstruction of these structures. Conclusions suggest that "the oldest investigated karez systems originated in the Uyghur Huihe Dynasty (790–1755, 回紇). A second phase may have occurred during late Huihe/Qing Dynasty (清) (after 1755 CE).[49] More precise: "[…] the oldest investigated karez systems originated in the early 15th century".[50] If these findings can be further substantiated, then the Turfan Basin can be interpreted with good reason as the missing link to the so far isolated occurrences of qanat-like structures in Korea and Japan the origins of which are so far unknown.

It is a fact that very similar structures to the Persian qanats are also reported from as far away as Japan. According to Shoko Okazaki, these horizontal channels with vertical shafts and underground tunnels strikingly resemble their Middle Eastern counterparts. Okazaki refers to more than 280 of these irrigation devices called mambo in the foothills of central Japan and mentions

46 Christoph Jentsch, "Die Kareze in Afghanistan," *Erdkunde* 24 (1970): 112–120.

47 Ahmad S. Khattak, *Mutual Sustainability of Tube-well Farming and Aquifers. Perspectives from Baluchistan, Pakistan* (Cham, Heidelberg, New York, Dordrecht: Springer, 2014), 168; Fred Scholz, "Die physisch- und sozialgeographischen Ursachen für die Aufgabe und den Erhalt der Kareze in Belutschistan," *Die Erde* 103 (1972): 302–315.

48 Bertil Mächtle et al., "The Age and Origin of Karez Systems of Silk Road Oases around Turpan, Xinjiang, P.R. of China," in *Socio-Environmental Dynamics along the Historical Silk Road* (Kiel: Graduate School at Kiel University—PAGES/Past Global Changes, Springer, 2019). https://doi.org/10.1007/978-3-030-00728-7_17.

49 Mächtle, "The Age and Origin of Karez Systems of Silk Road Oases around Turpan, Xinjiang, 359. https://doi.org/10.1007/978-3-030-00728-7_17.

50 Ibid., 368.

the existence of similar underground water channels in Korea. His conclusion reads as follows:

> I support the hypothesis that the origin of Japanese underground water channels ... can be traced to Korea, and that in all probability this technique was brought to Korea via Turfan, north-west China, from Iran.[51]

3 Qanats, Vernacular Architecture, and Urban Design: Making Deserts Liveable

Our survey of water and wind as hidden, yet highly efficient and sustainable 'purveyors' of renewable resources in extremely arid environments would not be complete without referring to their impact on the architectural culture and the traditions of urban design and layout.

In both rural and urban areas not only in Iran, but in the entire Gulf region as well as in neighbouring countries, there are remarkable architectural structures specifically designed to cope with the scarcity of water and the impacts of summer heat-waves and sandstorms. Focussing on their function and micro-climatic as well as physical effects on climate and culture, it must be noted that architecture here is an immediate response to the adversity of their natural environment. Both water and wind are determining factors to their function and even here, qanat play a major role.

The combination of wind and water in an extremely hot environment causes a physical reaction: water evaporates, thus increasing the humidity of the surrounding hot air resulting in a remarkable cooling effect, especially in closed or sheltered rooms. Desert people reacted very skilfully to the adversity of their environment. Over centuries, they developed passive cooling systems on the basis of the very limited natural resource water and the abundant availability of wind—an eminently sustainable approach. Among the most remarkable developments among these passive cooling systems are the underground water reservoirs and a kind of natural cold-storage facility: underground cisterns (*āb ānbār*) and ice houses (*yaḫčal*) (see fig. 5.3). Both structures were common in villages and urban centres; and they exist in some locations to this day, although mostly in the form of museum-like reminders

51 Shoko Okazaki, "The Japanese Qanat (Mambo): Its Technology and Origin. Beaumont," in *Qanat, Kariz and Khattara: Traditional Water Systems in the Middle East and North Africa*, ed. Peter Beaumont, Michael Bonine, and Keith McLachlan, 267–279 (London: University of London, Centre for Middle Eastern Studies, MENAS Press 1989), 278.

FIGURE 5.3 A reconstructed *yaḫčal* in Abarkuh, central Iran. © Eckart Ehlers 2018. The shadow in the foreground marks the shadow of the east-west-running adobe wall; the shallow basins in front of the *yaḫčal* hold the water for the freezing process in winter months.

of a great cultural past.[52] Perfectly embedded into the annual climatic cycle of their environments, the winter flow of qanat water that is not needed for irrigation purposes is used to replenish *āb-ambars*. Otherwise, they depend on occasional sheet floods as a consequence of rare heavy rainstorms, or winter precipitation. Cisterns are mostly small circular pools, excavated from the ground and covered by a dome in order to protect the water against evaporation and against pollution by desert storms. Whenever supplemented by a wind tower (*bādgir*)—and in central Iran that was very common—cisterns become perfect examples of ingenious passive and highly sustainable cooling systems. *Yaḫčal*, on the other hand, depend only partly on qanat water, more on rainfall and snow melt. Their water is collected in small ponds or shallow basins where it freezes in cold winter nights. Ponds and basins are often protected by up to ten metre-high adobe walls running in an east-west direction to reduce the impact of solar radiation, thus serving as an additional tool for the freezing process. Frozen ice is then stored in large-domed storage pits up to ten metres deep and tall. Protective straw mats reduce evaporation so that the ice can be preserved as a natural refrigerator for many months (fig. 5.3).

The combination of water and wind as an almost perfect human response to the challenges of nature has produced yet another remarkable invention closely connected to the afore-mentioned structures: wind-catchers or wind-towers (*bādgir*). In many desert cities, not only in Iran, *bādgir*s still dominate the skyline. In rural areas they are less common or are entirely replaced by small and simple openings on the roof of the houses. Their urban architecture is fascinating, their function no less. *Bādgir*s are mostly rectangular structures with open vents on either two or all four sides, depending on the prevailing winds. They stand above the roofs and vaults of the houses like chimneys and transport wind into their interior. Fresh air circulation within the dwellings is one benefit of this system. When the air is directed to water pools in the basement or where a qanat even provides running water in the subterraneous dwelling rooms of better-off houses, their cooling effects are remarkable. They enable families to spend the days of hot summer in the subterraneous basements of their premises. Their efficacy (fig. 5.4) in combination with Iran's vernacular traditional architecture are perfect examples of the close interaction between climate and skilful adaptations by human culture. Studies by

52 Alireza Dehgani-Sanij and Ali Sayigh, *Cisterns: Sustainable Development, Architecture and Energy* (Aalborg, Delft, River Publishers, 2014).

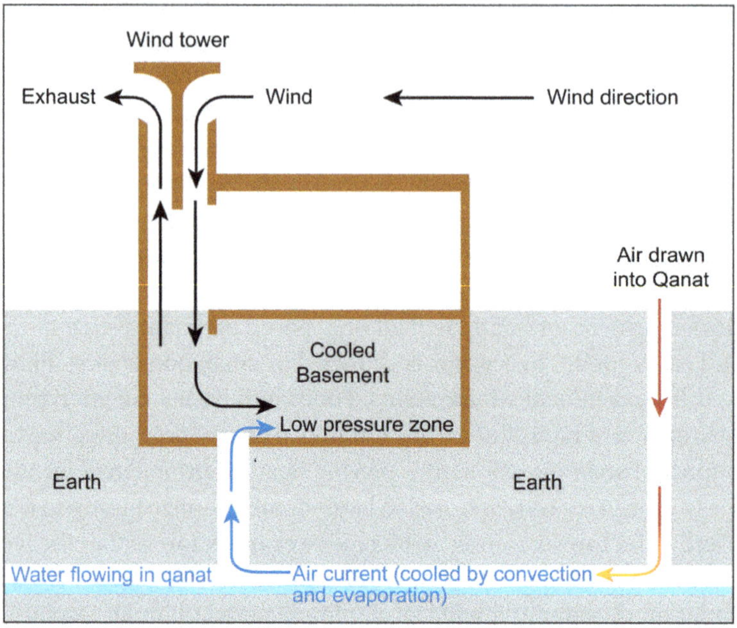

FIGURE 5.4 Passive cooling system through wind and water (modified after Mehdi N. Bahadori et al.)

Bahadori, Beazley, Bonine and more recent references by Semsar-Yazdi and Labbaf-Khaneiki or by Javadi pay appropriate tribute to these achievements.[53]

The impact of wind and water, however, is by no means limited to individual buildings or single objects. On the contrary: in the desert environments of Western Asia and North Africa, both rural and urban landscapes as a whole are influenced and shaped by these two elements. Especially in desert regions with a dense network of qanat-like structures, both rural and urban settlements have been comprehensively shaped by those downhill oriented artificial water-courses and their linear channels—and they are to this day.[54] At the same time, they also created social hierarchies and economically stratified societies at a small regional scale, with larger houses and gardens on the upper

53 Mehdi N. Bahadori, "Passive Cooling Systems in Iranian Architecture," *Scientific American* 238. 2 (1978): 144–154; Mehdi N. Bahadori and A. Dehgani Sanij, *Wind Towers—Architecture, Climate and Sustainability* (Basel: Springer International Publishing, 2014); Elizabeth Beazley, "Some Vernacular Buildings of the Iranian Plateau," *Iran* XV(1977): 89–102; Semsar-Yazdi, and Labaf-Khaneiki, ed., *Qanats in its Cradle*; Seyed M.R. Javadi, *La maîtrise millenaire de la terre, de l'eau et du vent en Iran. Yasouj/Iran* (Razbar, 2016), 135.

54 Michael E. Bonine, "The Morphogenesis of Iranian Cities," *Annals of the Association of American Geographers* 69 (1979): 208–224.

and well-watered reaches of the channels and smaller units (and sometimes less water) on the lower ends.

The conclusion of our deliberations on the great achievements of desert societies *vis-à-vis* the challenges of extremely harsh environments are the palaces, gardens, and,—above all—the old urban centres in the desert regions between Morocco in the west and Central Asia and the edge of the Indus valley in the east.[55] It is here that wind and water in combination with architecture and urban planning come together to create a specific urban culture based on the abovementioned creative adjustments to nature. Building materials, house designs, and street patterns create a cultural uniqueness that makes historical city centres of the Near and Middle Eastern city a very distinct type of its own.[56] Some authors go so far to say that urban building and planning principles are also and even decisively influenced by Islamic law and Quranic requirements.[57] The uniqueness of these historical and traditional centres is accentuated by the fact that houses are built around small inner courtyards. Wherever possible, small ponds surrounded by tiny arrangements of a few green plants or flowerpots simulate a kind of paradeisos in the dense urban fabric (fig. 5.5 and 5.6).

It would go beyond the limits of this paper to dive into the intricate details of Near and Middle Eastern urbanism. However, the unique combination of mud-brick architecture, water supply by 'hidden waters', underground channels and open irrigation canals and the catching of wind and air current circulation as part of a passive cooling system: that is a perfect match for the inseparable unity of climate and culture. It was again and above all

55 For a comprehensive survey of North Africa and the Near/Middle East cities up to the early 1990s see Michael E. Bonine et al., *The Middle Eastern City and Islamic Urbanism. An Annotated Bibliography of Western Literature* (Bonn: F. Dümmlers Verlag, 1994).

56 Leon C. Brown, ed., *From Madina to Metropolis. Heritage and Change in the Near Eastern City* (Princeton: The Darwin Press, 1973), 343; Abdulaziz Y. Saqqaf, ed., *The Middle East City. Ancient Traditions Confront a Modern World* (New York: Paragon House Publishers, 1987), 393; Albert H. Hourani and S.M. Stern, ed., *The Islamic City. A Colloquium* (Oxford: University of Pennsylvania Press, 1970), 222; Eugen Wirth, *Die orientalische Stadt im islamischen Vorderasien und Nordafrika. Städtische Bausubstanz und räumliche Ordnung, Wirtschaftsleben und soziale Organisation* (Mainz: Verlag Philipp von Zabern, 2000), 584; Ira M. Lapidus, ed., *Middle Eastern Cities. A Symposium on Ancient, Islamic, and Contemporary Middle Eastern Urbanism* (Berkeley, Los Angeles: Univ. of California Press, 1969), 206. For a comprehensive bibliography see Michael E., Bonine, Eckart Ehlers, Th. Krafft, G. Stöber, ed., *The Middle Eastern City and Islamic Urbanism. An Annotated Bibliography of Western Literature*. (Bonn: F. Dümmlers Verlag, 1994).

57 Besim S. Hakim, *Arabic-Islamic Cities. Building and Planning Principles* (London: KPI Limited, 1986).

Michael E. Bonine whose analysis of the morphogenesis of Iranian cities[58] and whose survey of adaptations of indigenous housing in Central Iran to aridity[59] unveiled the deep impact of climate on urban culture. Especially in Central Iran and its fringes, but also in neighbouring countries and in North Africa (esp. in the Maghreb/Morocco), many urban centres have preserved their traditional historical urban fabric. Often characterised as "desert cities", famous examples are Kerman,[60] Kashan,[61] or Yazd[62] in Iran. Other desert towns are e.g. Ghardaia in Algeria, Fes,[63] or Marrakesh[64] in Morocco, Muscat in Oman,[65] as well as Sanaa in Yemen[66]—to name just a few. Bianca has presented a comprehensive analysis and all-embracing symbiosis of Middle Eastern and North African urbanism; climate, water, and vegetation, urban design, and indigenous courthouse architecture are all parts of what he calls "life studies" in the Islamic world.[67]

---

58 Bonine, "The Morphogenesis of Iranian Cities," 208–224.

59 Michael E. Bonine, "Aridity and Structure: Adaptations of Indigenous Housing in Central Iran," in *Desert Housing: Balancing Experience and Technology for Dwelling in Hot Arid Zones*, ed. Kenneth N. Clark and P. Paylore, 193–219 (Tuscon: University of Arizona, 1980).

60 English, *City and Village in Iran: Settlement and Economy in the Kirman Basin.*

61 Vincent F. Costello, *Keshan—A City and Region in Fars* (London, New York: Bowker Publications, 1976); Hengameh Dowlatshahi, *Kashan—The Great Pearl* (Tehran: Gooya Art House, 2006), 168.

62 Bonine, *Yazd and its Hinterland.*

63 Anton Escher and Eugen Wirth, ed., "Die Medina von Fes. Geographische Beiträge zur Persistenz und Dynamik, Verfall und Erneuerung einer traditionellen islamischen Stadt," *Erlanger Geographische Arbeiten* 53 (1992).

64 Gaston Duverdun, *Marrakech. Des origines à 1912*, 2 vols (Rabat, Editions Techniques Nord-Africaines, 1959/1966).

65 Fred Scholz, *Muscat—Sultanat Oman. Geographische Skizze einer einzigartigen arabischen Stadt,* 2 vols (Berlin, Das Arabische Buch,1990).

66 Robert B. Serjeant, and Ronald Lewcock, *Sana. An Arabian Islamic City* (Cambridge: World of Islam Fetival Trust, 1983).

67 Bianca, *Hofhaus und Paradiesgarten: Architektur und Lebensformen in der islamischen Welt.* Beyond those aforementioned publications there are invaluable documentations of the great traditions of North African architectures in the following books: Herbert Popp and A. Kassah, *Les ksour du sud Tunisien. Atlas illustré d'un patrimoine culturel* (Bayreuth; Naturwissenschaftliche Gesellschaft, 2010); Herbert Popp, M. Aït Hamfa, and Brahim El Fasskaoui, *Les agadirs de l'Anti-Atlas occidental. Atlas illustré d'un patrimoine culturel du Sud marocain* (Bayreuth: Naturwissenschaftliche Gesellschaft, 2011); Herbert Popp and Brahim El Fasskaoui, *Die Straße der Kasbahs—einst und jetzt. Le route des Kasbahs—jadis et maintenant—The Road of the Kasbahs—Past and Present* (Bayreuth: Michael Imhof Verlag, 2017); See also Thomas M. Schmitt, "Protection du patrimoine culturel et transformation socioculturelle: la « vallée du M'zab ». Un site classé patrimoine mondial en Algérie," in *Les pays du Maghreb. Contributions de la géographie humaine allemande*, ed. Herbert Popp, 38–49 (Bayreuth: University of Bayreuth, 2009).

FIGURE 5.5 View of part of the ancient city centre of Yazd, central Iran. © ECKART EHLERS 1998. SEE TEXT FOR MORE DETAILS.

FIGURE 5.6 Typical elements of a courtyard house in central Iran. © ECKART EHLERS 2012. CONSTRUCTED IN THE 19TH AND EARLY 20TH CENTURIES BY RICH MERCHANTS AND LANDLORDS, THESE MANSIONS WERE BUILT AROUND A CENTRAL COURTYARD WITH A WATER POOL, FLOWERBEDS, AND FRUIT TREES. BUILDING MATERIALS OF THE HOUSES ARE SUN-DRIED ADOBE AND BAKED CLAY BRICKS; WIND TOWERS AND OPEN BALCONIES PROVIDE PLEASANT TEMPERATURES IN THE SUMMER MONTHS. COMMON URBAN HOUSES (FIG. 5.5) ARE EQUIPPED WITH SMALLER VERSIONS OF THESE EARTHLY PARADISES.

Figure 5.5 summarises the very essence of a desert town, its layout and architectural design. Representing a typical section of the old city centre of Yazd, the photo shows characteristic components of the material culture of urbanism in extremely arid climatic environments: horizontal extension of the urban fabric (the wide use of locally manufactured adobe and clay bricks), the privacy of the domestic architectural design of houses (the courtyard structure of many buildings around small green spaces), the seemingly chaotic ground-plan with narrow alleys, cul-de-sacs, and shady protective walls. Figure 5.6 offers more detail through a closer look into the interior structure of an upper-class mansion typical of cities like Kashan, Yazd, Meybod, Esfahan, and many other places in the dry belt of the MENA.

4 Qanats and the Decay of the Qanat System: Ecology vs. Economy

One of the most remarkable features of qanats and their equivalents in the Near and Middle East and beyond is their ecological modernity: they represent sustainability in an almost classical sense. Although based on a millennia-old technology, the mining and provision of irrigation water was based almost exclusively on the exploitation of subterraneous and hidden renewable aquifers until quite recently. Casual rainfall and winter snowfall form the fragile basis of their annual replenishment, more or less dependent on the meteorological conditions and on the seasonality of climate. Climate, however, is not the only variable in the enduring successes of qanat irrigation. Due to the fact that the water-bearing aquifers and artificial galleries were excavated in the foothills of mountains, i.e. in the comparatively loose gravel and sand of their alluvial pediments (cf. map 5.1[68] and fig. 5.1), the geological setting was and still is a source of permanent uncertainty for the regular production and flow of the water. Water erosion of the flanks of the subterranean channels as well as falling debris require permanent cleaning and repair. Earthquakes especially repeatedly cause devastating damage so that many qanats collapsed, fell dry or were given up—not to speak of the disastrous consequences of warfare the wanton destruction of qanats as part of military action or strategic decisions in historical times.

One serious ecological setback of qanat hydrology, however, is the fact that water flows all year round. Thus, water is potentially wasted at times when climate prohibits agricultural production and irrigation is not needed. *Āb ānbār*s

---

68 Mohammad R. Sarvati, *Geomorphologische Untersuchungen in der Region Kashmar NE/Iran* (Würzburg: Würzburger geographische Arbeiten, 1986).

offer only limited storage capacity. Open pools in which qanat water could be stored are nor usable because of the high seepage and potentially high evaporation rates. A certain waste of water is therefore inevitable. Yet in spite of these obvious deficiencies, qanat technology in Iran and elsewhere has survived until today and follows—as mentioned before—almost the same principles documented 1000 years ago by Karadji.[69]

What then are the reasons for the rapid decline of qanat irrigation everywhere it was in use? In the middle of the 20th century, new technologies arose by which water could be produced in greater quantities and water-bearing geological strata at greater depth could be exploited. At the same time the demand for water increased. Population growth was one of the driving forces, the demands and needs of improved and increased agricultural production another. Together with the growing demand of the urban population, the rapid development of the industrial sector, and changing overall consumption patterns, these made traditional techniques of 'water production' obsolete, especially in the extremely arid parts of the MENA regions.

This is not the place to refer to the high-tech endeavours by oil-rich states like Saudi Arabia, the Gulf Emirates, or Libya and Algeria to tap the seemingly vast fossil water reservoirs in deeper geological strata. Aside from the fact that these resources are non-renewable and therefore limited, their 'mining' is costly and not sustainable. The same holds true for desalination technology for seawater, widespread again in the oil-rich coastal states of the Near and Middle East. Our focus on the traditional mining techniques of subterraneous water close to the surface and its decline is connected with the devastating effects of pump irrigation and the abundance of motor-driven wells.

Scientific analyses of the damaging effects of pump irrigation are very sparse. However, one early study of their negative impact on qanats was undertaken by Saidi, based on a regionally limited case study of the Assad'abad Basin near Hamadan/Iran. Over seventy villages with a total of 109 qanats in the year 1950 were analysed.[70] However, within a few years the number of qanats decreased rapidly due to the development of other forms of irrigation, i.e. pumps and wells. The drilling of shallow wells with normal diesel pumps (Pers. *chah-ye nim-e-amigh*) and of deep tube wells (Pers. *chah-ye amigh*) lowered the water table and caused qanats to dry up rapidly. *Chah* constructions increased from 39 shallow wells in 1962 to 439 in 1969 and almost 1,400 in 1979

69 Nadji, "Karadjis 'Erschließung verborgener Gewässer.'"

70 Abbas Saidi, "Agrarwirtschaftlicher Wandel in Iran unter besonderer Berücksichtigung der Bewässerungswirtschaft. Am Beispiel des Gebietes Assad'abad (West Iran)" (unpublished Ph.D. diss., Universität Marburg), 197.

in the Assad'abad Basin alone.[71] It can well be argued that Assad'abad is typical for many other regions not only in Iran—the development of modern tube wells and more effective pumps have accelerated the rapidly increasing deterioration of qanat culture.

While—to the best knowledge of the author—we do not have similarly detailed analyses of the deterioration of qanat irrigation in Iran or anywhere else, we fortunately have two recent surveys of the distribution and present status of qanats and qanat irrigation on a global scale.[72] Rough and partly incomplete as these surveys may be, they nevertheless provide deep insights into the general situation of this rapidly vanishing technology and its decreasing application. The overall trend of decline and abandonment is the same everywhere:

(a) overpumping of the groundwater reservoirs and lowering of the groundwater table due to the use of pumping devices;
(b) high costs of maintenance of the traditional qanat/karez/foggara structures and rapid decline of the knowledge and skill base required for their construction and maintenance;
(c) climate change and its effects on the climate regime in the Near and Middle East: less precipitation, higher temperatures and evaporation.

These and other trends can be observed in all parts of MENA. The number of observed and documented qanats/karez/rhettaras is dropping rapidly. And yet, the number of existing qanats is still remarkable.

According to Semsar Yazdi and Labbaf Khaneiki, Iran had a total of 34,355 qanats by the beginning of our century[73] followed by Afghanistan with 9,370 karez, of which 36 per cent had dried up,[74] and Oman with 4,112 falaj, 1004 of which were out of operation.[75] As to for Azerbaijan, the survey reports 885 karez. More scanty information refers to parts of Iraq and especially to the

71 Eckart Ehlers and Abbas Saidi, "Qanats and Pumped Wells—The Case of Assad'abad, Hamadan," in *Qanat, Kariz and Khattara: Traditional Water Systems in the Middle East and North Africa*, ed. P. Beaumont, Michael Bonine, and Kaith McLachlan (SOAS: The Middle East Centre, School of Oriental and African Studies. University of London, 1989), 89–112.

72 Semsar-Yazdi, and Labaf-Khaneiki, ed., *Qanats in its Cradle*; Semsar-Yazdi, and Labaf-Khaneiki, ed., *Veins of Desert*.

73 Semsar-Yazdi, and Labaf-Khaneiki, ed., *Qanats in its Cradle*, 88.

74 Sayed S. Shobair "Afghanistan," in *Qanats in its Cradle. Qanats in its Cradle. Situation of Qanat (Kariz, Karez, Falaj) in the World*, ed. Ali A. Semsar-Yazdi and Majid Labaf-Khaneiki (Yazd: International Center on Qanats & Hydraulic Structures, 2012), 22.

75 Abdullah al-Ghafri, "Sultanate of Oman," in *Qanats in its Cradle. Qanats in its Cradle. Situation of Qanat (Kariz, Karez, Falaj) in the World*, ed. Ali A. Semsar-Yazdi, and Majid Labaf-Khaneiki (Yazd: International Center on Qanats & Hydraulic Structures, 2012), 196.

province of Sulaimanya with 380 karez, many of them inactive[76] and to the Baluchi part of Pakistan with an aggregate total of over eight hundred qanat.[77] Additional figures are presented by Semsar Yazdi and Labbaf Khaneiki for China's Xinjiang Uighur AR with 404 karez, and for Algeria's Adrar region with 1,416 foggaras in total, 855 of which seem to be in operation.[78] For Morocco, the authors refer to approximately six hundred khettaras in the Haouz of Marrakesh with many of them "abandoned since the early 1970s as they have dried up", while in the Tafilalt "half of the 400 are still in use".[79] Scanty references to Spain and Italy should be seen as a tribute to the medieval control of the Iberian Peninsula and Sicily by the Arabs and their role as mediators of Near and Middle Eastern culture and historical know-how.

5 Summary and Conclusions

Our review of the role of water and wind for the development of a specific 'desert culture' in the dry belt of North Africa and the Near and Middle East has demonstrated very clearly that qanats and their linguistic synonyms are a specific cultural phenomenon of MENA. As such, they are characterised not only by their origin in the highlands of present-day northwestern Iran, but also by their perseverance over several thousand years and by their successful application in the human conquest of extremely harsh natural environments. Referring to our introductory remarks, they are striking examples of the 'creative adjustment' of people and societies to the challenges of nature. They have led to the development of a specific "genre de vie" (Ger. *Lebensform*) with equally specific forms of material culture and socio-economic structures. The 'mining' of subterranean waters and their combination with ubiquitous wind power not only enabled agriculture to flourish, even if it was and is on a small scale, but also allowed the creation of permanent settlements with climate-adapted architecture specifically designed to cope with the climatic adversity of their environment.

---

76 Dale Lightfoot, "Iraq,"in *Qanats in its Cradle. Qanats in its Cradle. Situation of Qanat (Kariz, Karez, Falaj) in the World*, ed. Ali A. Semsar-Yazdi and Majid Labaf-Khaneiki (Yazd: International Center on Qanats & Hydraulic Structures, 2012), 154.

77 Muhammad A. Kahlown, "Pakistan," in *Qanats in its Cradle. Qanats in its Cradle. Situation of Qanat (Kariz, Karez, Falaj) in the World*, ed. Ali A. Semsar-Yazdi and Majid Labaf-Khaneiki (Yazd: International Center on Qanats & Hydraulic Structures, 2012), 295–296.

78 Semsar-Yazdi and Labaf-Khaneiki, ed., *Veins of Desert*.

79 Ibid., 166.

It speaks to the ingenuity of these innovative technologies that in the course of history, qanat technology was transferred to other parts of our globe. The diffusion within MENA and its periphery has already been mentioned. In the Iberian Peninsula, the Arabs applied qanat irrigation as far north as Madrid which, according to Braun, was retained by the Spaniards after the Reconquista and even extended. Qanat technology experienced its heyday in the 19th century. It were also most likely the Spanish conquerors who transferred this knowledge to Mexico and to some smaller areas in Peru and northern Chile.

The very fact that qanat culture is under severe stress and obviously coming to an end does not diminish its importance as a great cultural achievement and as a remarkable facilitator of human wellbeing under adverse climatic and environmental conditions of desert regions on an almost global scale. It is fair to say that the development of this technology in the Middle East almost three millennia ago is not only the foundation of specific desert cultures in MENA, but in a broader sense also a contribution of this region and the ingenuity of its inhabitants to the overall history of mankind. Therefore, it is more than justified to term "qanat culture" and "desert culture" as synonyms. As early as 1967, the German geographer Carl Troll declared qanats part of a "histoire de l'humanité", mankind's universal history. While some of these evaluations may seem somewhat exaggerated—especially in comparison to the hydraulic civilisations of the Nile valley, of Mesopotamia or the Indus valley, their impact on human history, their cultures and their civilisations—qanat technology itself is by no means less spectacular.

In view of all these considerations, it is not surprising that the existence of this great cultural tradition and the maintenance of its rapidly vanishing remainder has been acknowledged by the United Nations Educational, Scientific, and Cultural Organisation UNESCO. In 2016, qanats were included in the UNESCO World Heritage Programme highlighting them "as an exceptional testimony to the cultural traditions and civilisations in desert areas with an arid climate". Is there any better characterisation of the close interaction between climate and culture?

PART 3

Present

CHAPTER 6

The Vanishing of Iran's Lake Urmia

Matthias Schmidt

Abstract

The Middle East is characterised by high water stress. Current climate change processes and increasing demand are leading to even higher water stress in the future. A drastic example not only of the mismatch between limited water availability and high water consumption, but also of intense human intervention in hydrology, is the desiccation of Lake Urmia, Iran's largest inland lake and one of the most substantial hypersaline lakes in the world. This massive environmental crisis which has long gone unnoticed by the world is by no means a natural phenomenon but a human-made environmental and social disaster which can also be seen as a disastrous feature of the Anthropocene. This article deals with the causes and consequences of the water crisis surrounding Lake Urmia.

1 Introduction

Aridity and desert-like landscapes are widespread images connected with the Middle East. In fact, large parts of the Middle East are characterised by irregular and low precipitation patterns and by sparse vegetation cover, steppes and deserts. In order to live in such environmental conditions, humans have had to develop specific adaptations. Over the centuries, societies in the Middle East have evolved sophisticated forms of water management to match the spatial and seasonal limited water availability. The famous qanats as well as distinctive systems of water procurement and distribution[1] are striking indicators of these

1 Hans E. Wulff, "The Qanats of Iran," *Scientific American* 218 (1968): 94–105; Peter Beaumont, "Qanat Systems in Iran," *Hydrological Science Journal* 16 (1971): 39–50; Peter Beaumont, and Michael E. Bonine, *Qanat, Kariz and Khattara: Traditional Water Systems in the Middle East and North Africa* (London: University of London, 1989); Mahmoud Jomehpour, "Qanat Irrigation Systems as Important and Ingenious Agricultural Heritage: Case Study of the Qanats of Kashan, Iran," *International Journal of Environmental Studies* 66 (2009): 297–315; Asghar Tahmasebi, "Indigenous Knowledge for Water Management in Iran's Dry Land—Siraf," *International Journal of Environmental Studies* 66.3 (2009): 317–325; Fatemeh Zafarnejad, "The Contribution of Dams to Iran's Desertification," *International Journal of*

 | DOI:10.1163/9789004444973_008

efforts, established thousands of years ago (see also Eckart Ehlers, "Deserts—Wind—Water," Chapter Five in this volume). These physical structures and institutional arrangements were well adapted to the climatic conditions of the region and allowed the evolution of advanced civilisations. Human creativity and agency were the central factors in using limited natural resources, most importantly water, and creating the ability to live in such environments.

Over the past centuries, however, it has become more and more obvious that human beings not only have the ability to establish and evolve advanced civilisations including the creation of sophisticated technologies and institutions, but also that human agency is a destructive force leading to extensive changes and severe damage to the natural environment more often than we would wish. Human beings have not only transformed the land cover in almost all corners of the planet by removing forests, planting crops, digging canals or founding settlements, but their construction works and exploitation efforts have also led to widespread erosion, interference in the ecology through the extinction of species, and the burning of fossil fuels on a vast scale leading to pollution and, finally, global climate change.

The Middle East is particularly exposed to climate change since already low and decreasing precipitation rates can quickly lead to severe droughts. Prolonged droughts, desertification, and landscape degradation are already widespread in the Middle East;[2] in fact, the water supply is running short in many places through diminishing groundwater tables or the springs drying up. These obvious and highly problematic examples of environmental change and water shortage are often explained by climate change, though a closer look very quickly shows that these effects are in fact the result of a number of factors.

Today, the Middle East is experiencing severe water scarcity which is projected to get even worse in the near future.[3] As a consequence, there is high

Environmental Studies 66.3 (2009): 327–341; Yazdi Semsar, Ali Asghar, and Majid Labbaf Khaneiki, *Qanat Knowledge: Construction and Maintenance* (Dordrecht: Springer, 2016).

2 Madjid Abbaspour, and Alieh Sabetraftar, "Review of Cycles and Indices of Drought and their Effect on Water Resources, Ecological, Biological, Agricultural, Social and Economical Issues in Iran," *International Journal of Environmental Studies* 62.6 (2005); Zafarnejad, "The Contribution,"; Amiraslani, Farshad, and Deirdre Dragovich, "Combating Desertification in Iran over the Last 50 Years: An Overview of Changing Approaches," *Journal of Environmental Management* 92 (2011): 1–13; Iraj Emadodin, and Hans Rudolf Bork, "Degradation of Soils as a Result of Long-term Human-induced Transformation of the Environment in Iran: An Overview," *Journal of Land Use Science* 7.2 (2012): 203–219.

3 United Nations World Water Assessment Programme (WWAP)/UN-Water, *The United Nations World Water Development Report 2018: Nature-Based Solutions for Water* (Paris: UNESCO, 2018), 12. Regions are considered severely water scarce when total annual withdrawals for human use exceed forty per cent of the total available renewable surface water resources.

water stress,[4] which means an unfavourable water withdrawal-to-availability ratio.[5] Thus, water scarcity is not only a function of precipitation and evaporation, but it must also be seen in connection with human water consumption. Population growth aligned with increasing food demand, economic development, and modernisation efforts in the Middle East have all led to growing water demand, resulting in increasing water stress and even water crises in the region.[6]

A particularly drastic example of one such water crisis is the desiccation of Lake Urmia, Iran's largest lake. Lake Urmia is located in a highly dynamic region in which population numbers and settlements have scaled up significantly over the last few decades and, following the paradigm of technocentric modernisation, where several dams have been built for power generation and to feed irrigation systems, leading to an enormous expansion of irrigated farmlands. As a result, the inflow from several rivers into Lake Urmia has decreased dramatically.

No other lake of this magnitude has dried out so quickly in such a short time. Parallels are evident when comparisons are made to the once much larger, now largely dried-out Aral Sea.[7] These processes became well-known in science and are discussed as the so-called Aral Sea syndrome.[8] Worldwide attention to the disastrous developments at Lake Urmia, however, came relatively late, mainly due to the political isolation of Iran and the restrictive information policies of its government. A major characteristic of the Aral Sea syndrome is the restructuring of landscapes by centrally planned large-scale projects including the construction of dams, canals, and irrigation systems to achieve strategic goals without considering ecological and social consequences and ignoring

4 WWAP, *The United Nations World Water Development Report 2016: Water and Jobs* (Paris: UNESCO, 2016), 17.

5 Karen Frenken, *Irrigation in the Middle East Region in Figures. AQUASTAT Survey—2008* (Rome: Food and Agriculture Organization of the United Nations, 2009); Petra Hellegers, Walter Immerzeel, and Peter Droogers, "Economic Concepts to Address Future Water Supply-Demand Imbalances in Iran, Morocco and Saudi Arabia," *Journal of Hydrology* 502 (2013): 62–67.

6 Richard Foltz, "Iran's Water Crisis: Cultural, Political, and Ethical Dimensions." *Journal of Agricultural and Environmental Ethics* 15 (2002): 357–359.

7 Philip Micklin, "The Aral Sea Disaster," *Annual Review of Earth and Planetary Sciences* 35 (2007): 47–72; Philip Micklin, Nikolai V. Aladin, and Igor Plotnikov, *The Aral Sea: The Devastation and Partial Rehabilitation of a Great Lake* (Heidelberg: Springer, 2014); Amir AghaKouchak et al., "Aral Sea Syndrome."

8 Wissenschaftlicher Beirat der Bundesregierung Globale Umweltveränderungen, *Welt im Wandel: Herausforderungen für die deutsche Wissenschaft*. Jahresgutachten 1996 (Bremerhaven: WBGU, 1996), 125.

local conditions. Its consequences include large-scale environmental degradation, biodiversity loss, local climate change, deficient fresh water supply, soil degradation, forced resettlement of local populations, and water conflicts between political entities.

This paper analyses and discusses the main reasons for and consequences of the desiccation of Lake Urmia as an example of human destructive effects on the water cycle in the Middle East.

2 Desiccation of Lake Urmia

Lake Urmia is located at an altitude of 1276 metres at the edge of the Zagros Mountains in northwestern Iran and is the country's largest inland body of water (map 6.1). Due to its unique ecology as a wetland in a desert-like landscape, it was declared a protected wetland area by the Ramsar Convention in 1975 and a UNESCO Biosphere Reserve in 1976. At the time, despite its high salt content, it had a remarkably high level of biodiversity[9] and was the largest habitat of the endemic *Artemia urmiana*, a brine shrimp species that adapted very well to the hypersaline environment. *Artemia urmiana* was also an important element in the food chain for more than two hundred species of migratory birds such as ibis, storks, geese, and gulls that rested on or near the lake during their annual migration.[10] In addition, river mouths and brackish water swamps represented important biodiversity hotspots.

Twenty years ago, Lake Urmia covered an area of about 5,200 square kilometres. Since then, it has lost about eighty per cent of its water volume, causing the lake's surface to shrink to fewer than 2,000 square kilometres and the overall level to fall by more than six metres[11] (fig. 6.1 and 6.2). At the same time,

---

9 Alireza Asem et al., "Biodiversity of the Hypersaline Urmia Lake National Park (NW Iran)," *Diversity* 6 (2014): 126.
Alireza Asem, Amin Eimanifar, and Michale Wink, "Update of 'Biodiversity of the Hypersaline Urmia Lake National Park (NW Iran)'," *Diversity* 8.6 (2016): 7.

10 UNEP-GEAS, "The Drying of Iran's Lake Urmia and its Environmental Consequences," *Environmental Development* 2 (2012): 129; Amin Eimanifar, and Feridon Mohebbi, "Urmia Lake (Northwest Iran): A Brief Review," *Saline Systems* 3.5 (2007): 5–6.

11 Iran Environment and Wildlife Watch (IEW), accessed May 18, 2018. http://www.iew.ir/1396/10/19/57250; Somayeh Sima and Mohammad Tajrishy, "Using Satellite Data to Extract Volume—Area—Elevation Relationships for Urmia Lake, Iran," *Journal of Great Lakes Research* 39.1 (2013): 94; Amir AghaKouchak et al., "Aral Sea Syndrome Desiccates Lake Urmia: Call for Action," *Journal of Great Lakes Research* 41.1 (2015): 308; Mohammad J. Tourianet al., "A Spaceborne Multisensor Approach to Monitor the Desiccation of Lake Urmia in Iran," *Remote Sensing of Environment* 156 (2015): 349;

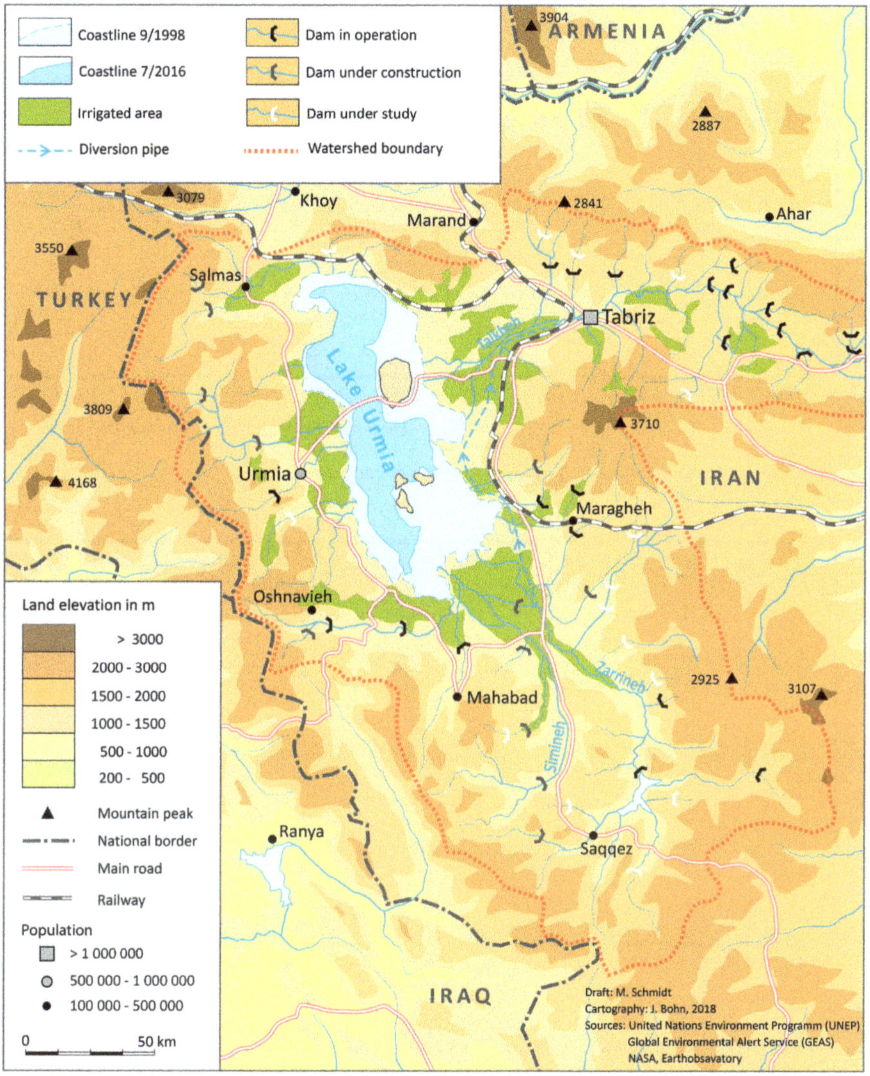

MAP 6.1 Lake Urmia watershed in Northwest Iran with irrigated areas and water dams.

Somayeh Shadkam et al., "Impacts of Climate Change and Water Resources Development on the Declining Inflow into Iran's Urmia Lake," *Journal of Great Lakes Research* 42 (2016): 943; Somayeh Shadkam et al., "Water-saving Interventions Assessment Framework: an Application for the Urmia Lake Restoration Program." *Geophysical Research Abstracts* 19 (2017): 1; Urmia Lake Restoration Program (ULRP), "Urmia Lake Restoration Program: Brief Report and Projects Outline," last modified October 2, 2015, accessed May 18, 2018. https://usu.instructure.com/files/61057414.

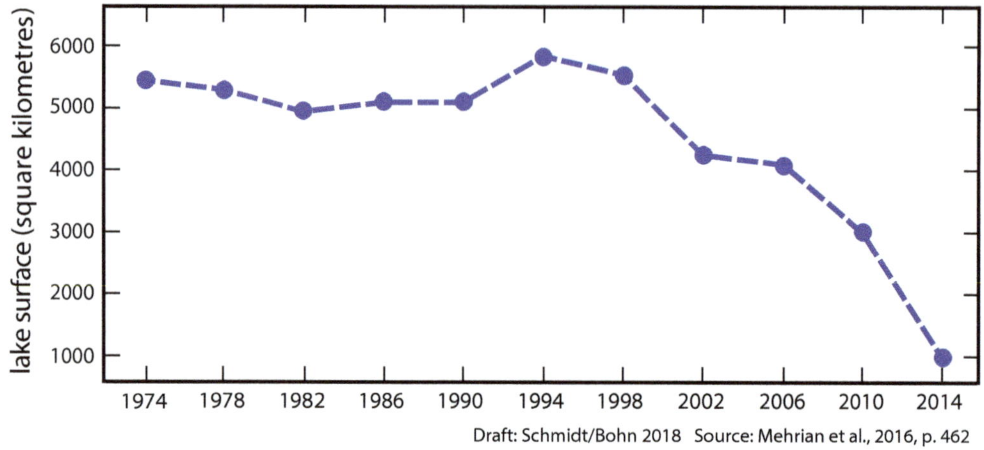

FIGURE 6.1 Lake Urmia's surface area changes between 1974 and 2014.

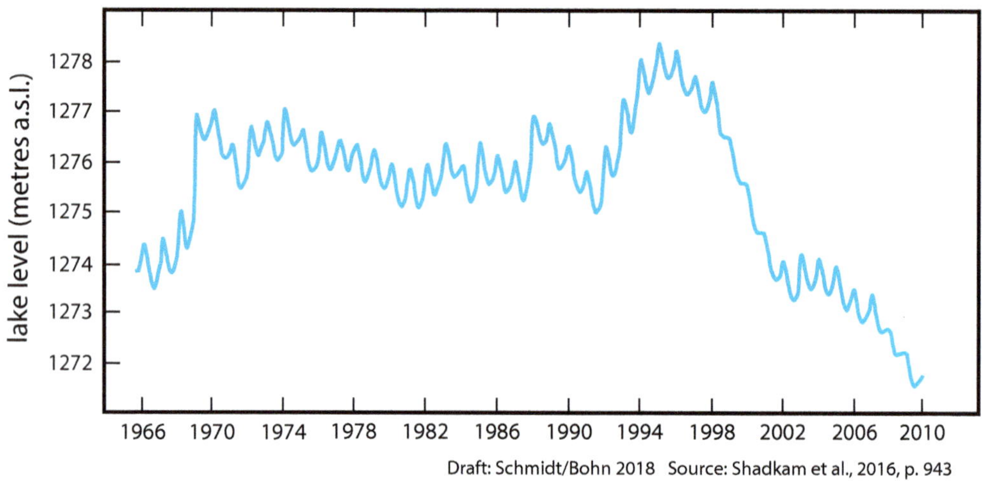

FIGURE 6.2 Lake Urmia's surface elevation changes between 1966 and 2010.

the salt content in the remaining lake water has risen to such an extent that most organisms cannot survive.

Significant fluctuations in water levels and shorelines are not unusual for endorheic lakes, as can be seen, for example, at Great Salt Lake in the USA or Lake Chad in Central Africa. Accordingly, water volume—and thus the surface area of Lake Urmia—has always varied considerably due to climate change processes and weather variability. However, the current water loss is unparalleled in terms of its sheer speed, causing a historical low the causes of which cannot be traced back to natural phenomena alone. The recent development of Lake Urmia is a rather extreme example of human intervention in a fragile ecosystem.

The reasons for this dramatic water loss are diverse and of intense interest to the scientific community.[12] Desiccation is blamed consistently on a combination of climate change and direct human interference in the water cycle,[13] but in either case, one may safely suggest that it has been influenced by humans since the present climate change is human-induced, and can thus be seen as a warning sign which could presage future events and hazards.

3 Climatic Conditions and Recent Changes

Water is an essential natural resource and is widely available in many areas of the world, but in short supply in others. Although the Middle East is commonly seen as an arid region with very limited water resources, there are areas with high precipitation rates and water abundance such as Eastern Anatolia, the Lebanon Mountains, or the Elburz Mountains. However, the prevalent image of the Middle East as a desert or steppe region is not wrong in principle since large areas are subject to arid to semi-arid climatic conditions.

This holds particularly true for Iran, where the climate in most parts of the country is characterised by low and irregular rainfall and high temperatures, and thus enormous evaporation potential. Significantly higher precipitation

12 Alizadeh-Choobari, Omid, and Mohammad S. Najafi, "Extreme Weather Events in Iran under a Changing Climate," *Climate Dynamics* 50.1–2 (2018); Hassanzadeh, Elmira, Mahdi Zarghami, and Yousef Hassanzadeh, "Determining the Main Factors in Declining the Urmia Lake Level by Using System Dynamics Modeling." *Water Resource Management* 26.1 (2012): 129–145; Sheida Jalili, Sajad Ahmad Hamidi, and Reza Namdar Ghanbari, "Climate Variability and Anthropogenic Effects on Lake Urmia Water Level Fluctuations, Northwestern Iran," *Hydrological Sciences Journal* 61.10 (2016); Shadkam et al., "Impacts of Climate,"; Shadkam, Somayeh et al., "Preserving the World Second Largest Hypersaline Lake under Future Irrigation and Climate Change," *Science of the Total Environment* 559 (2016); Taravat Talebiet al., "The Late-Holocene Climate Change, Vegetation Dynamics, Lake-level Changes and Anthropogenic Impacts in the Lake Urmia Region, NW Iran." *Quaternary International* 408 (2016).

13 AghaKouchak et al., "Aral Sea Syndrome," 309–310; Farshad Fathian, Zohreh Dehghan, and Saeid Eslamian, "Analysis of Water Level Changes in Lake Urmia Based on Data Characteristics and Non-parametric Test," *International Journal of Hydrology Science and Technology* 4.1 (2014): 35–36; Fatemeh G. Hamzekhani, Bahram Saghafian, and Shahab Araghinejad, "Environmental Management in Urmia Lake: Thresholds Approach." *International Journal of Water Resources Development* 32.1 (2016): 84–86; Hassanzadeh, Zarghami, and Hassanzadeh, "Determining,"; Jalili, Hamidi, and Ghanbari, "Climate Variability," 139–143.

rates can only be found in the Zagros and the Elburz Mountains where water generation exceeds evaporation and water consumption so that these areas serve as the main water sources for other regions of Iran.

Various studies as well as the daily experience of the local population demonstrate that the Middle East is highly affected by present and future climate change which is already one of the main reasons for worsening extreme heat, drought, and aridity.[14] Climate projections for the Middle East predict a high increase in average temperatures and a strong decrease in precipitation.[15] Iran already warmed by nearly 1.3°C between 1951 and 2013 and is showing an increase in the frequency of heat extremes, while annual precipitation has decreased by eight millimetres per decade, causing an expansion of the country's dry zones.[16]

Moreover, these climatic and hydrological changes, paired with population growth and modernisation efforts, and thus increasing water consumption, will intensify the already severe levels of water stress in most regions.[17] The agricultural sector in particular is highly exposed to changing climatic conditions,[18] because in rain-fed areas, lower precipitation and more frequent droughts will lead to decreasing yields[19] and more crop failures while reduced

---

14 Katharina Waha et al., "Climate Change Impacts in the Middle East and Northern Africa (MENA) Region and their Implications for Vulnerable Population Groups," *Regional Environmental Change* 17 (2017).

15 Jason P. Evans, "21st Century Climate Change in the Middle East," *Climatic Change* 92 (2009): 431; Hossein Tabari and Parisa Hosseinzadeh Talaee, "Temporal Variability of Precipitation over Iran: 1966–2005," *Journal of Hydrology* 396 (2011): 313; Mohammad R. Kousari, Hossein Ahani, and Razieh Hendizadeh, "Temporal and Spatial Trend Detection of Maximum Air Temperature in Iran during 1960–2005," *Global and Planetary Change* 111 (2013): 97; Edoardo Bucchignani et al., "Climate Change Projections for the Middle East-North Africa Domain with COSMO-CLM at Different Spatial Resolutions," *Advances in Climate Change Research* 9 (2018): 66; Tugba Ozturk et al., "Future Projections of Temperature and Precipitation Climatology for CORDEX-MENA Domain Using RegCM4.4," *Atmospheric Research* 206 (2018): 102–104.

16 Alizadeh-Choobari, and Najafi, "Extreme Weather," 249.

17 Kaveh Madani, "Water Management in Iran: What is Causing the Looming Crisis?" *Journal of Environmental Studies and Sciences* 4 (2014): 315; Hossein Hashemi, "Climate Change and the Future of Water Management in Iran," *Middle East Critique* 24.3 (2015): 320–321.

18 Hamideh Maleksaeidi, and Ezatollah Karami, "Social-Ecological Resilience and Sustainable Agriculture under Water Scarcity," *Agroecology and Sustainable Food Systems* 37.3 (2013): 262–290.

19 Sarvenaz Farhangfar et al., "Vulnerability Assessment of Wheat and Maize Production Affected by Drought and Climate Change," *International Journal of Disaster Risk Reduction* 13 (2015): 37; Hassan M. Mosammam et al., "Analyzing the Potential Impacts of Climate

water discharge,[20] already critically low, will affect most irrigated lands with negative consequences for food production.[21] Additionally, higher average temperatures and more heat extremes are increasing the evaporation potential, resulting in higher water demand for crop fields. Furthermore, from a livestock point of view, prolonged dry seasons reduce the length of time that rangelands can be grazed.[22]

In semi-arid areas such as northwestern Iran, precipitation is irregular and long dry periods are not unusual, which has led to considerable fluctuations in the level of Lake Urmia in the past,[23] reaching its peak between 1994 and 1996. The climate development of the last decades, however, has engendered a significant trend towards higher air temperatures, decreasing precipitation, and increasing drought periods.[24] For instance, the average annual precipitation in the Urmia basin decreased by 14.8 millimetres per decade from 1968 to 2011 while the average temperature increased by 0.18°C per decade in the same period.[25]

This not only reduced direct inflow into the lake,[26] but also increased the evaporation rate and subsequently the water demand of the agricultural sector.[27] A climate change scenario indicates a further increase in temperature by +0.88°C and a precipitation loss equal to 0.37 millimetre per decade by the year 2100.[28] It is clear that these trends are consequences of human-induced

---

Change on Rainfed Wheat Production in Hamedan Province, Iran, via Generalized Additive Models," *Journal of Water and Climate Change* 7.1 (2016), 212.

20 Mahdi Zarghami, et al., "Impacts of Climate Change on Runoffs in East Azerbaijan, Iran," *Global and Planetary Change* 78 (2011): 137.

21 Alireza Gohari et al., "Climate Change Impacts on Crop Production in Iran's Zayandeh-Rud River Basin," *Science of the Total Environment* 442 (2013): 405; Waha et al., "Climate Change Impacts," 1623–1624.

22 Evans, "21st Century," 417; Asghar Tahmasebi, Eckart Ehlers, and Conrad Schetter, "Climate Change and Mountain Pastoralism—The Shahsevan of Northwest Iran," *Erdkunde* 67.4 (2013): 320.

23 Shadkam et al., "Impacts of Climate Change," 943; Jalili, Hamidi, and Ghanbari, "Climate Variability," 1766.

24 Tabari, and Talaee, "Temporal Variability," 313; Amir H. Delju et al., "Observed Climate Variability and Change in Urmia Lake Basin, Iran," *Theoretical and Applied Climatology* 111.1–2 (2013): 295; Jamileh Farajzadeh, Ahmad Fakheri Fard, and Saeed Lotfi, "Modeling of Monthly Rainfall and Runoff of Urmia Lake Basin Using 'Feed-Forward Neural Network' and 'Time Series Analysis' Model," *Water Resources and Industry* 7–8 (2014): 46–47; Alireza Taravat et al., "A Spaceborne Multisensory, Multitemporal Approach to Monitor Water Level and Storage Variations of Lakes," *Water* 8.11 (2016): 478.

25 Zarrineh, and Abad, "Integrated," 43.

26 Zarghami et al., "Impacts of Climate Change," 137.

27 ULRP, "Urmia Lake Restoration Program."

28 Zarrineh, and Abad, "Integrated," 43.

climate change, and so the increased air temperatures and aridity that appear to be natural processes must in fact be counted as anthropogenic causes.

It seems obvious that climate change has already influenced and will continue to transform hydrology in many areas of Iran, and the desiccation of Lake Urmia is an extreme example of this. Moreover, it has the potential to put additional ecological, economic, and societal pressure on the already existing high water stress as it poses serious challenges to populations whose livelihoods depend principally on water and natural resources.[29] Paired with a lack of resilience, this will cause high levels of vulnerability within the agricultural sector. However, a closer look at Lake Urmia shows that an explanation based purely on climatic arguments is insufficient since other factors are more relevant to this severe ecological disaster.

4 Human Hydrological Interference

In addition to climatic factors, various direct effects of human agency are even more relevant to the desiccation of Lake Urmia. The socioeconomic development in Northwest Iran in the form of high population growth, increasing urbanisation, and extensive infrastructure expansion has been accompanied by massive intervention in the water cycle.[30]

Notably, 48 dams have been built in the Lake Urmia catchment area over the last few decades. These are used for power generation, flood protection, and the expansion of irrigation areas. Another 24 dams have been under construction or were planned until recently[31] (see map 6.1). In principle, hydro-

29 Marzieh Keshavarz, Ezatollah Karami, and Mansoor Zibaei, "Adaptation of Iranian Farmers to Climate Variability and Change," *Regional Environmental Change* 14 (2014): 1163.

30 Shahram Khalighi Sigaroodi and Shiva Ebrahimi, "Effects of Land Use Change on Surface Water Regime (Case Study Orumieh Lake of Iran)," *Procedia Environmental Sciences* 2 (2010); Delju et al., "Observed Climate"; Farajzadeh, Fakheri F., and Lotfi, "Modeling of Monthly"; Hassanzadeh, Zarghami, and Hassanzadeh, "Determining"; Henareh Khalyani, Mayer, and Norman, "Water Flows"; Somayeh Shadkam et al., "Preserving the World Second Largest Hypersaline Lake under Future Irrigation and Climate Change," *Science of the Total Environment* 559 (2016); Talebi et al., "The Late-Holocene"; Mostafa Zeinoddini, Arash Bakhtiari, and Majid Ehteshami, "Long-term Impacts from Damming and Water Level Manipulation on Flow and Salinity Regimes in Lake Urmia, Iran," *Water and Environment Journal* 29 (2015); Alizadeh-Choobari and, Njafi, "Extreme Weather,"; Jalili, Hamidi, and Ghanbari, "Climate Variability."

31 Azad Henareh Khalyani, Audrey L. Mayer, and Emma S. Norman, "Water Flows toward Power: Socioecological Degradation of Lake Urmia, Iran," *Society & Natural Resources* 27.7 (2014): 762.

logical electricity generation does not lead to a reduction in available water volume because only the kinetic energy of the water is used and not the water itself. Nevertheless, with the extension of open water bodies under arid climatic conditions, large amounts of water are lost as flow velocities are lowered and the water surface massively enlarged, which results in significantly higher evaporation.

An even more significant intervention in the water cycle is the function of the reservoirs for irrigation purposes. Agriculture is the largest water consumer in Iran, accounting for 92 per cent of the total demand.[32] In the Lake Urmia catchment area, irrigated farmland has increased in size by 67.5 per cent, from 3,035 square kilometres in 1984 to 5,086 square kilometres in 2014.[33] Overall, the total size of cultivated land has tripled in recent times,[34] while a great deal of previously solely rain-fed agricultural land has become irrigated farmland. This major increase in farmland is the result of demographic and political developments: on the one hand, the high population growth in Iran—the number of inhabitants rose from 61 to 82 million between 1996 and 2018—and the changed consumption patterns that go along with ongoing modernisation processes, and on the other hand, the international trade embargo imposed on Iran that forced the nation's government to increase food production to secure the food supply of its own population.

In addition to the discharge of surface water from tributaries, groundwater is increasingly used intensively in the Lake Urmia catchment area. Almost 90,000 deep wells,[35] most of which were drilled without permission, provide groundwater for agricultural irrigation and drinking water use. The strong increase in the number of deep-water wells is linked to population growth as well as to inheritance laws. Islamic law provides for a division in the case of land inheritance according to which the land is divided equally among the heirs on the principle that male descendants receive twice the share as females. This form of inheritance law leads to the division of farms and thus to the reduction of landed property, and since most farms look to their own water supply, more and more deep wells are being drilled. Due to the high extraction rate of groundwater, levels in some places have fallen by up to 16 metres, which in the

32 Madani, "Water Management," 320.

33 Majid Ramezani Mehrian et al., "Investigating the Causality of Changes in the Landscape Pattern of Lake Urmia Basin, Iran, Using Remote Sensing and Time Series Analysis," *Environmental Monitoring and Assessment* 188 (2016): 462.

34 Simon Dalby and Zahra Moussavi, "Environmental Security, Geopolitics and the Case of Lake Urmia's Disappearance," *Global Change, Peace & Security* 29.1 (2017): 48.

35 "Urmia Lake Restoration Program."

vicinity of Lake Urmia has also increased the risk of salt water intrusion into the groundwater body.[36]

In addition to the expansion of irrigated farmland, there has also been a major shift in cultivation patterns over recent years towards economically more valuable, but thirstier crops such as sugar beets and apples.[37] About seven per cent of the irrigated arable land in the Urmia area has been converted into fruit gardens,[38] and many vineyards are replaced by water-intensive apple plantations.

Two construction projects, affecting the lake ecology directly and indirectly are also highly controversial. First, since 2000, the road between the cities of Tabriz and Urmia has been shortened greatly by means of a 15 kilometre-long causeway across Lake Urmia. Due to this massive structural intervention, the lake is now divided into two parts connected to each other by a 1.25 kilometre-wide passage which severely hampers circulation and water exchange between the northern and southern parts of the lake.[39] As a result, sedimentation behaviour has changed significantly and the lake water now heats up faster, which in turn leads to higher evaporation rates.

The second construction project is a water pipeline that has channelled about three billion cubic metres of water annually from the Zarrineh River into the metropolitan region of Tabriz since 1999.[40] This in turn has significantly reduced the runoff of one of the most important inlets into the lake. In this context, the disproportionately high demand for water in the rapidly growing urban agglomerations of Tabriz (about 2.2 million inhabitants) and Urmia (about 0.7 million inhabitants) is a major cause of increased water consumption. Undoubtedly, the cities in the catchment area of Lake Urmia show a marked increase in water utilisation. As a matter of fact, however, the water is used but not necessarily consumed in the cities because it diffuses back into the water cycle via the sewage system or other diffusers—undoubtedly in a much reduced quality—and is therefore only partly lost to the water cycle at Lake Urmia.

---

36 Mahdi Zarghami, "Effective Watershed Management. Case Study of Urmia Lake, Iran," *Lake and Reservoir Management* 27.1 (2011): 90.

37 Dalby, and Moussavi, "Environmental," 47; Ercan Ayboğa and Akgün Ilhan, "Iran's Dam Policy and the Case of the Lake Urmia," last modified July 31, 2012, accessed May 18, 2018. http://www.academia.edu/9613644.

38 Sigarodi, and Ebrahimi, "Effects of Land Use," 256.

39 Aliasghar Marjani and Mirmosadegh Jamali, "Role of Exchange Flow in Salt Water Balance of Urmia Lake," *Dynamics of Atmospheres and Oceans* 65 (2014): 1; Zeinoddini, Bakhtiari, and Ehteshami, "Long-Term Impacts," 85–86.

40 Henareh Khalyani, Mayer, and Norman, "Water Flows," 763.

The construction boom in luxury holiday homes on the outskirts of urban agglomerations and in rural areas has been largely neglected in scientific debates, but it, too, is associated with high water consumption.[41] In many places, arable, pasture, and fallow lands have been transformed into building land where wealthy city dwellers construct weekend cottages, often with large ornamental gardens. Finally, aquaculture farms, which have been set up in recent years, must also be regarded as unfavourable to the water supply since they require a lot of water and are associated with high evaporation.[42]

Currently, approximately seventy per cent of renewable water resources in the Lake Urmia basin are consumed, mainly by the agricultural sector, instead of being discharged into the lake. According to ULRP, the runoff amount feeding the lake has decreased by fifty per cent during the past two decades, amounting to a total of thirty billion cubic metres of 'lost water.'[43] Therefore, human intervention in the expanding agricultural sector is considered the major contributor to Lake Urmia's water crisis as feeding insufficient water into the terminal lake is the main cause of its drying out. Consequently, scientists refer to this phenomenon as a "socioeconomic drought"[44] or an "Anthropocene disaster."[45]

5 Ecological and Socio-economic Consequences of the Urmia Disaster

The drying out of large parts of Lake Urmia and the increase in the salinity of the remaining water have had a destructive effect on the aquatic ecosystem, leading to the interruption of food chains and threatening biodiversity. Due to high evaporation and the accompanying accumulation of salt in the lake's residual water, the saline content of Lake Urmia has almost doubled from 166

41 Sigaroodi and Ebrahimi, "Effects of Land-use," 256.

42 Nina Zarrineh and Mohammad Azari Najaf Abad, "Integrated Water Resources Management in Iran: Environmental, Socio-economic and Political Review of Drought in Lake Urmia," *International Journal of Water Resources and Environmental Engineering* 6.1 (2014).

43 As of January 2018, the lakes total water volume stood at 1.1 billion cubic metres, around 15.17 billion cubic metres below its long-term average volume. "Urmia Lake Restoration Program."

44 Kaveh Madani, Amir AghaKouchak, and Ali Mirchi, "Iran's Socio-economic Drought: Challenges of a Water-Bankrupt Nation," *Iranian Studies* 49.6 (2016).

45 Matthias Schmidt, "Wasserkrise am Urmiasee im Iran: Eine Umwelt- und Sozialkatastrophe des Anthropozäns," *Geographische Rundschau* 70.1–2 (2018).

to over 300 g/l since 1995.[46] This has led to the subsequent extinction of many species, and even the saltwater brine shrimp *Artemia urmia*, adapted to the hypersaline environment, looks to be becoming extinct soon, thus eliminating the most important food source for migratory birds.

Apart from the destructive consequences for the ecosystem, the desiccation of Lake Urmia has also had massive negative impacts on the lives and economies of the two provinces of West and East Azerbaijan. About 6.5 million people live in the catchment area of Lake Urmia.[47] Accompanying the drying out of the lake bottom (fig. 6.3), there has been an increase in wind erosion and the occurrence of dust storms in recent years which threatens surrounding farmlands, livestock and livelihoods through salt deposits.[48] The aeolian dust load may transport high concentrations of fine-grain saline or alkaline materials and other potentially toxic components, and it is thus a health hazard that could lead to an increase in respiratory and immune diseases.[49] A further depletion of the lake will most certainly amplify the health problems of the local population.

In the past, Lake Urmia was known all over Iran as a popular area for relaxation that allowed various bathing opportunities and water sports. Spa tourism in the form of mud applications was widespread and led to large investments in tourist infrastructure, and numerous holiday facilities, tourist resorts, and marinas for water sports were built up to the turn of the millennium, when a tipping point was reached and the number of visitors who benefitted from medicinal mud applications steadily decreased.[50] With the desiccation of Lake Urmia, tourism-dependent businesses such as the production of clay for healing purposes, tourist shops, and boat rentals, were no longer viable,[51] and hotels, holiday homes, and boat jetties were now far away from the water's edge (fig. 6.4). As a result, many households that relied directly on tourism needed to look for other income opportunities.

---

46 Marjani, and Jamali, "Role of," 2.

47 SEDAC—Socioeconomic Data and Applications Centre, "Gridded Population of the World: Future Estimates," accessed May 18, 2018. http://sedac.ciesin.columbia.edu/mapping/popest/gpw-v4/.

48 Vahid Garousi et al. "Environmental Crisis in Lake Urmia, Iran: A Systematic Review of Causes, Negative Consequences and Possible Solutions," in *Proceedings of the 6th International Perspective on Water Resources & the Environment* (*IPWE*), (Izmir,2013).

49 Henareh Khalyani, Mayer, and Norman, "Water flows," 762; Zarrineh, and Abad, "Integrated," 45.

50 Zarrineh, and Abad, "Integrated," 45.

51 Farhudy, and Hajilou, "Strategic Evaluation of Tourism Industry Development Pattern with Emphasis of Ecotourism (Sample is Urmia Lake)" (paper presented at the "International Conference on Lake Urmia," Urmia, December 8–12, 2012).

FIGURE 6.3 Desiccated lake bottom of Lake Urmia.
© SCHMIDT 2015.

FIGURE 6.4 Former shore of Lake Urmia.
© SCHMIDT 2014.

The reduction of Lake Urmia's inflows, caused by upstream water extraction and the draining of groundwater levels due to the overexploitation of groundwater resources, has resulted in major challenges for the region's agriculture.[52] Deep-water wells have to be drilled deeper and water shortages have become an everyday problem. Furthermore, water quality is decreasing throughout the basin due to fertiliser and pesticide deposits as well as the overuse of local groundwater which has led to salt-water intrusions into aquifers.[53] Currently, farmers do not have to pay for water, but investment costs are rising and with a further depression of groundwater levels, many farming households could lose their economic basis which would result in unemployment, emigration from rural regions, and the disintegration of rural settlements and infrastructure.

It is obvious that the ecological disaster at Lake Urmia is creating increasingly difficult socio-economic conditions for millions of households living in the basin's villages and cities. Local reports have indicated that many people around the lake have already abandoned the area, and it is believed that a further escalation in environmental degradation will result in many more people choosing to migrate either temporarily or permanently as their livelihoods and their economic and social well-being are undermined.[54]

6 Perception of the Urmia Disaster and Prospects

The obvious ecological crisis at Lake Urmia, resulting in negative socio-economic impacts, is increasing the sense of resentment and resistance in the affected population of the region; for instance, demonstrations and riots took place in Tabriz in 2011 and 2012, addressing the deterioration of the environmental situation in the Urmia area and calling on the government to act.[55] In addition, the 2009 film *Keshtzar haye sepid* (The White Meadows) which impressively presents the Urmia disaster as an apocalypse can be understood

52 Arezoo Anvari, and Mohammad Valaie, "Evaluation of the Effects of Water Level Decline of Urmia Lake in Sustainable Rural Development: Case Study: Central Marhamat Abad Rural District, Miandoab County," *Bulletin of Environment, Pharmacology and Life Sciences* 4.7 (2015): 70.

53 Dalby, and Moussavi, "Environmental," 46.

54 Elham Torabian, "Exploring Social Vulnerability and Environmental Migration in Urmia Lake of Iran: Comparative Insights from the Aral Sea," in *The State of Environmental Migration 2014*, ed. François Gemenne, Pauline Brücker, and Dina Ionesco (Geneva: IOM, 2014).

55 Henareh Khalyani, Mayer, and Norman, "Water Flows," 764.

as an attack on the current ignorance of most politicians in relation to the unfolding catastrophe.

Although the Iranian government has now recognised the high socio-economic costs of the environmental crisis and promised to secure the water supply of the population living in the Urmia catchment area and to save Lake Urmia in 2012, the situation has not changed greatly. According to the affected population, bewilderment and frustration still dominate in this regard.

In the context of a study project at the University of Hannover in 2015, assessments and perceptions of the Urmia crisis were surveyed by interviewing residents in the area as well as various functionaries. According to the respondents, the drying out of the lake is perceived as depressing and catastrophic. Their statements reflect lost pride of this once-largest lake in Iran as well as frustration and profound grief over the socio-ecological disaster:

> "The lake is everything to us."
>
> "We are children of the lake. The loss is great and painful; the lake was our pride."
>
> "When I look at the lake, I feel very bad, my thoughts are confused, I am sad."

In 2014, Iran's President Rouhani declared the saving of Lake Urmia a national priority. Since then, an energetic campaign has been put into action in politics and science, for instance by holding conferences with the aim of developing solutions. Various ways out of the crisis have been discussed, most of which technological in nature. One of the most questionable proposals was the transfer of water from the Caspian Sea to Lake Urmia. To this end vast amounts of water would have to be pumped over an altitude difference more than 1,300 metres, which would mean enormous energy requirements over a long period of time. Proposals to discharge water from Lake Van in Turkey or from Armenia must also be filed in the category of unrealistic utopia. Up to the present day, technological measures such as discharging water from remote rivers into the Urmia Basin[56] or chemical desalination plants[57] have been discussed. However, these 'solutions' would lead to undetermined environmental impacts as well as unpredictable economic and social consequences, and this certainly would

56 Eimanifar, and Mohebbi, "Urmia Lake," 3; Zarghami, "Effective Watershed Management," 91.

57 Abdolreza Karbassi, Gholamreza Nabi Bidhendi, Amirhossein Pejman, Mehdi Esmaeili Bidhendi, "Environmental Impacts of Desalination on the Ecology of Lake Urmia," *Journal of Great Lakes Research* 36 (2010): 423. https://doi.org/10.1016/j.jglr.2010.06.004.

not be the way out of the current situation.[58] In fact, the Urmia disaster is less a purely ecological problem that can be solved technologically but rather a complex mixture of political and socioeconomic causes, all of which require political and institutional resolutions.[59]

Agriculture is the largest water consumer in the Urmia catchment area. At the same time, it offers the greatest potential for a turnaround in fortunes because water use efficiency is poor. Water loss in reservoirs and canals through evaporation and infiltration is very high, and a lot of irrigation water is wasted through the common practice of flood irrigation. There is a great saving potential here, for example through the use of water-saving technologies such as drop irrigation, grey water recycling, or the cultivation of less water-intensive crops.

Water shortages, in quantitative and qualitative terms, along with ecological degradation, high unemployment, and the massive demographic shifts caused by rapid urban growth on the one hand, and migration from rural areas and the resulting ageing of the rural population on the other, pose great potential for conflict.

Climate models suggest that temperatures will continue to rise and droughts will become more likely,[60] resulting in an expected increase in evaporation rates which will further reduce the lake level.[61] Local climatic conditions will also change with further water surface decrease since the balancing effect of the water body will be reduced and thus hotter and drier weather conditions can be expected in the summer. State water use plans until recently envisaged the construction of additional dams and an extension of irrigated agricultural land in the Urmia catchment area by 25 per cent. In the course of the recently adopted Urmia Lake Restoration Program, these projects were halted and a forty per cent reduction target in water consumption set up. However, according to model calculations, the lake level will continue to drop even with the successful implementation and realisation of this goal, due to worsening climate change.[62]

58 UNEP-GEAS, „The Drying," 136; AghaKouchak et al., "Aral Sea Syndrome,"310–311.

59 Habib Alipour and Hossein Ghasemi Tangal Olya, "Sustainable Planning Model toward Reviving Lake Urmia," *International Journal of Water Resources Development* 31.4 (2015): 535–536.

60 Farajzadeh, Fakheri Fard, and Lotfi, "Modeling of Monthly," 38.

61 Sima, and Tajrishy, "Using Satellite Data," 97–98.

62 Shadkam et al., "Preserving," 324.

7 Conclusion

The environmental and societal disaster witnessed at Lake Urmia is the result of various processes. These include anthropogenic climate change along with reduced precipitation and higher air temperatures (and thus increased evaporation), the construction of small and large dams to discharge water for agricultural purposes, the expansion of irrigated farmland, the excessive abstraction of groundwater, and increasing water requirements in urban agglomerations and for holiday home gardens. The result is an ecological disaster with political and socioeconomic consequences which must currently be considered just the tip of the iceberg. In political and scientific discussions on the Urmia disaster, technological solutions are still being discussed while institutional and socioeconomic aspects are mostly neglected. Instead, agro-economic practices, water governance, clientelism and patronage, as well as the increasing vulnerability of certain population groups need to be put to the test.

The Urmia tragedy represents a general water crisis in the Middle East where water is widely treated as an unlimited resource that can be exploited without much hesitation. Interference with the water cycle is implemented on a breathtaking scale, including the construction of numerous reservoirs, massive water redirection projects, and the expansion of irrigated areas. In contrast to traditional water management systems such as qanats, which are well adapted to the environmental conditions, the application of technology with high energy inputs has led to the exploitation of water resources far beyond their recovery rate. Water utilisation in the Middle East is far from being sustainable; groundwater levels are falling all over the region, and water has to be pumped from ever-deeper levels.

The major reason for the paucity of water is the government's policy of supporting and subsidising the agricultural sector to match growing food demand. The unstable political situation in the entire region has also meant that every country is trying to secure its food requirements through its own agricultural sector since they cannot depend on food imports. As a result, water is treated as an open access, free of charge resource, leading to very wasteful water use. Additionally, all governments in the Middle East follow a modernisation paradigm at any cost, and the erection of massive and impressive dams as a matter of national pride showing the technological and economic potency of the national economy. Dams continue to serve as symbols of development, progress, and modernity.

In view of the major ongoing desiccation processes in many parts of the Middle East, Lake Urmia is a vivid example of how ecological and societal systems are intertwined and how human interference influences hydrology and the ecosystem of a whole region, resulting in unpredictable consequences for the population. In this sense, Lake Urmia is a striking, disastrous testimony to the so-called Aral Sea syndrome. Understanding the complex set of causes and the development of a sustainable resource management programme requires a multidisciplinary approach and must include social science perspectives.

CHAPTER 7

Water in the Middle East: Hotspots of Scarcity and Conflict

Tobias von Lossow

Abstract

The "water question" is one of the key challenges for human and socio-economic development in the Middle East. The region is in the midst of a severe water crisis, scarcity continues to intensify since demand is increasing while water supply is decreasing. Water scarcity has the potential to threaten the region's stability since increasing competition over scarce resources contributes to other political, economic or social tensions and intensifies conflicts over the use and distribution of water. Given its importance, water also plays an important role as a target, tool and weapon in the region's fragile and conflict settings. Today's water challenges require technical and political responses as well as a broader behavioural change. Touching upon some of today's most pressing issues related to water and conflict, this overview outlines some of the most important water challenges in the Middle East.

1 Introduction

The importance of water for the Middle East is reflected in prominent, repeatedly cited, and often exalted images, metaphors, legends, and myths. Water has always been acknowledged as key for regional socio-economic development, valued as pre-condition of irrigated agriculture and food production, and equated with life and survival. For thousands of years, the advanced utilization of water and investment and innovation in water management guaranteed the continued existence of civilisations and cultures.[1]

Mesopotamia—"the land between the rivers"—for instance is referred to as the cradle of civilisation, based on organised irrigated agriculture and hydraulic engineering. As they did more than 6000 years ago, far-reaching systems of irrigation canals, water reservoirs and drainage systems build the infrastructural

1 World Bank, *Beyond Scarcity: Water Security in the Middle East and North Africa* (Washington, DC. World Bank, 2017), xv.

 | DOI:10.1163/9789004444973_009

backbone of today's agriculture in Iraq[2] (see also Chapters Three and Four of this volume).

Another example is the myth of the "Greening (or blooming) of the desert" based on extraordinary water engineering that brought about man-made rivers, canal systems and storage facilities. Several civilisations, societies, and modern states such as in Egypt or in Israel, equate achievements in this field with their broader human and socio-economic development. Agricultural production along the Nile still plays an important role in constituting Egyptian identity and also affects the conflict with upstream Ethiopia over the Grand Ethiopian Renaissance Dam (GERD). For Israel, the utilisation of land and water and agricultural production in difficult climatic and hydrological circumstances have historically been the existential basis guaranteeing the state's economic development in a politically hostile environment.[3]

Such images and myths are deeply rooted in the Middle East and substantially shape the identity of the societies, cultures, nations, and nation states in the region. They have influenced cultural, social, and political narratives to this day and, as a consequence, continue to play an important role in water-related politics and policies in the region—at the local, national and regional level alike.

Against the background of the outstanding importance of water in the Middle East and the exacerbating water crisis, this contribution sheds light on some of the most pressing water issues and trends in the region by navigating the field of water and conflict in a cursory manner. Following this introduction, some of the main features of water scarcity and water conflict will be outlined. In the third part, the role of water in armed conflict in the region will be analysed with a particular focus on the use of water as a weapon. The final part outlines solutions that help to address central challenges related to water scarcity and conflict in the Middle East. Comprehensive country-by-country analysis would exceed the scope of this contribution, but specific examples and case studies are referred to throughout this chapter to illustrate the broader, region-wide phenomena relevant for this volume.

---

2 Tobias von Lossow, *More than Infrastructures: Water Challenges in Iraq* (Den Haag: Clingendael, July 2018), accessed December 20, 2018. https://www.clingendael.org/sites/default/files/2018-07/PB_PSI_water_challenges_Iraq.pdf.

3 Mark Zeitoun, *Power and Water in the Middle East: The Hidden Politics of the Palestinian-Israeli Water Conflict* (London: I.B. Tauris, 2011), 65–66.

2 Water Scarcity, Competition, and Conflict

The Middle East and North Africa (MENA) is the most water scarce region in the world facing a broad variety of water-related challenges. While the region covers 4.9 per cent of the global landmass and hosts 4.4 per cent of world's population, it provides only 1.1 per cent of the globe's total renewable freshwater resources.[4] The water challenges are not only in the provision of drinking water, but decisively affect agriculture and food production as well as hydroelectricity generation. Moreover, the quality and quantity of water supplies impacts public health and economic development and plays an important role for social, cultural, and religious practices.

Despite the fact that the region is home to some of the world's wealthiest as well as some of its poorest nations with highly diverse socio-economic and political characteristics and parameters, such as Egypt, Iran, Israel, Saudi Arabia or Yemen, they all face similar water-related challenges. The lack of water resources and water scarcity pose a significant threat for the whole region and its social, political, and economic as well as environmental development. Increasing and intensifying competition over freshwater resources among states, sectors, groups, and individuals has the potential to threaten stability, security, and peace on the local, national, and regional level.[5]

In ancient times, the Middle East was relatively water rich and home to impressive agricultural production. The so-called "Fertile Crescent" (see map 0.2 in the Introduction) was the bread basket of the whole region, where predominantly irrigated agriculture even allowed exporting agricultural goods and commodities. Over the last centuries, though, water became more and more of a challenge as scarcity and shortages led to a dramatic decrease in agricultural production. The Middle East ran out of water in the 1970s already, and since then the rate of food imports has increased constantly. The region is no longer able to meet all of its water needs including water for food production, for industrial and municipal needs, and household needs.[6] Today the Middle East imports more than fifty per cent of its caloric demand, about sixty

4 Aysegül Kibaroglu, *Natural Cooperation. Facing Water Challenges in the Middle East* (Washington, D.C.: Middle East Institute, 2016), 1.

5 See for example Neda A. Zawahri, *Water Security in the Middle East and North Africa* (Durham: Duke University, 2017), accessed December 20, 2018. https://humanitiesfutures.org/papers/water-security-middle-east-north-africa/.

6 Tony Allan (J.A.), *The Middle East Water Question. Hydroploitics and the Global Economy* (I.B. Tauris, London, New York, 2002), 5–6.

per cent of its wheat—there is no other region as dependent on food imports in the world.[7]

The Middle East suffers heavily from water scarcity, insufficient or inadequate access to water and sanitation which also threatens human security and sustainable development. More than sixty per cent of the population in MENA live under conditions of high or very high surface water stress, particularly in remote and rural areas, compared to the global average of 35 per cent.[8] While the global average of annual per capita water availability is about 7,000 m^3, in MENA it is only 1,200 m^3 which is below the minimum requirement of 1,700 m^3 for an active and healthy life.[9] And it is only slightly above the threshold of water scarcity of 1,000 m^3, below which individuals are affected by severe consequences for their health and wellbeing.[10] With a few exceptions, most of the countries in the region already suffer from water scarcity (fig. 7.1) while some even face serious shortages, as for instance Iran and Iraq. Of course this indicator varies markedly throughout the region. A division into sub-regions such as the Gulf or the Maghreb indicates (groups of) countries that face similar hydrological and water conditions. Turkey, for instance, has more than 2,500 m^3 water per person and year available while several states on the Arabian Peninsula have less than 250 m^3 (see fig. 7.1).

The importance of water, the challenging hydrological and climatic conditions, and the resulting conflict potential are nothing new to the region. But several parameters and drivers of water scarcity have increased or intensified over the recent years, posing unprecedented challenges in scale and scope today. Rapid growth in the region with some dramatic social, economic, political, environmental, and climatic developments outpaced necessary, often postponed adjustments and reforms in the water sector. The aim of all of today's water-related efforts is to increase water security in the region, which can be defined as the "availability of an acceptable quantity and quality of water for health, livelihoods, ecosystems and production, coupled with an acceptable level of water related risks to people, environments, and economies".[11] Various

7 World Bank Group, *Agriculture and Rural Development in MENA*, World Bank Group (2008): 1.
8 World Bank, *Beyond Scarcity*, xxv.
9 UNESCO, *Water for People, Water for Life*, The United Nations World Water Development Report (UNESCO, 2003), 10.
10 See for instance Stefan Deconinck, *Security as a Threat to Development: The Geopolitics of Water Scarcity in the Nile River Basin*, Royal High Institute of Defence (2009/updated: 2017): 3.
11 David Grey and Claudia W. Sadoff, "Sink or Swim? Water Security for Growth and Development," *Water Policy* 9 (2007): 545.

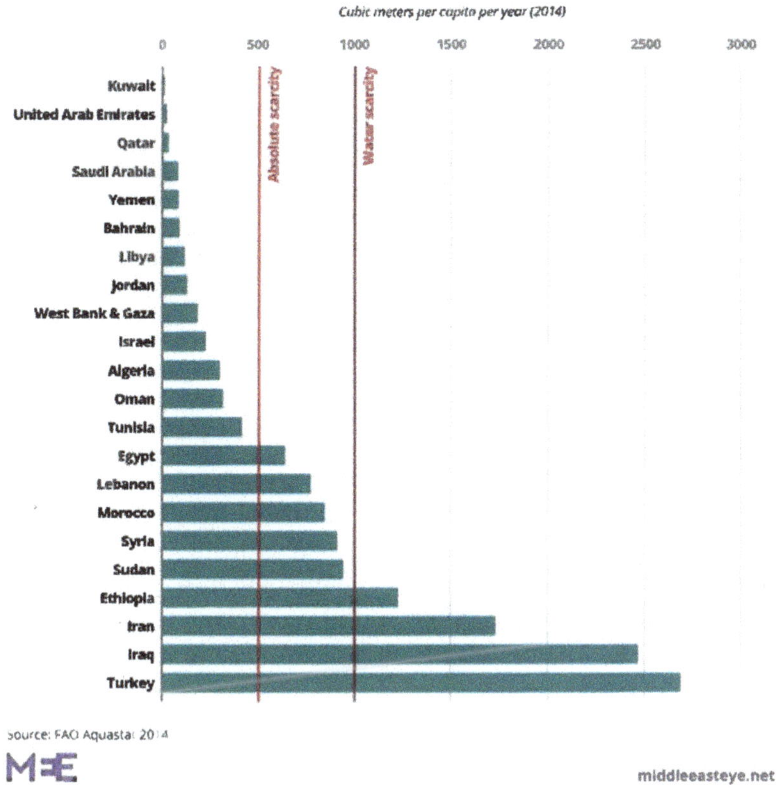

FIGURE 7.1 Water scarcity in the MENA region. From *Middle East Eye* (2018).

developments and indicators suggest that this will be more and more challenging in the future and require tremendous technical, political, and social efforts.

2.1 *The Gap: Increasing Demand and Decreasing Supply*

The pressure on the region's water resources is reflected and driven by an increasing water consumption while water supplies have been constantly decreasing over the past decades. On the demand side, population growth is one of the key drivers. With an annual rate of 1.7–2.0 per cent, the Middle East expects its population to grow from 300 million today to 500 million by 2025.[12] While some countries in the region such as Oman (4.7 per cent, 2017) have fast growing populations, others such as Iran (1.1 per cent, 2017) have rather more

12 Kibaroglu, *Natural Cooperation*, 1.

moderate rates, and some such as Syria even have negative population growth (-0.9 per cent, 2017).[13]

Another driver is socio-economic development in the region over the last fifty years. Increasing economic and industrial activities lead to higher water consumption. While about a quarter of the region's population, predominantly those living in rural areas, are poor, the Middle East is also home to the two wealthiest nations in the world, Qatar and Saudi Arabia. With a few exceptions such as Yemen, most of the countries in the region are middle-income countries, some of which benefit from their oil and gas reserves. Related to socio-economic development, rising individual incomes also result in more water-intensive lifestyles. Changing water consumption, for instance through dietary preferences for water-intensive products such as beef, are among the most important drivers.[14] Qatar, for instance, has the highest per capita water consumption with five hundred litres per day—a figure which is even more striking as the country basically has no freshwater resources, but almost completely depends on desalination, wastewater treatment, and fresh water imports. There are no river basins, almost no rainfall, and the groundwater resources are nearly depleted.[15] Even if climate change is far more important on the supply side, it does play a role here, too, since increasing temperatures and more frequent and more extreme heatwaves increase water demand for cooling systems, drinking water, and hydroelectricity generation. During the heatwave of 2018, Iran increased its water consumption at the tributaries of the Tigris and reduced the water flow to Iraq further downstream.

Parallel with the increasing demand, water supplies in the region have shrunk dramatically over the past decades mainly due to over-consumption, pollution, and climate change. Overall water availability in the region is likely to halve by 2050.[16] The Middle East is highly dependent on its few renewable water resources, in particular the large river basins, the Euphrates and Tigris, the Jordan and the Nile—literally blue lifelines in a largely arid region. Just a few areas have abundant rainfall or even snow, such as parts of Iraq, Iran or Lebanon. The other important source of supply is groundwater, largely fossil

---

13 Word Bank, "World Bank Indicators," accessed December 20, 2018. https://data.worldbank.org/indicator/SP.POP.GROW?end=2017&locations=ZQ&start=2016.

14 See for example Tobias von Lossow, "Klimawandel und Wassermangel in Afrika: Möglichkeiten der Kooperation am Beispiel Nil," in *Klimawandel und Sicherheit. Herausforderungen, Reaktionen und Handlungsmöglichkeiten*, ed. Steffen Angenendt, Susanne Dröge, and Jörn Richert (Baden-Baden: Nomos, 2011).

15 Husam M. Baalousha and Omar K.M. Ouda, "Domestic Water Demand Challenges in Qatar," *Arabian Journal of Geosciences* 10 (2017): 1–4.

16 Kibaroglu, Natural Cooperation, 1.

aquifers. Against the background of increasing water demand, these scarce resources have been overused. Over the past decades, several fossil aquifers have been rapidly depleted while the level of consumption of renewable groundwater and surface waters has exceeded the pace of their replenishment. In addition, water resources have been increasingly polluted by untreated household sewage, industrial waste water, herbicides, and fertilisers.[17]

One of the key drivers of decreasing supplies is climate change hitting the region in multiple ways and additionally aggravating the water situation. The Middle East is one of the most arid regions in the world, with high natural variability in precipitation. Climate change leads to less and less predictable rainfall, irregular snow melt, and more frequent and prolonged extreme weather events, particularly droughts and floods, but also sand and dust storms, that all hamper water supplies. Increasing temperatures and more frequent and prolonged heatwaves result in increasing evaporation. Sea-level rise favours saltwater-intrusion into groundwater aquifers such as in Gaza, as well as surface water systems affecting rivers dozens of kilometres upstream, as for example in Iraq along the Shatt al-Arab as high as Basra about 65 kilometres inland.[18] Climate change also leads to more frequent and more severe extreme weather events such as floods, droughts, heatwaves, sandstorms, and dust storms that worsen the water situation and threaten supply systems.

2.2 *Governance and Fragility: Structural, Systemic, and Contextual Drivers*

Alongside accelerating socio-economic developments that dramatically increase water demand and diminish supply, the Middle East features a whole set of interlinked structural, contextual, and systemic drivers that additionally exacerbate the situation, for instance by favouring inappropriate water use practices and high levels of consumption. At the centre of this problem is poor or insufficient water governance that largely fails to either develop or implement appropriate water (-related) policies and necessary reforms. Weak, ineffective, and inefficient institutions have simply not been able to adjust to the rapidly changing conditions in the water sector. To adequately respond to the immense challenges of today would require flexible and participatory mechanisms and accountability structures.[19] Instead, in many cases institutions fail

---

17 Håkan Tropp and Anders Jägerskog, *Water Scarcity Challenges in the Middle East and North Africa* (MENA) (UNDP, 2006), 3.

18 von Lossow, "More than Infrastructures", 6.

19 Christopher Ward, Sabine Beddies, and Sandra Ruckstuhl, *Managing Water Scarcity to Reduce Conflict Risks in the Middle East and North Africa* (Washington, World Bank, 2011), 9.

to develop and implement sustainable water management and lack political power and autonomy as well as appropriate funding. Moreover, the institutions are not able or willing to implement and enforce water law.[20]

Another central problem is that water is undervalued. Governments in the region promoted water development over decades through subsidies and low water prices.[21] This, ironically, encourages high levels of water consumption under conditions of heavy shortage and widespread scarcity. Widespread and generous water subsidies in the Middle East are among the highest in the world, also for the agricultural sector, the largest water user. Such systems favour users with high water consumption since they benefit comparatively more.[22] With water prices kept artificially cheap and water service fees low, the subsidies de facto promote further resource degradation instead of encouraging reasonable and efficient use of the scarce water available. At the same time, incentives for water saving and water conservation are largely absent. As a consequence, large amounts of water are wasted or even lost. Leakage of thirty to forty per cent in the supply network and irrigation systems are the rule rather than the exception.

On top of the hydrological consequences, these policies also aggravate the financial deficit in the water sector. This compounds financial constraints in some countries and in turn prevents from investments in water infrastructure and services, for example in Yemen, Lebanon, or Iraq. Postponed investments and the slow adoption of technological innovation have also contributed to the scarcity situation in the region.[23]

Moreover, the subsidies compound vulnerabilities and can reaffirm existing political imbalances among user groups when it comes to the allocation of scarce resources. While water for agriculture is largely subsidised, industries and households have to pay regular prices. Instead of balancing out water asymmetries, the power structures in place often contribute to reaffirming and fostering inequitable access.[24] Since water is framed as a national priority, it is centralised, highly politicised, and often dealt with by the heads of state rather than the respective ministries. Therefore, water policies and investment decisions often focus on prestigious large-scale infrastructure projects at the cost of smaller, decentralised and more sustainable solutions. In particular large-scale

20 Ibid.

21 Ward, Beddies, and Ruckstuhl, *Managing Water Scarcity*, 8.

22 World Bank, *Beyond Scarcity*.

23 Ibid.

24 Ward, Beddies, and Ruckstuhl, *Managing Water Scarcity*, 9.

hydro-electricity projects, desalination plants, or irrigation schemes follow this approach, as for instance the GERD in Ethiopia.

Another important contextual factor that aggravates water-related problems and challenges are political tensions, violence, armed conflicts, and war. Many of the most fragile countries in the region also suffer from high levels of water stress.[25] The fragile situation over decades in countries such as Iraq, Syria or Yemen, prevented them from investing in, (re-) building, and appropriately maintaining water infrastructure and supply systems and often left them physically damaged and neglected. Protracted conflicts and crises as well as less predictable shocks additionally increase the vulnerability of water management, resulting a downward spiral of insufficient water management and deteriorating conditions.[26] In Iraq, the recurring cycles of violent conflict or war and post-conflict settings left physically damaged water infrastructures. They have largely been left untouched and continuously deteriorated further over 25 years—without major modernisation measures.[27]

2.3 *From Scarcity to Conflict*

The increasing resource scarcity creates or intensifies tensions over the distribution and utilisation of water. Competing interests between water users at the local, national, and regional level result in various kinds of water-related disputes and conflicts. Competition between the sectors is visible across the different levels, with agriculture as the largest water consumer with approximately seventy per cent on global average, while industrial needs account for twenty per cent and households for ten per cent of all water consumption. In some countries, the figure for the agricultural sector can be above ninety per cent.[28] But conflicts also arise within the sectors, for example between agro-enterprises that produce crops for export and small-scale farming that produces goods for the local market.

At the domestic level, local tensions over water occur between or within communities, between various socio-economic groups along centre and periphery, between urban and rural consumers, or in the context of the privatisation

25 World Bank, *Beyond Scarcity*.

26 International Committee of the Red Cross, Geneva (ICRC), *Urban Services During Protracted Armed Conflict: A Call for a Better Approach to Assisting Affected People* (Geneva ICRC, 2015).

27 von Lossow, "More than Infrastructures", 5.

28 Eran Feitelson and Amit Tubi, "A Main Driver or an Intermediate Variable? Climate Change, Water and Security in the Middle East," *Global Environmental Change* 44 (2017): 39–48.

of water supply services. Sometimes violent protests against the state or the government erupt over insufficient water supply and sanitation services. Such conflicts over water allocation can also contribute to broader anti-government protests: In the summer of 2018, for instance, the violent protests in Iraq that erupted in Basra and spread all over the country featured a water component. Despite these protests being driven by broader discontent with the government, the protestors repeatedly stressed the lack of basic water services and electricity supplies as a key driver that motivated their protests.[29] All kinds of interventions in the existing water regime resulting in reduced supplies can cause protests and conflict. Attempts to re-allocate water to higher value use in urban areas—also in the light of rapid urbanisation trends—often lead to conflict as they did in Jordan or Iran, where farmers protested against these policies. Another conflict issue is the degradation of environmental conditions due to industrial pollution such as heavy metal contaminants and household sewage. In Lebanon, for example, only eight per cent of the wastewater is treated, most of the sewage is discharged into the ocean or back into the water systems.[30]

A different source of tension are trans-boundary water conflicts. About sixty per cent of all surface waters in MENA are shared across boundaries, among them the region's major trans-boundary river basins. In addition, huge and largely fossil aquifers lie beneath several countries. Conflicts over such trans-boundary waters have specific characteristics. In trans-boundary water conflicts, water is rarely the one and only reason for an increasing escalation, but rather one element of a larger set of conflict drivers. Like other environmental and resource factors, water and water scarcity have played a role, influenced or contributed to violent conflicts in the region throughout history.[31] But while domestic water conflicts have the potential to escalate violently, there is no empirical evidence for a violent inter-state water conflict or war over trans-boundary waters. Instead, states tend to cooperate over water even if the competition over the resource is increasing.[32] A prominent example is the Red Sea-Dead Sea project involving Jordan, Palestine, and Israel along

29 See for example Zeit Online, "*Tote bei Protesten im Irak*," 23 July, 2018. https://www.zeit.de/politik/ausland/2018-07/irak-sueden-demonstrationen-korruption-arbeitslosigkeit-staatliche-dienstleistungen.

30 Ward et al., *Managing Water Scarcity*, 8.

31 Peter Gleick, "Water, Drought, Climate Change, and Conflict in Syria," *Weather, Climate and Society* 6.3 (2014): 332.

32 Aaron T. Wolf, "Conflict and Cooperation along International Waterways," *Water Policy* 1 (1998): 251–265.

the Jordan River. Nonetheless, conflicts over water along the Euphrates and Tigris, the Nile, the Jordan and the Orontes, or between Israel and Palestine over groundwater resources have been intensifying over the last decades, in particular along the Nile and along Euphrates and Tigris.[33]

Since the distribution and allocation of water resources is based on political decisions, water conflicts are themselves genuinely political conflicts that are intrinsically linked with the legitimacy of the state. This is less a matter of scarcity or abundancy, but at a certain point and level political institutions get involved in the question on how available water is used. This is the more relevant and politically sensitive in water scarce contexts than under circumstances of abundance of water. The failure to provide basic services such as water and electricity supplies can erode the legitimacy of the state, accelerate socio-economic tensions and political discontent, and thus compound social and political fragility. The 2018 anti-government protests that started in Basra and later spread to other Iraqi cities were based on broader political discontent, but had been triggered by the lack of sufficient water and electricity supplies when a heatwave hit the region.[34]

3 Water during Conflicts

Besides the scarcity-related challenges and conflicts over resources, water plays a central role in the region's political conflicts and even more during violent conflicts and (civil) wars both as a target and a weapon.[35] The outstanding role of water in the societies and politics in the Middle East, the hydrological and climatic conditions as well as centralised often autocratic decision-making processes make water a powerful political instrument and military tool. This became particularly obvious in various conflicts over the past decades.

3.1 *Water as Tool, Target and Weapon*

At the domestic level, water has widely been used as a political instrument throughout history. On the one hand, access to water has been used to deprivilege and marginalise certain geographic areas, ethnic groups, religious and political minorities, or socio-economic groups. In the Middle East, this practice

---

33 See for instance Tobias von Lossow, "The Multiple Crisis. Perspectives on Water Scarcity in the Euphrates and Tigris Basin," *ORIENT* 58 (2017).

34 Zeit online, Tote bei Protesten im Irak.

35 Tobias von Lossow, "The Rebirth of Water as a Weapon: IS in Syria and Iraq," *The International Spectator* 51.3 (2016): 82–99.

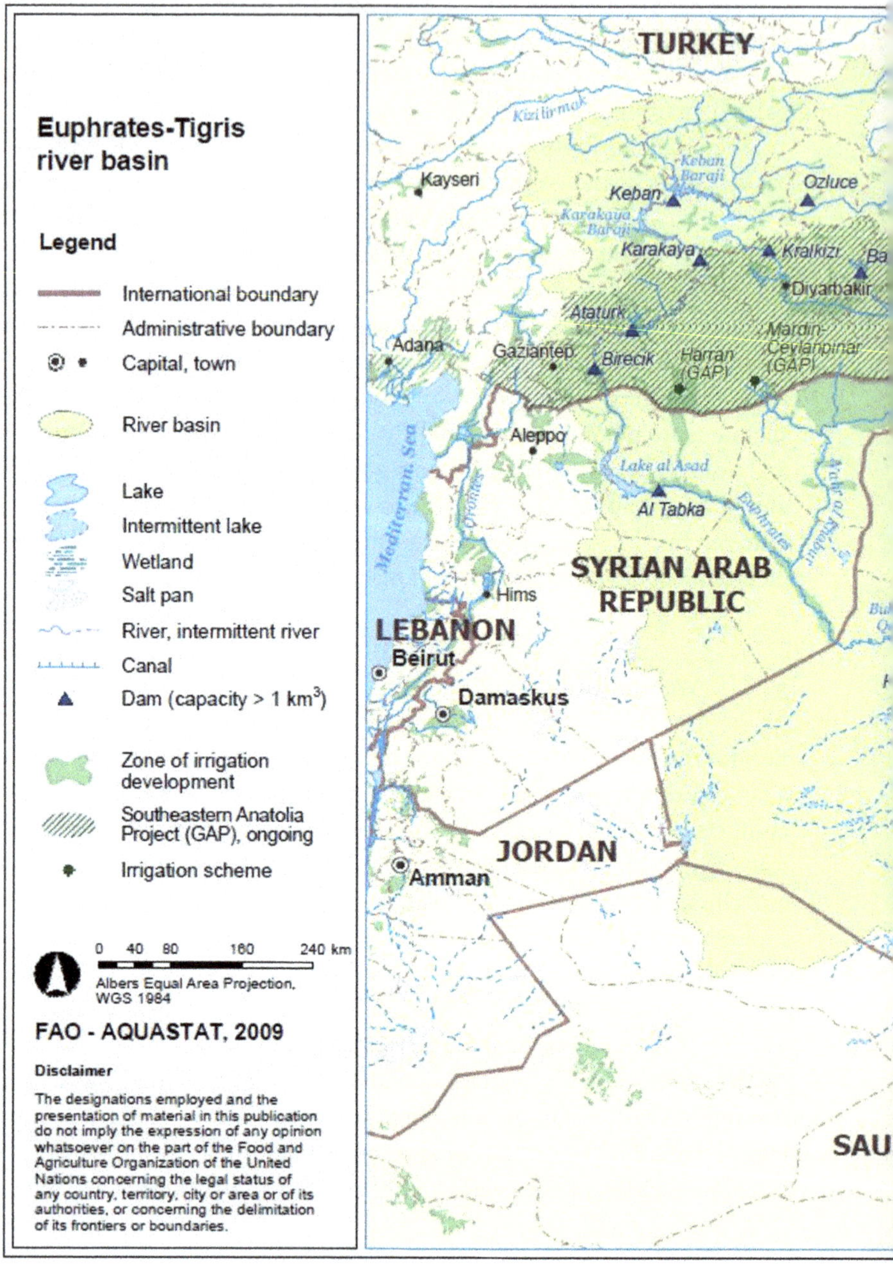

MAP 7.1 Euphrates-Tigris river basin.

ARMENIA
AZERBAIJAN
Yerevan
Palandoken
Boundary of former Nagorno-Karabakh Autonomous Oblast
Lake Van
Tabriz
Orumiyeh
Lake Orumieh
Greater Zab
Aras
TURKEY
ARM.
AZER.
TURKM
SYRIA
LEBANON
ISRAEL
IRAQ
IRAN
JORDAN
KUW.
EGYPT
SAUDI ARABIA
UAE
SUDAN
OMAN
ERITREA
YEMEN
IRAN
(ISLAMIC REPUBLIC OF)
Tigris
East Al-Jazeera
Mosul
Mosul
Arbil
Karaj
Tehran
Lesser Zab
Dokan
Dibbis
Kirkuk
As Sulaymahiyah
Dasht-e Kavir (salt desert)
Derbendi Khan
Diyala
Qom
Tharthar Lake
Samarra-Tharthar
Kermanshah
Hamrin
Baghdad
Ramadi-Habbaniya
Karkheh
Esfahan
Razzaza Lake
Karkheh
Dez
Karbala
Dez
Karun 1
Karun 3
Karkheh-Shavoor
Gotvand
Karun
An Najaf
Ahvaz
Euphrates
IRAQ
Al Basrah
Hawr al Hammar
Shatt Al-Arab
Persian Gulf
Kuwait
KUWAIT
DI ARABIA

has repeatedly been observed. One of the outstanding examples was in Iraq in the 1990s, when Saddam Hussein had the Marshes in Southern Iraq drained as a revenge act against the local population for rising up against his regime.[36] Diverting the water from the Marshlands had dramatic socio-economic and environmental consequences. Thanks to tremendous restoration efforts, about fifty per cent of the wetland area could be recovered in the late 2000s. Yet about ten years later, the area once more suffers from the diversion of water (see below), droughts, heatwaves, and decreasing water discharge into the Euphrates and Tigris. On the other hand, water supply can also work as reward for certain parts of the population favouring certain groups and areas over others in order to gain political support and thus strengthen the power base of the state.

During various armed conflicts over the past decades both within and between state entities, water resources and infrastructures have repeatedly become targets of military violence. Water installations such as dams, sewage or treatment plants, pumping stations or piping systems have regularly been damaged or destroyed intentionally as well as unintentionally.

Moreover, water has also been used as a weapon, and "directly (ab)used as in instrument of war, as a tool in order to achieve strategic political and tactical military goals".[37] Control of water resources and infrastructure allows the manipulation of water flows and supplies and thus the targeting of specific populations, the opponent's political leadership, and military forces. The recent conflicts and wars in Syria and Iraq, but also in Yemen showed that this practice has gained new prominence in the region, even though it is banned by international humanitarian law and the law of war.[38] In the Syrian war, for example, nearly every faction has been weaponising water—various rebel groups, the so-called Islamic State (IS), government troops, and pro-regime militias as well as Russia through airstrikes. Damascus has also been cut-off from the springs at Ain al-Fijah, the capital's most important water supplies, several times.[39]

3.2 *Water as a Weapon: The Case of IS*

Among the most recent and prominent examples for the weaponisation of water has been the IS. Even if IS was not the only actor, it took this practice to a

36 von Lossow, "More than Infrastructures," 7.
37 von Lossow, "Rebirth of Water as a Weapon," 84.
38 Tobias von Lossow, "Weaponization of Water in the Middle East—'Lessons Learned' from IS", in *Water and Conflict in the Middle East,* ed. King, Marcus D. (Hurst Publishers, 2019).
39 Alisa Reznick, "Weaponizing Syria's Water."

new level and instrumentalised these war crimes for its propaganda purposes.[40] The militia has strategically used water as a weapon—in a more targeted, systematic, consistent and at the same time flexible manner addressing the local, the regional and the national level. IS repeatedly weaponised water in all its various facets in the portfolio of its day-to-day operations. Particularly during its expansion phase 2012–2015, jihadist groups that in 2014 became or joined the IS captured important water infrastructure. IS used water as a weapon in three different ways—by making sure that there is either too little or too much water available, or that water is rendered unusable.[41]

During the period of the militia's territorial expansion, water was primarily weaponised by denying supplies or flooding certain swathes of land. Major dams in the Euphrates and Tigris basin played a significant role and allowed them to intervene in the rivers' water flow and water and electricity supply systems. With the exception of the Haditha dam which was defended by the Iraqi Army with close air support by the US, all major dams and barrages upstream of Baghdad were under IS control in 2015 at least for a certain period of time (see map 7.2). Water resources in Iraq—as well as land—were increasingly contaminated when IS came under more and more military pressure from 2016 onwards until its official military defeat in Iraq in December 2017.[42]

The case of Fallujah dam in April 2014 illustrates in a nutshell the broad spectrum, the variety, multiplicity, simultaneity, and gravity of how IS weaponised water when the militia gained control of the dam and immediately proceeded to weaponise water. Even if this example was exceptional for the IS in its frequency and impact, it illustrates and the logic and mechanisms of the water weapon.[43] After capturing the dam, the militia first closed the gates in order to withhold the water of the Euphrates and to threaten and deliberately deprive the largely Shiʿi population in the lower reaches of the river basin of water supplies. In doing so, the militia threatened Iraq's most important agricultural centers and thus the food security of the country. By retaining the water, IS at the same time submerged government and military facilities along the banks since the reservoir's water table behind the barrage quickly rose.[44]

40 von Lossow, "Weaponization of Water in the Middle East."

41 von Lossow, "The Rebirth of Water as a Weapon," 87.

42 von Lossow, "Lessons Learned."

43 Ibid.

44 See al-Akhbar, "ISIS Captures Iraqi Dam, Floods Areas," April 11, 2014, https://english.al-akhbar.com/node/19385; H. Ali Mohanad, "ISIS' Path of Destruction Drains Iraq and Syria's Water Supplies," *Al Arabiya*, June 21, 2014, http://english.alarabiya.net/en/perspective/analysis/2014/06/21/ISIS-path-of-destruction-drains-Iraq-and-Syria-s-water-supplies.html.

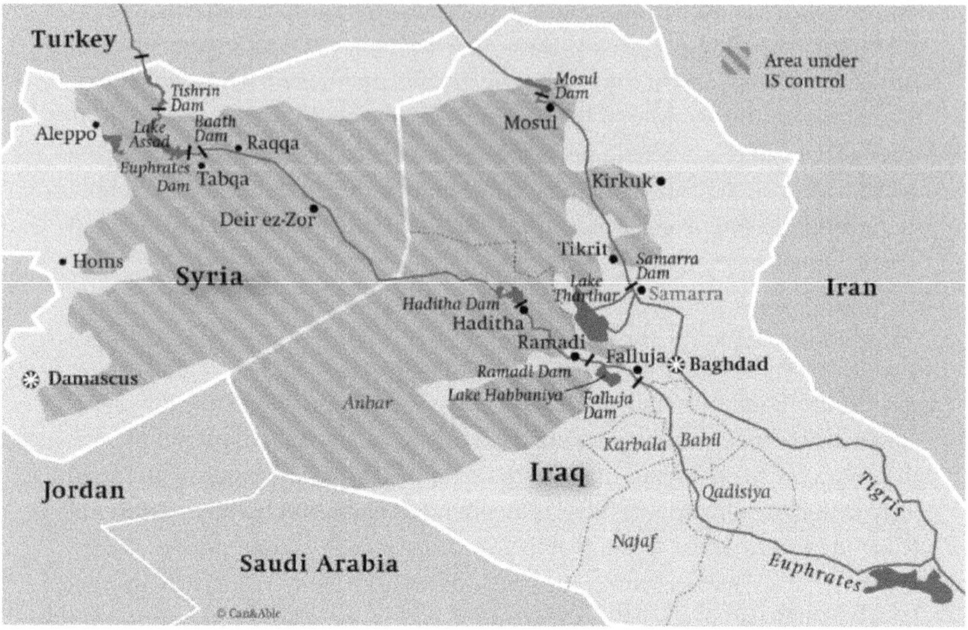

MAP 7.2 Areas controlled by the IS in December 2015: Water as a Weapon: IS on Euphrates and Tigris, SWP Comments.
© CAN AND ABLE 2016

TABLE 7.1 Overview of major dams controlled by IS, adjusted from von Lossow, 2019 (forthcoming)

Dam	Under IS control
Syria	
Tishrin Dam	11/2012–12/2015
Euphrates Dam / Tabqa Dam	02/2013–05/2017
Baath Dam	02/2013–06/2017
Iraq	
Mosul Dam	07–18 August 2014
Haditha Dam	—
Samarra Dam	04/2014–10/2015
Ramadi Dam	05/2015–01/2016
Fallujah Dam / Nuaimiya Dam	02/2014–06/2016

Two days later, the militia decided to divert these retained waters through an irrigation canal into a side valley, flooding lands up to one hundred kilometres away. The city of Abu Ghraib was up to four metres under water, 10,000 houses and two hundred square kilometres of fertile farmland were destroyed, the harvest for 2014 was lost, livestock killed, and about 60,000 people were forced to flee.[45] By submerging these territories, the IS also prevented the rapid advance of follow-on units of the Iraqi army from approaching IS positions near Fallujah. In addition, the militia also hampered the parliamentary elections held on 30 April, 2014. As a consequence of the flooding, only a third of the polling stations in Anbar province—the largest by area—were able to open on election day.[46]

Fallujah, and in particular the flooding, was among the most dramatic examples of how the IS used water as a weapon. The flooding had particularly far-reaching consequences. Yet the event also illustrated how political and military purposes can overlap. It showed further how intended as well as unintended consequences can serve the aims of a conflict party.

4 Approaches and Solutions

The current water crisis in the Middle East is unprecedented in scale and scope and will require a new, innovative and comprehensive response to address the great challenges associated with it. Efforts to tackle these challenges can only be successful if they are embedded in broader national and regional approaches that include technological innovation as well as considerable political and socio-economic adjustments.

4.1 *Technology: Desalination*

Today's complex water challenges in the region require technological innovation, for instance, to increase water productivity or to produce nonconventional water by wastewater recycling or desalination. The technological

45 Debora MacKenzie, "Extremists in Iraq Now Control the Country's Rivers," *New Scientist*, June 12, 2014, www.newscientist.com/article/dn25722-extremists-in-iraq-now-control-the-countrys-rivers/; Cameron Harrington and Schuyler Null, "What Can Iraq's Fight over the Mosul Dam Tell Us about Water Security?," *NewSecurityBeat Wilson Center*, August 20, 2014, www.newsecuritybeat.org/2014/08/iraqs-fight-mosul-dam-water-security/.

46 Birgit Svensson, "Flood Disaster in Iraq: Water as an Instrument of War," *Qantara.de*, June 13, 2014, https://en.qantara.de/content/flood-disaster-in-iraq-water-as-an-instrument-of-war.

options and their cost-effectiveness are rapidly improving, which will allow for further new opportunities in the future.[47] Further investment in research and development is necessary to better address the constantly growing gap between water demand and supply. It needs smart, tailored and comprehensive innovation to enhance sustainable water use and to achieve water security.

While waste water treatment, water reuse, or irrigation technologies are common sense, desalination of sea-water on a larger scale is more controversially discussed but at the same time an extremely prominent approach in the region. Israel is among the pioneering countries in desalination technology worldwide and has turned its water situation upside down: in 2014 the country produced more water than it consumed for the first time ever.[48] Five major and two minor desalination plants along the Mediterranean coast are one of the core elements of the country's larger water strategy—alongside wastewater treatment, innovation in agricultural irrigation, and efforts to reduce water losses. Also, other countries in the region, particularly the oil-rich Gulf States, have increasingly been focusing on desalination in recent years. Saudi Arabia and Qatar, for instance, have invested in their own research centres to develop tailored desalination approaches and technologies for the countries' specific desalination needs. Qatar already relies more or less completely on desalination today since the last of the country's tiny freshwater resources are almost depleted.[49]

The massive energy need is still challenging, even if the advance of renewable energies and new desalination technologies—most prominently reverse osmosis technology—have made desalination more efficient. The potentially bigger problem is environmental: it is the vast amounts of salt which are left as the other end product alongside desalinated water. Today, the brine is usually dumped back into the sea right after the desalination process, which threatens and demonstrably destroys ecosystems and marine wildlife along with fish stocks near the coast. The dumping of brine also increases the salinity level in the sea water, which in turn challenges the desalination process even more since more salt needs to be filtered out. Further risks and problems are

47 World Bank, *Beyond Scarcity*.

48 Tobias von Lossow, "Blaues Wunder. Wasser war für Israel stets von existenzieller Bedeutung—gesellschaftlich, politisch, wirtschaftlich," *Internationale Politik* (*IP*) 2 (2016): 53.

49 A central argument for desalination, see Mannan Mehzabeen et al., "Examining the Life-cycle Environmental Impacts of Desalination: A Case Study in the State of Qatar," in *Desalination* 452 (2019): 238–246.

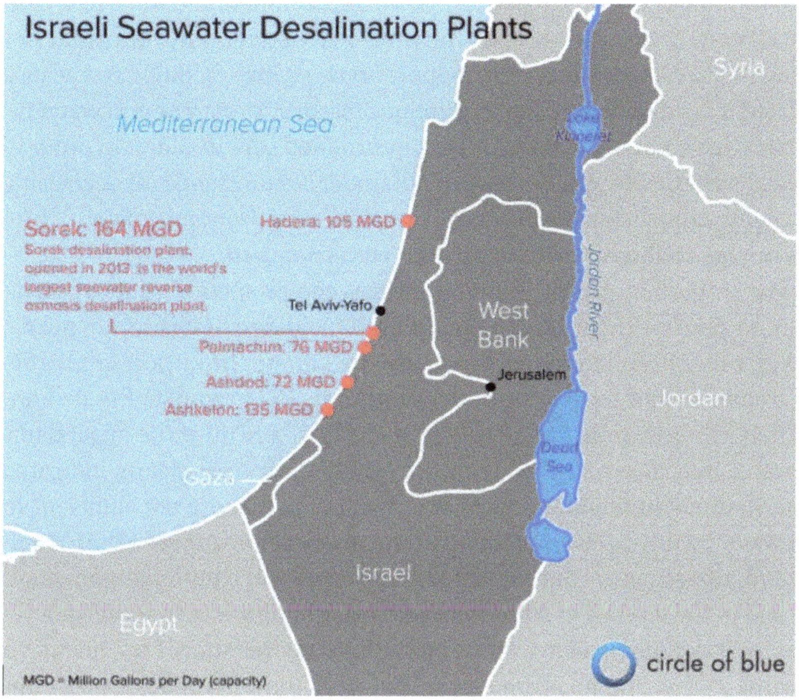

MAP 7.3 Israeli seawater desalination plants
Note: https://www.circleofblue.org/2016/middle-east/israels-mediterranean-desalination-plants-shift-regional-water-balance/attachment/israeli-desalination-plants-02/.

pharmaceutical remnants and the lack of minerals and trace elements in the drinking water as all of the latter are removed.[50]

Given the hydrological situation in some areas in the Middle East, desalination is a central pillar for the several countries' water security. Given the massive challenges that come with it and remain to be resolved, desalination should be used within reason, to a limited amount and for certain purposes only. For drinking water, desalination might provide a realistic, ecologically acceptable solution in the future, but it cannot provide the large amounts needed for agriculture and food production without environmental damage.

4.2 *Investment in and Protection of Infrastructure*

Another entry point is water infrastructure. In many countries in the region, the water infrastructure is outdated, often has been neglected over decades

50 Von Lossow, "Blaues Wunder."

and heavily damaged or destroyed during violent conflicts and wars. As a consequence, many countries lose up to forty per cent of their water due to lack of pipelines, high evaporation rates in inappropriate storage facilities, or because of outdated irrigation technology. For various reasons, investments in water infrastructure have not been made or been postponed over decades. In order to implement sustainable water management under circumstances of decreasing and changing rainfall, increasing temperatures, etc. appropriate infrastructure such as storage facilities, treatment plants, etc. is required.

Another important aspect in this context is the protection of water infrastructure to prevent them from being utilised by armed actors to weaponise water. Key water installations such as dams are among the critical infrastructure of a country and need to be better protected. The example of IS demonstrated that a militia could literally overrun and take control of the major dams in Syria and Iraq. The recent wars in the Middle East have clearly indicated that international humanitarian law and the laws of war are not sufficient to protect water facilities. The protection of the physical integrity of water installations and infrastructure during armed could represent a minimal agreement among and in the interest of all conflict parties. In addition, it could act as a vehicle for trust-building and to foster water-related cooperation. This includes a technical as well as a political dimension, but it circumvents delicate questions of water control, quantity, and quality.[51]

4.3 *Politics and Reform*

Besides technological innovation and investment in infrastructure, political intervention and reform in water-related sectors are crucial. Addressing the various shortcomings in national water governance is key, particularly the establishment of efficient, transparent and participatory institutions. Of course, this implies several sensitive, politically and financially costly reforms in the water sector—but also reforms in other fields, first and foremost in agriculture. In nearly all of the countries in the region, water for irrigation has been subsidised—either directly through water prices as in Egypt or indirectly on diesel for groundwater pumps as it was in Syria before the war.[52] Despite the harmful impact, it is politically highly sensitive to remove the subsidies since the agricultural sector employs large parts of the population and has substantial influence.

---

51 von Lossow, "More than Infrastructures," 10.

52 Francesca de Châtel, "The Role of Drought and Climate Change in the Syrian Uprising: Untangling the Triggers of the Revolution," *Middle Eastern Studies*, 50 (2014): 526.

The driver behind the subsidies is the high degree of dependence on food imports. All countries in the Middle East are net food importers, Egypt for instance is the largest wheat importer in the world. The lack of food security makes a reasonable argument to expand the capacity of (often subsidised) agricultural production—despite low crop yields, high water losses due to dilapidated water infrastructure and evaporation, and decreasing water resources. Food security and water security are in the vital interest of a sovereign state, so it is imperative to decrease the dependence on food imports. In combination with the image and importance of water in society and the political weight of the agricultural sector, necessary water-related reforms are extremely sensitive and were, as a consequence, often not properly addressed in the past.

One exception is Saudi Arabia which implemented remarkable reforms in the agricultural sector in the 2000s. Riyadh largely stopped the large-scale wheat production they began in the 1990s after a few years as groundwater resources had dramatically diminished. The country turned away from water-intensive products and shifted towards more drought-resilient and heat-resistant crops such as dates. These crops were often of higher value and grown for export. As a consequence, Saudi Arabia has increasingly invested in arable agricultural land elsewhere, for example in Sudan or Ethiopia, to guarantee its food security through imports of wheat and other water-intensive crops produced outside the country. Adjusting its own agricultural sector in this way and also counting on the import of virtual water to decrease the pressure on domestic resources is a smart, highly rational, hydrologically wise and economically feasible decision. But it also externalises the pressure on water resources which might increase and intensify water- and food-related problems and conflicts in other countries and regions. In Sudan, for example, such land grabs are not only accelerating domestic competition over water, but has the potential to intensify the conflict over Nile water between Egypt and the upstream riparian states on the regional level.

5 Time for a Revolution

Water is absolutely fundamental for the development and stability of the region. The scale and scope of today's water challenges in the Middle East require new ways of looking at and addressing them. The myths and images related to water highlight the pioneering spirit in the Middle East which had always been able to develop new approaches to address water-related challenges. For centuries, this adaptation to often harsh environmental, hydrologic and climatic conditions provided the basis for livelihoods, socio-economic development, and the survival of civilisations. However, recent decades clearly demonstrated that the current socio-economic, hydrological, and climatic dynamics are extraordinarily complex and urgent while necessary political adjustments or corrections have been politically ignored or de-emphasised for decades.

Current water-related challenges and ubiquitous scarcity require a whole set of new technical as well as political measures since water management, irrigation technologies, agricultural policies, etc. are not able to meet current water needs. If water governance and institutional capacities are not ready and able to meet these challenges, the region will experience more severe scarcity situations, even more frequent supply interruptions, the consequences of floods, and droughts. In addition, countries that fail to achieve or at least to improve water security are threatened in their economic growth and social and political stability through increased vulnerability to hydrological shocks.

It will require another, different water revolution in the Middle East to respond to rapidly changing conditions on a scale and scope more drastic and extensive than ever before. Alongside technological innovations, investment in infrastructure and political effort, this endeavor calls for a broader behavioural change in society and culture. Water affects many fields of policy inside and outside the water sector and, to a greater or lesser extent, every individual. As a consequence, a shift towards promoting water saving practices, modest water use, adjusted lifestyles, etc. would have to be based on society as a whole supporting it.

It is a longer process that involves tremendous and deeply-rooted socio-economic and cultural changes which are everything but simple to achieve after the casual use of water has been accepted and enabled over decades. In order to successfully tackle today's and the future's water challenges, the region has to develop a new image or narrative for the water question over time—one that imagines a region, nations, states, and individuals that are determined to overcome the business-as-usual of the last fifty years in order to sustain a hydrological future.

CHAPTER 8

Climate Change and the Emerging Information Societies in the Arab Region

Kerstin Fritzsche

Abstract

In recent decades, the countries in the Arab region have made efforts to develop towards information societies and to transform their mainly fossil-fuel-based and poorly diversified economies into knowledge economies. Modern information and communication technologies (ICTs) play a central role in this transformation process. Besides, the Arab region is also one of the most vulnerable to the adverse impact of climate change. Against this backdrop, this chapter explores the potential direct and indirect implications of digital change processes and the emergence of information societies in the Arab countries for climate change mitigation and adaptation.

1 Introduction

The term 'information society' first appeared in 1960s Japan to describe the societal and economic change processes experienced at the dawn of the computer age.[1] In the 1960s and 1970s, a vivid academic debate emerged about the nature of a new type of society that evolved with the increasing spread of information and communication technologies (ICTs). With the decline of traditional industries and the rise of white-collar sectors, industrial societies were believed to be transforming into societies that would build on information, knowledge, and innovation as central economic factors. Drucker, rather than focusing on information, discussed the emergence of knowledge as "the very foundation of modern economy and modern society and the very principle of social action",[2] thus coining the terms 'knowledge society' and 'knowledge economy'. Castells rejected both the terms information society and

1 Alistair S. Duff, David Craig, and David A. McNeill, "A Note on the Origins of the 'Information Society'", *Journal of Information Science* 22.2 (2016): 119.

2 Peter F. Drucker, *The Age of Discontinuity: Guidelines to Our Changing Society* (New Brunswick, London: Transaction Publishers, 2011), 371.

DOI:10.1163/9789004444973_010

knowledge society, and developed the idea of a 'network society',[3] underscoring the interconnectedness and interdependencies of a world penetrated by ICTs. Aside from these examples, a multitude of approaches[4] were developed to describe this new type of society, rendering the term information society a generic concept[5] or a 'metaphor'[6] that subsumes different social, cultural, and economic phenomena in the broader context of the emergence and spread of ICTs. In this broad sense, the term 'information society' is applied in this chapter to describe the complex social, cultural, and economic change processes happening throughout the Arab region in connection with the spread and increasing relevance of modern ICTs in numerous areas of life.

At the international level, the discourse on the question of information society only emerged a few years after the United Nations Earth Summit in Rio in 1992 had emphasised the importance of environment and sustainable development for humanity.[7] In particular, the widespread use of ICTs in the information society and their potential environmental benefits and challenges became major areas of concern. Starting in the late 1990s, research increasingly explored the role of ICTs in the broader context of sustainability.[8] The extent of this research, however, remained limited in relation to the anticipated impacts of ICTs and it was not until the 2000s that ICTs and their potential contributions to climate change mitigation and adaptation received more attention.[9] More recently, under the impression of a worldwide 'digital transformation' and major international policy processes such as the United Nations' Agenda 2030 for Sustainable Development and the adoption of the Paris Agreement, research on the climate change implications of ICTs has gained momentum and become more differentiated.

3 Manuel Castells, *The Rise of the Network Society: The Information Age: Economy, Society, and Culture* (Hoboken: John Wiley & Sons, Ltd., 2009).

4 Webster provides five different definitions of the information society: technology-centred, economic, occupational, spatial, and cultural. See: Frank Webster, *Theories of the Information Society* (London, New York: Routledge, 2007), 8–21.

5 Webster, *Theories of the Information Society*, 263.

6 Ljerka Luic and Dejan Glumac, "The Role of ICT Technology in the Knowledge Society," (paper presented at the conferences "TELSIKS 2009–9th International Conference on Telecommunications in Modern Satellite, Cable, and Broadcasting Services," 2009), 310.

7 European Information Technology Observatory (EITO), "The Impact of ICT on Sustainable Development, 2002," accessed July 25, 2017. http://homepage.cs.latrobe.edu.au/sloke/greenIT/eito_forum_2002.pdf, 251.

8 Angelica V. Ospina and Richard Heeks, *Unveiling the Links between ICTs & Climate Change in Developing Countries: A Scoping Study* (Manchester, Institute for Development Policy and Management, SED, 2010), 5, accessed July 24, 2017. http://citeseerx.ist.psu.edu/viewdoc/download?doi=10.1.1.841.822&rep=rep1&type=pdf.

9 Ibid.

In this light, this chapter reflects on the transformation of the Arab countries[10] towards information societies from a climate change perspective. Given the ubiquity and variety of ICTs, the complex climate change challenges in the Arab region, and the heterogeneity of the Arab countries, this endeavour is bound to be incomplete and cursory. However, while the transformation of the Arab countries towards information societies and knowledge economies has received wider attention, especially with regard to challenges in the areas of infrastructure, economy, education, innovation, and regulation, the environmental dimension[11]—more specifically, the aspect of climate change—has so far been a minor subject. This chapter therefore aims to encourage and contribute to a debate on climate change and the emerging information societies in the Arab region by providing an overview of some key issues, a glimpse of current developments in the region, and food for thought concerning future research on the issue of digital transformation and climate change in the Arab region.

The chapter is structured as follows: It first provides an overview of the state of the information society in the Arab countries. Then, an analytical framework is presented to explore how ICTs relate to climate change adaptation and mitigation in the Arab region. Subsequently, the different levels of analysis, namely, first-order (or direct) and second-order (or indirect) effects of ICTs,

---

10 In this chapter, the Arab countries are understood as the 22 members of the League of Arab States (LAS): Algeria, Bahrain, the Comoros, Djibouti, Egypt, Iraq, Jordan, Kuwait, Lebanon, Libya, Mauretania, Morocco, Oman, Palestine, Qatar, Saudi Arabia, Somalia, Sudan, Syria, Tunisia, United Arab Emirates and Yemen. This chapter will, however, particularly focus on the North African and Middle Eastern countries as well as the Arab Gulf states.

11 In particular the Technology for Development Division of UN ESCWA has issued several studies over the past years that address ICTs or related topics in the context of e.g. innovation, employment, education, internet and ICT sector governance, etc., as for example: Economic and Social Commission for Western Asia (ESCWA), *Perspectives on the Digital Economy in the Arab Region* (Beirut: United Nations Economic and Social Commission for Western Asia, 2018); ESCWA, *Innovation Policy for Inclusive Sustainable Development in the Arab Region* (Beirut: United Nations Economic and Social Commission for Western Asia, 2017); ESCWA, *The Innovation Landscape in Arab Countries: A Critical Analysis* (Beirut: United Nations Economic and Social Commission for Western Asia, 2017); ESCWA, *Regional Profile of the Information Society in the Arab Region* (Beirut: United Nations Economic and Social Commission for Western Asia, 2016); ESCWA, *Impact of ICT on Arab Youth: Employment, Education and Social Change*, (Beirut: United Nations Economic and Social Commission for Western Asia, 2013). While some of these studies relate directly to the UN Sustainable Development Goals, none of these publications explores environmental or climate change issues more in-depth. Likewise, academic literature dealing with digital transformation in the Arab countries and environmental implications is still sparse.

are explored with a particular emphasis on how climate change mitigation and adaptation could be supported by the application of ICTs in the region. The chapter concludes with a brief summary of findings and suggestions for further research on the theme of ICTs and climate change in the Arab countries.

2 State of the Information Society in the Arab Region

In the early and mid-2000s, Arab countries started to develop ICT policies and strategies and invested in telecommunication infrastructures to promote the transformation towards information societies.[12] This development took place against the backdrop of not only the increasing importance of ICTs in the global economy, but also the deepening technological and digital divide between so-called developed and developing countries. Moreover, many Arab countries recognised that their present fossil-dependent and poorly diversified economies would fall short in providing sustainable economic growth and in creating employment opportunities for their young and growing populations. Since 1990, the population of the Arab region has increased from 225 million to more than four hundred million.[13] Approximately sixty per cent of the Arab populations are younger than thirty years.[14] Against this backdrop, a study conducted by the World Economic Forum (WEF) suggests that by 2020, more than eighty million additional jobs would be required to accommodate the large annual increases in workforce and to maintain youth unemployment at tolerable levels.[15] Today, the Arab countries already suffer above-average youth unemployment (19 per cent in 2013 compared to 13 per cent worldwide).[16] The transition to diversified economies building on knowledge and modern

---

12 ESCWA, "Regional Profile of the Information Society in the Arab Region," 15.

13 United Nations Department of Economic and Social Affairs (UN DESA), *World Population Prospects 2017* (New York: United Nations Department of Economic and Social Affairs, Population Division, 2017).

14 UNDP, *Arab Human Development Report 2016: Youth and the Prospects for Human Development in a Changing Reality* (New York: United Nations Development Programme Regional Bureau for Arab States, 2016), 22.

15 World Economic Forum, *Addressing the 100 Million Youth Challenge. Perspectives on Youth Employment in the Arab World in 2012* (Geneva: World Economic Forum, 2012), 15.

16 UNDP, *Arab Human Development Report 2016*, 31, citing World Bank, "World Development Indicators 2015," accessed December 16, 2012. http://data.worldbank.org/data-catalog/world-development-indicators; ILO, *Global Employment Trends for Youth 2015: Scaling Up Investments in Decent Jobs for Youth* (Geneva: International Labour Organization, 2015). http://www.ilo.org/wcmsp5/groups/public/---dgreports/---dcomm/---publ/documents/publication/wcms_412015.pdf.

technologies is therefore perceived as one important measure to address increasing societal and economic pressures and social friction that, among other factors, spurred the popular uprisings of the Arab Spring in early 2011.

However, there are marked differences between the Arab countries with regard to the implementation and success of digital development efforts due to major differences in political stability, financial resources, and political will as well as know-how.[17] Countries such as the United Arab Emirates, Bahrain, and Qatar have some of the most tech-savvy governments in the world.[18] Not surprisingly, they score highly on different digital development indices, such as for example the ICT Development Index[19] of the International Telecommunication Union (ITU) and the World Economic Forum's Networked Readiness Index,[20] which evaluate a country's ability to exploit the opportunities of ICTs on the basis of different sets of indicators. Internet infrastructures in the region are relatively good, but maturity levels vary across the region and are particularly low in its least developed countries (LDCs)[21] and conflict-ridden countries according to an assessment conducted by ESCWA.[22] Similarly, internet use, which is one important indicator to measure the state of the information society, differs significantly between the Arab countries. As table 8.1[23] shows, a large percentage of the populations of the Arab Gulf states in particular, but also of Lebanon, use the internet. Countries in the Maghreb and the Mashreq such as Morocco, Jordan and Tunisia have rates of internet use (as a proportion of population) that exceed the global average, while Egypt and Algeria remain below the world average—as does the Arab region overall. Conflict-ridden countries such as Iraq, Libya and Syria, and the Arab states belonging to the group of LDCs, show significantly low rates of internet use among their populations.

17 ESCWA, "Regional Profile of the Information Society in the Arab Region," 15–16.

18 World Economic Forum, "Networked Readiness Index 2016. Infographics and Shareables," accessed July 27, 2017. http://reports.weforum.org/global-information-technology-report-2016/infographics-and-shareables/.

19 ITU, "ICT Development Index 2016," accessed July 27, 2017. http://www.itu.int/net4/ITU-D/idi/2016/.

20 World Economic Forum, "Networked Readiness Index 2016," accessed July 27, 2017.

21 The group of the least developed countries (LDCs) is identified by the Committee for Development Policy (CDP), which is mandated by the UN General Assembly (GA) and Economic and Social Council (ECOSOC) to review the list of LDCs every three years. The list of LDCs as of March 2018 can be found here: https://www.un.org/development/desa/dpad/wp-content/uploads/sites/45/publication/ldc_list.pdf, accessed March 25, 2018.

22 ESCWA, "Regional Profile of the Information Society in the Arab Region," 23–33.

23 World Bank, World Development Indicators; International Telecommunication Union, World Telecommunication/ICT Development Report and database.

TABLE 8.1 Individuals using the internet (per cent of population)

	2000	2005	2010	2015
Algeria	0.5	5.8	12.5	38.2
Bahrain	6.2	21.3	55.0	93.5
Comoros	0.3	2.0	5.1	7.5
Djibouti	0.2	1.0	6.5	11.9
Egypt	0.6	12.8	21.6	37.8
Iraq	..	0.9	2.5	17.2
Jordan	2.6	12.9	27.2	53.4
Kuwait	6.7	25.9	61.4	82.1
Lebanon	8.0	10.1	43.7	74.0
Libya	0.2	3.9	14.0	19.0
Mauritania	0.2	0.7	4.0	15.2
Morocco	0.7	15.1	52.0	57.1
Oman	3.5	6.7	35.8	74.2
Qatar	4.9	24.7	69.0	92.9
Saudi Arabia	2.2	12.7	41.0	69.6
Somalia	0.0	1.1	..	1.8
Sudan	0.0	1.3	16.7	26.6
Syria	0.2	5.6	20.7	30.0
Tunisia	2.8	9.7	36.8	48.5
United Arab Emirates	23.6	40.0	68.0	91.2
Palestine	1.1	16.0	37.4	57.4
Yemen	0.1	1.0	12.4	25.1
Arab World*	1.1	8.4	24.5	39.9
World	6.8	15.8	29.1	43.9

* Includes all members of the League of Arab States.

However, the development of information societies and thriving knowledge-based economies is not just about connecting to and using the internet. Aside from ICT infrastructure and regulations, education, research, and development as well as innovation form major pillars of knowledge economies.[24]

24 Ramona D. Leon, "The Pillars of the Knowledge Based Society," *SSRN Electronic Journal* (2011): 3, accessed July 27, 2017. doi:10.2139/ssrn.1907630.

With regard to education, the quality of learning and training in the Arab countries remains low overall by international standards.[25] Despite major efforts in many Arab countries to modernise and reform their education systems, the skills and competencies of young people are not adequately developed, and this poses a major hurdle for economic growth.[26] Given the need for highly skilled personnel in the areas of natural and computer sciences, engineering, and mathematics, much attention has been devoted to the improvement and extension of higher education in the Arab region. The region has also experienced a major expansion of private universities and branches of established Western universities. The model of Western research universities is on the rise, particularly among the Gulf Cooperation Council (GCC) states.[27] In Saudi Arabia, Qatar, and Egypt, this development has translated into growing numbers of academic publications.[28] However, despite some exceptions, research at Arab universities remains generally weak, hampered by financial constraints, pressure from increasing numbers of students, and a widespread culture within Arab higher education that places little emphasis on research as an essential task of academic personnel.[29]

Furthermore, the UNESCO Science Report points out that science, technology, and innovation policies in Arab countries have so far had little success in catalysing knowledge production, especially because they failed to involve businesses and private sector actors in research and development (R&D) activities.[30] In 2014, the Council of Ministers of Higher Education and Scientific Research of the Arab League therefore stressed the role of the private sector in regional and interdisciplinary collaborations for R&D.[31] However, to date, enabling environments for private-sector companies in the Arab countries remain insufficient, restraining their potential to contribute to innovation and economic growth. In particular, small businesses and young

---

25 Roberta Gatti et al., *Jobs for Shared Prosperity: Time for Action in the Middle East and North Africa* (Washington, DC: World Bank, 2013), 24, accessed July 27, 2017, https://openknowledge.worldbank.org/handle/10986/13284.

26 Ibid.

27 Samar A. Abid al Tamimi, "Reshaping Higher Education in the Gulf States: Study Abroad Trends and Student Experiences," *Gulf Affairs* (2017): 10–13, accessed July 27, 2017. http://www.oxgaps.org/files/gulf_affairs_spring_2017_full_issue.pdf.

28 Moneef R. Zou'bi. "The Arab States," in *UNESCO Science Report: Towards 2030*, ed. UNESCO (Paris: UNESCO, 2015), accessed July 27, 2017. https://en.unesco.org/sites/default/files/usr15_the_arab_states.pdf, 444.

29 Ibid., 438.

30 Ibid., 435.

31 Ibid.

entrepreneurs struggle with regulations and practices that favour larger and better-established firms.[32]

In summary, the transformation of Arab countries to information societies and knowledge economies is most advanced among the GCC countries,[33] which are endowed with the necessary financial resources to upgrade ICT infrastructures, improve higher education, develop human capacities, and spur innovation. Countries such as Egypt, Morocco, Jordan and Tunisia have made significant progress in the past decade, particularly in providing internet access, but also with regard to developing innovation policies. However, many of these countries continue to suffer from a strong rural-urban divide, the challenge of restrained financial resources, the pressure of large numbers of young people, and overall socio-economic difficulties. However, modern ICTs continuously gain importance in the Arab countries, which is why it is worthwhile taking a closer look at their potential implications for climate change mitigation and adaptation.

3 Framework for Exploring ICTs from a Climate Change Perspective

Assessing the relevance of ICTs with regard to climate change is a complex endeavour since it is not only their material dimension and energy consumption that leaves a carbon footprint, but also the context in which they are applied, how and for what purpose ICTs are utilised, as well as the social innovations and new practices to which they give birth. Many assessments of environmental and climate change implications of ICTs draw on taxonomies that suggest three levels of analysis:[34]

1) First-order or direct effects stemming from the design, production, deployment and disposal of ICTs;
2) second-order or indirect effects related to the use of ICTs; and
3) third-order effects, usually considered as more profound and often structural changes related to ICTs.

---

32 UNDP, "Arab Human Development Report", 30.

33 These are: Bahrain, Kuwait, Oman, Qatar, Saudi Arabia and the United Arab Emirates.

34 See for example: EITO, "The Impact of ICT on Sustainable Development" and; Frans Berkhout and Julia Hertin, *Impacts of Information and Communication Technologies on Environmental Sustainability: Speculations and Evidence. Report to the OECD* (Brighton: Science and Technology Policy Research Unit, University of Sussex, 2001), accessed October 19, 2017. http://www.oecd.org/sti/inno/1897156.pdf. A useful overview of taxonomies of different authors is further provided by Nathaniel C. Horner, Arman Shehabi, and Inês L. Azevedo, "Known Unknowns: Indirect Energy Effects of Information and Communication Technology," *Environmental Research Letters* 11.10 (2016): 4.

The latter naturally builds on assumptions about digital transformations in a more or less distant future that encompass countless variables. A scenario analysis of potential systemic effects of ICT deployment in the Arab region would exceed the scope of this chapter. Therefore, although Erdmann and Hilty rightly point out that "it seems indispensable to account for third-order effects—in particular rebound effects—in any macroeconomic assessment of ICT impacts on GHG emissions",[35] the following sections focus on exploring the direct and indirect implications of ICTs for climate change in the Arab region. This includes ICTs' contribution to greenhouse gas (GHG) emissions, but also opportunities for climate change mitigation and adaptation that could be exploited through the use of digital technologies. The analysis focuses on trends and applications that are either emerging or already manifest in the Arab region or that appear relevant against the backdrop of major transitions, e.g., in the energy sector. Table 8.2 summarises how first- and second-order effects of ICTs are understood in the context of this chapter, and provides examples:

TABLE 8.2 Understanding of the two dimensions of ICTs relevant to climate change adaptation and mitigation

	Explanation	Examples
First-order effects	Effects relevant to climate change and related directly to the production, deployment and disposal of ICTs	CO_2 emissions: related to the production process of ICTs; the energy consumption of a device over its lifespan; or stemming from recycling or disposal of a device
Second-order effects	Effects relevant to climate change and related indirectly to the deployment of ICTs	Decreasing energy consumption in private households due to smart energy management; increasing water resource efficiency

35 Lorenz Erdmann and Lorenz M. Hilty, "Scenario Analysis: Exploring the Macroeconomic Impacts of Information and Communication Technologies on Greenhouse Gas Emissions," *Journal of Industrial Ecology* (2010): 827, accessed March 22, 2018. https://onlinelibrary.wiley.com/doi/epdf/10.1111/j.1530-9290.2010.00277.x.

4 First-order Climate Change Effects of ICTs in the Arab Region

4.1 *GHG Emissions of ICT Production*

The term ICTs encompasses a wide range of different technological products manufactured along global supply chains.[36] Numerous factors, such as the design of a product, materials, the mode of production, and the sources of energy used in the production process influence the environmental and carbon footprint of the production stage of ICTs. A study conducted by Manhart et al. points out that the production of electronic devices has a major impact on the overall GHG emissions of a device, with emissions during production largely stemming from the extraction and processing of resources, manufacturing, and the assembly of components as well as testing.[37] The study further reveals that the shares of different components in the GHG emissions of the production stage vary significantly and that very few components often account for the largest amounts of GHG emissions.[38]

Over the past years, the Arab region has seen a rapid adoption of modern information and communication technologies such as smartphones due to falling prices of both devices and mobile internet connections, and this trend will likely continue.[39] With the number of smart devices increasing and with new applications they offer, e.g. streaming of videos, the need for infrastructure and data centres will also grow. Lange and Santarius argue that while the resource and energy use for manufacturing a single smartphone may appear comparatively low, it is exactly this dynamic and the increasing number of devices that would lead to a mounting overall environmental footprint of ICT production.[40] Material and energy consumption for the production of ICTs deployed in the Arab region will therefore likely rise. However, most ICTs used in the region are imported. Consequently, the GHG emissions related to the production of ICTs for the Arab market mainly occurs outside the region. However, many Arab countries aspire to further develop their own ICT sectors, not only in terms

36 Berkhout, and Hertin, *Impacts of Information and Communication Technologies on Environmental Sustainability: Speculations and Evidence*, 8.

37 Andreas Manhart, et al., *Resource Efficiency in the ICT Sector. Final Report, November 2016* (Freiburg: Öko-Institut, 2016), accessed March 22, 2018. https://www.greenpeace.de/sites/www.greenpeace.de/files/publications/20161109_oeko_resource_efficency_final_full-report.pdf, 21.

38 Ibid., 28–30.

39 GSMA, "The Mobile Economy Middle East and North Africa 2017," accessed March 22, 2018. https://www.gsmaintelligence.com/research/?file=84935f5774975f3d35c8ed9a41b9c1a4&download, 18.

40 Steffen Lange, Tilman Santarius, *Smarte grüne Welt? Digitalisierung zwischen Überwachung, Konsum und Nachhaltigkeit* (München: oekom, 2018).

of ICT services, but also with regard to developing and producing components and devices. The carbon footprint of ICT production in Arab countries will therefore depend on the size and production of this industry, the kind of materials and energy used, how efficiently they are applied in the production process, and the environmental management schemes implemented.

4.2 *CO_2 Emissions from ICT Operation*

Electricity consumption during operation contributes considerably to the carbon footprint of ICTs, with the source for electricity generation being an important factor. Global electricity consumption from ICTs is expected to rise over the coming years. A study by Andrae and Edler suggests that by 2030, communication technologies could account for eight per cent of global electricity usage in a best case scenario, 21 per cent in the expected scenario, and up to 51 per cent in the worst case scenario.[41] Some authors argue for a rather moderate overall increase in energy demand from ICT usage due to increasing energy efficiency.[42] With regard to GHG emission from ICTs, a study conducted by the Global e-Sustainability Initiative (GeSI) and Accenture came to the conclusions that, despite its rapid growth, the ICT sector would decrease its share to approximately two per cent of global carbon dioxide emissions by 2030.[43] This will particularly be due to investments made in the sector to reduce emissions and the increasing energy efficiency of ICTs.

Reflecting these numbers in the context of the Arab countries, it should first be noted that electricity consumption has risen steeply in the region, mainly due to economic development and population growth. However, in most Arab countries, in particular the LDCs, electricity consumption currently ranges below the world average of approximately three MWh/capita (in 2014); conversely, per capita energy consumption among the GCC countries far exceeds the global average, ranging from 6.13 MWh/capita in Oman to 16.7 MWh/capita in Qatar.[44] The contribution of ICT to electricity consumption in the Arab countries is difficult to determine. However, as already mentioned above, the Arab region has shown significant growth in ownership of mobile phones,

41 Anders Andrae and Tomas Edler, "On Global Electricity Usage of Communication Technology: Trends to 2030," *Challenges* 6.1 (2015): 117–157.

42 Erol Gelenbe and Yves Caseau, "The Impact of Information Technology on Energy Consumption and Carbon Emissions," *Ubiquity* (2015): 1–15, accessed July 20, 2017. doi:10.1145/2755977, 3.

43 GeSI, and Accenture, "#SMARTer 2030: ICT Solutions for 21st Century Challenges," accessed July 25, 2017. http://smarter2030.gesi.org/downloads/Full_report.pdf, 10.

44 International Energy Agency, "Statistics," accessed July 27, 2017. http://www.iea.org/statistics/statisticssearch/.

computers, laptops, and other devices in recent years. Along with increasing internet connectivity, it is likely that electricity consumption by ICTs will therefore increase throughout the region. Currently, the majority of electricity generation in the Arab countries depends on natural gas and oil (approximately 58 and 37 per cent, respectively, in 2014).[45] If the region manages a transition toward high shares of renewable energies, exploiting in particular the vast solar energy potentials of the region as envisaged in the renewable energy plans of many countries in the region, this could contribute significantly to lowering carbon emissions from ICT usage in the Arab countries.

4.3 *The Carbon Footprint of ICTs' Afterlife*

With the increasing spread of ICTs, electronic waste (e-waste) has become a mounting challenge. The latest Global E-Waste Monitor estimates that in 2016, 44.7 million metric tonnes of e-waste were generated worldwide, an amount equivalent to 4,500 Eiffel Towers, which is expected to rise to 52.2 million metric tonnes by 2021.[46] The increasing affordability of ICTs that also allows ownership of multiple devices, ever shorter replacement cycles, electrification of non-electrical equipment, and increasing demand for cloud computing services and data centres are some of the key drivers for e-waste generation.[47] Only twenty per cent of global e-waste is documentably collected and recycled, which in turn means that large amounts of energy and valuable materials used in the production process are lost if devices are not reused or recycled.[48] Where e-waste is not properly disposed of, hazardous materials contained in electronic devices could be released into the environment e.g. through dumping or burning. The latter in particular contributes to atmospheric emissions of CO_2 and other climate pollutants.

Arab states increasingly contribute to the amount of e-waste. However, there are large differences between the countries of the region: For 2016, Bahrain (15.5 kg/inh.), Kuwait (15.8 kg/inh.), and Saudi Arabia (15.9 kg/inh.) showed some of the highest numbers of e-waste generated per inhabitant; Morocco (3.7 kg/inh.), Algeria (6.2 kg/inh.), and Tunisia (5.6 kg/inh.) rank considerably lower, while the Arab LDCs range between 0.8 (Comoros) and 1.5 (Yemen) kg/

45 Reem Nejdawi et al., *Arab Sustainable Development Report* (Beirut: ESCWA, 2015), 102.

46 Kees Baldé et al., *The Global E-waste Monitor* (Bonn, Geneva, Vienna: United Nations University (UNU), International Telecommunication Union (ITU), and International Solid Waste Association (ISWA), 2017), 4, accessed March 22, 2018. https://www.itu.int/en/ITU-D/Climate-Change/Documents/GEM%202017/Global-E-waste%20Monitor%202017%20.pdf.

47 Ibid., 19.

48 Ibid., 5–7.

inhabitant.[49] A 2009 study concluded that in the Arab countries, awareness of the topic of e-waste—among both governments and the general public—remains compartively low.[50] More recent figures are difficult to obtain, but an overview provided by the Global E-Waste Monitor shows that none of the Arab countries had national e-waste regulations in place in January 2017.[51] However, several Arab countries have initiated steps to deal with the increasing amounts of e-waste management. For example, the Green ICT strategy of the Egyptian Government explicitly includes the development of a national e-waste management scheme.[52] The carbon and environmental footprints of ICTs in the Arab region will therefore also depend on the determination of Arab countries to implement effective measures for dealing with the e-waste arising from their emerging information societies.

5 Second-order Effects of ICTs from a Climate Change Perspective

5.1 *ICTs and Climate Change Mitigation*

While the previous sections dealt with the direct climate change implications of ICTs, especially with regard to CO_2 emissions, the following sections are dedicated to exploring the more indirect effects of deploying ICTs, starting with the question of how ICTs might contribute to mitigating climate change.

The role of second-order effects of ICTs for climate change mitigation is ambiguous. While the energy consumption and CO_2 emissions directly associated with ICTs are rising, a 2015 study by the Global e-Sustainability Initiative (GeSI) and Accenture argues that the application of ICTs could lead to a twenty per cent reduction of global CO_2 emissions in other sectors by 2030.[53] Moreover, the study suggests that the deployment of ICTs could decouple economic growth from carbon emissions and contribute significantly to sustainable, low-carbon development.[54] These statements are based on models and projections, but empirical evidence on the reduction of greenhouse gas (GHG)

49 Ibid., 102–108.

50 Hossam Allam and Simon Inauen, *E-Waste Management Practices in the Arab Region* (Cairo: CEDARE, 2009), accessed October 19, 2017. http://citeseerx.ist.psu.edu/viewdoc/download?doi=10.1.1.625.2567&rep=rep1&type=pdf.

51 Baldé, et al., *The Global E-waste Monitor*, 102.

52 Ministry of Communications and Information Technologies, "Egyptian Green ICT Strategy: Pillars—Partners—Programs 2010–2018," accessed October 19, 2017. https://www.epa.gov/sites/production/files/2014-08/documents/egypt-_green_ict.pdf.

53 GeSI, and Accenture, "#SMARTer 2030: ICT Solutions for 21st Century Challenges," accessed July 25, 2017. http://smarter2030.gesi.org/downloads/Full_report.pdf.

54 Ibid., 92.

emissions from the application of ICTs is still rare. This is certainly also due to the fact that many digital innovations that could have major impacts on GHG emissions (for example, by leveraging the 'internet of things' to contribute to energy saving by improving industrial and commercial energy efficiency) have not yet been implemented on a larger scale. In particular the effect of the 'internet of things' remains "the big unknown".[55]

Furthermore, there is increasing concern that the positive effects of improved energy efficiency through ICTs could be negated by the so-called rebound effect.[56] This could manifest in multiple ways: For example, ICTs might enable more consumer-centred and individualised products, thus contributing to more efficient production using fewer resources and less energy. However, along with reduced costs, this might encourage greater consumption, thereby increasing net demand for resources and energy and exacerbating overall GHG emissions. Similarly, companies or individuals that improve their energy efficiency through digital technologies might ultimately neglect further energy-saving measures and thus not fully exploit their potential to reduce emissions of climate pollutants.

While the impact of ICTs on energy and resource efficiency is unclear, their role in the transition towards renewable energy systems appears decisive. In order to meet increasing electricity demands and diversify their national energy mix, many Arab countries have made strong efforts in recent years to exploit their renewable energy capacities. Almost all countries, even the Arab LDCs, have set ambitious renewable energy targets. In particular, Morocco is leading the way in the development of solar and wind energy, aiming to reach more than fifty per cent of installed capacity with renewables by 2030.[57]

---

55 Caseau Gelenbe, "The Impact of Information Technology," 6.

56 See for example: Cédric Gossart, "Rebound Effects and ICT: A Review of the Literature," in *ICT Innovations for Sustainability. Advances in Intelligent Systems and Computing*, vol. 310, ed Lorenz M. Hilty, and Bernard Aebischer (Berlin: Springer, 2015), 435–448; Nathaniel C. Horner, Arman Shehabi, and Inês L. Azevedo, "Known Unknowns: Indirect Energy Effects of Information and Communication Technology," *Environmental Research Letters* 11.10 (2016): 103001, accessed July 26, 2017, doi:10.1088/1748–9326/11/10/103001; Tilman Santarius, "Digitalization, Efficiency and the Rebound Effect," 2017, accessed July 26, 2017, https://www.degrowth.info/en/2017/02/digitalization-efficiency-and-the-rebound-effect/.

57 IRENA, *Renewable Energy in the Arab Region. Overview of Developments* (Abu Dhabi, International Renewable Energy Agency, 2016), 14, accessed July 26, 2017. http://www.irena.org/DocumentDownloads/Publications/IRENA_Arab_Region_Overview_2016.pdf. Also includes a comprehensive overview of the renewable energy targets of most Arab League member states.

Wind and solar power currently only account for a small share of electricity generation in Arab countries. However, their remarkable growth may eventually pose significant challenges to the stability of Arab electricity systems. As Brand[58] points out, there are three main issues to be solved:

(1) The handling of vast amounts of data that are produced by digitalised electricity systems, with the Arab Gulf countries already developing in this direction by rolling out smart meters;
(2) Managing the intermittency of solar and wind generation and dealing with an increasing number of electricity producers as renewable energies allow more decentralised and small-scale energy production, and;
(3) The emergence of new market structures and actors such as private and commercial 'prosumers' due to this decentralised production.

With renewable energies gaining larger shares of national energy mixes, the development of 'smart grids' in the Arab countries appears decisive. Since national electricity markets in the Arab region are relatively small, regional cooperation would be central to the development of stable smart grids. Grid interconnections already exist between the countries of the Maghreb and the Mashreq, including Turkey. The lack of harmonised regulations has so far, however, been a major obstacle to expanding the trade in electricity within these regions, which therefore remains at modest levels.[59] In the countries of the GCC, grid connections and an agreement signed in 2009 allow regional electricity exchange.[60] In particular in the Maghreb and Mashreq countries, strong efforts will be needed to develop suitable and smart regional power networks and electricity markets to accommodate expanding shares of renewable energy.

Furthermore, the heavily regulated electricity markets of the Arab countries currently provide little opportunity for private actors or even households to produce and sell electricity from private solar energy devices. This poses an obstacle to the development of new technologies and energy-related business models in the region because, for example, energy start-ups often lack enabling

58 Bernhard Brand, *Les énergies renouvelables et la digitalisation des systèmes électriques : Tendances et défis pour les pays de la région MENA* (working paper February 2017), 5, accessed July 27, 2017. http://www.enerpirica.com/download/Enerpirica_MENA_Digitalisation_14-02-2017.pdf.

59 World Bank, "Middle East and North Africa. Integration of Electricity Networks in the Arab World: Regional Market Structure and Design," accessed July 28, 2017. https://openknowledge.worldbank.org/bitstream/handle/10986/19329/ACS71240ESW0WH0I0and0II000Final0PDF.pdf?sequence=1&isAllowed=y, 2.

60 Ibid., 1.

regulatory environments and the financial and market incentives required to develop smart, renewable energy and ICT solutions in and for the region.

5.2 *ICTs and Climate Change Adaptation*

There is a wide range of possibilities for how ICTs could support adaptation to climate change and contribute to managing disaster risks, whether related to climate change or not. ICTs assist the collection of data for climate change adaptation and its analysis and thereby support informed decision-making, adaptive planning, and disaster risk-management. Furthermore, ICTs support monitoring and evaluation of adaptation measures and enable better documentation and knowledge exchange, thereby becoming essential parts of strategies and practical solutions for climate change challenges in a variety of sectors such as water, food and agriculture, energy, biodiversity, human habitats, and many more. However, the application of ICTs is not without risks. As Ospina and Heeks point out, the use of ICT for adaptation and disaster risk-management needs to be sensitive to the livelihoods and power relations of specific contexts in order to avoid negative consequences for already disadvantaged groups.[61] Furthermore, one would need to be aware that ICTs might divert resources from other important activities, and could also provide unreliable or misinterpreted data that may lead to maladaptation.[62]

One exemplary area that offers a range of possibilities for applying ICTs to climate change adaptation in the Arab countries is the agricultural sector. This remains a major employer in many Arab states and plays an important role in food security especially in the lower-income countries of the region. As a backbone of the economy and labour market in rural areas, the well-being of the agricultural sector influences rural-urban migration as well as social and political stability, as demonstrated by the unrest in many Arab countries in 2008 and 2010/2011, when rising food prices among other factors played an important role. Agricultural production in the Arab countries is constrained by scarcity of arable land and water resources[63] and highly vulnerable to climate change effects such as rising temperatures, shifting seasons, and decreasing precipitation. Water lies at the heart of the climate challenge for Arab

61 Angelica V. Ospina and Richard Heeks, "Linking ICTs and Climate Change Adaptation: A Conceptual Framework for e-Resilience and e-Adaptation," in *ICTs, Climate Change and Development: Themes and Strategic Actions*, ed. Richard Heeks and Angelica V. Ospina (Manchester: University of Manchester Press, 2012), 67–68.

62 Ibid.

63 Eckart Woertz, "Agriculture and Development in the Wake of the Arab Spring," *International Development Policy / Revue internationale de politique de développement* 7.7 (2017), accessed March 25, 2018. http://journals.openedition.org/poldev/2274.

countries' agricultural sector, as climate change will likely exacerbate water scarcity while at the same time increasing the demand for water.

ICTs could contribute to water-related climate change adaptation in the agricultural sector in several ways, for example by enabling data collection on soil moisture and crop water demand and analysing the need for irrigation against the backdrop of current weather information, therefore helping to design and manage optimised irrigation systems.[64] Software could monitor flows in water pipes and detect and assess losses.[65] ICT solutions could further provide farmers with information on rainfall forecasts and crop health. Many of these approaches are already being tested and used in Arab countries. In Egypt, ICTs support the Integrated Water Resource Management Action Plan of the Ministry of Water Resources and Irrigation in the Nile Delta by improving irrigation and drainage management to increase water efficiency.[66] Smart irrigation systems using soil moisture sensors have increased the water efficiency of greenhouses growing cucumbers and tomatoes in Abu Dhabi.[67] In Tunisia, farmers receive weather information via their mobile phones to make better-informed decisions on irrigating their crops.[68]

However, despite the increasing awareness of the need for climate-smart agricultural systems, the application of ICT and the availability of more and better data will only be part of the solution. Along with water-efficient and sustainable practices in agriculture, skills and knowledge need to be developed in agricultural businesses and relevant institutions to make purposeful use of the opportunities ICT can offer—alongside traditional knowledge and practices in the agricultural sector. This also underscores that developing infrastructures, competencies, and capabilities to benefit from modern digital technologies in Arab countries should not stop at the edge of cities and towns, but needs to

64 International Telecommunication Union, *ICT as an Enabler for Smart Water Management ITU-T Technology Watch Report October 2010* (Geneva: ITU, 2011), 8, accessed March 26, 2018. https://www.itu.int/dms_pub/itu-t/oth/23/01/T23010000100003PDFE.pdf.

65 Ibid., 8–9.

66 World Bank, and African Development Bank, *ICTs for Agriculture in Africa* (World Bank, 2012), 19, accessed July 28, 2017. http://siteresources.worldbank.org/EXTINFORMATIONANDCOMMUNICATIONANDTECHNOLOGIES/Resources/282822-1346223280837/Agriculture.pdf.

67 ITU, "Smart Irrigation for Sustainable Agriculture in Abu Dhabi: Case Study—United Arab Emirates," accessed July 28, 2017. https://itunews.itu.int/en/4853-Smart-irrigation-for-sustainable-agriculture-in-Abu-Dhabi.note.aspx.

68 Clare Pedrick, "A Future for Farming in the Dry Lands: Smart Solutions for Climate Change," accessed July 28, 2017. https://ccafs.cgiar.org/blog/future-farming-dry-lands-smart-solutions-climate-change#.WXtwb7ZpxLM.

include rural areas in order to ensure rural livelihoods and to bridge the technological rural-urban divide.

6 Conclusions

ICTs are spreading throughout the Arab region, accompanied by new challenges, but also opportunities for climate change mitigation and adaptation. However, as shown in the examples provided above, ICT is not a solution in itself, but requires conscious and deliberate deployment in order to achieve specific benefits and adequate framework conditions and regulations to avoid environmental risks. As shown by the exploration of the potential direct and indirect effects of ICT on climate change considerations in the Arab region, the transition towards renewable energy sources and the implementation of e-waste management schemes will be important pillars for a climate-friendly digital development in the Arab countries. Both require not only major investment, but—first and foremost—awareness of the implications of ICTs not only with regard to climate change, but also environmental sustainability. Furthermore, careful consideration should be given to trade-offs such as potential rebound effects, but also to issues related to data security and privacy.

The scope of this chapter is limited to only a fraction of the issues surrounding climate change and the environmental effects of ICTs in the Arab region. Nevertheless, these topics deserve further attention and should be explored in greater depth. For example, future studies could shed light on the potential of ICTs for energy and resource efficiency in Arab countries as well as opportunities and challenges of ICTs and digital platforms to foster renewable energy integration in the Arab region. Local case studies shedding light concrete applications of ICTs for adaptation e.g. in the agricultural sector or in urban areas could provide valuable lessons-learned and support informed decision-making. Research should further pay attention to the potential societal implications of ICTs in the Arab region and how they may relate to climate change and other environmental challenges. For example, consumer behavior and awareness of ICT-related environmental topics (including e-waste) in the Arab region could provide interesting insights, as could research on sharing economy or alternative forms of digitally enabled consumption and production in the Arab region. Finally, as mentioned above, third-order effects of ICTs on climate change have not been addressed in this chapter. However, exploring this topic using inter- and transdisciplinary methods that involve expertise from science, politics, economics, and civil society would be a worthwhile effort to better understand the potential implications of the emerging information societies in the Arab countries.

PART 4

Prospects

CHAPTER 9

DESERTEC: Europe—Middle East—North Africa Cooperation for Sustainable Energy

Franz Trieb

Abstract

The DESERTEC concept for energy, water, and climate security was published by the Trans-Mediterranean Renewable Energy Cooperation (TREC) a decade ago, in late 2007, in the form of a whitebook presented to the European Parliament. Being as fascinating as it was resilient, it created great expectations and considerable hype that was tangible in many international public and political discussions. Nevertheless, the original concept was lost along the way when a multitude of experts and consultants started to speculate about a futuristic infrastructure called supergrid instead of concentrating on a concrete first point-to-point power line interconnecting a flexible solar power plant in the desert (provider) with a demand centre in Europe (offtaker). The article at hand explains the original concept, the diversions that lead to its failure, and options to restart a concrete business case for a solar energy partnership in the EUMENA region.

1 Introduction

About 2000 years ago, the Roman Empire, by then ranging all around the Mediterranean Sea and north up to the British Islands, tried to build a wall across Europe—the limes—in order to keep out the barbarians that were trying to enter from Eastern-Central Europe. History tells us that this venture failed, and a little later, the Roman Empire vanished due to increasing social unrest caused by the greed and decadence of its leading class.

Although it is theoretically possible to learn from history, it is difficult in practice, especially if actors change roles: today, the former 'barbarians' are trying to build walls across Europe in order to keep out the former 'Romans' from all around the Mediterranean that are now coming to the heart of Europe in the course of the present refugee crisis. In the meantime, the leading class may also have changed, although perhaps not as much as would be desirable.

 | DOI:10.1163/9789004444973_011

In this context, the title of this paper looks extremely ambitious in several senses: firstly, it treats Europe, Middle East, and North Africa (EUMENA) as if it was one region, secondly, it re-starts thinking about the venture of energy cooperation within this region and thirdly, it aims at sustainable energy supply. In times of flourishing populism, separatism, and nationalism, each one of these innovations alone looks like an unattainable fiction, and all the more in combination.

Furthermore, the title above is very similar to that of a whitebook published over a decade ago, in 2007, by the Trans-Mediterranean Renewable Energy Cooperation[1] and a paper published by the German Aerospace Center DLR.[2] After being presented at the European Parliament, this paper started an unequaled hype about the so-called DESERTEC Initiative that sought to reach the above-mentioned goals, but came to a sudden end in late 2014. So recent history even tells us that our title story already had its chance, but failed to become reality.

Why then should we write (or read) another paper about this issue? Perhaps, in order to try to learn from history?

2 The Vision of Cooperation for Sustainable Energy

One major lesson we can learn from the DESERTEC history is to not confuse first steps with long-term visions. Map 9.1[3] illustrates the concept of EUMENA solar energy cooperation developed by the German Aerospace Center (DLR) and the Trans-Mediterranean Renewable Energy Cooperation (TREC) and later followed up by the DESERTEC Foundation that is based on producing flexible power on demand in concentrating solar thermal power stations (CSP) in the desert regions of MENA and sending it to European demand centers via

1 DESERTEC Foundation, *Clean Power from Deserts—The Desertec Concept for Energy, Water and Climate Security* (Bonn: Protext Verlag, 2009), accessed March 23, 2018. http://www.dun-eumena.com/sites/default/files/files/doc/trec_white_paper.pdf.

2 Franz Trieb and Hans Müller-Steinhagen, "Europe—Middle East—North Africa Cooperation for Sustainable Electricity and Water," *Sustainability Science* 2.2 (2007): 205–219. http://link.springer.com/article/10.1007%2Fs11625-007-0025-x#/page-1.

3 Franz Trieb et al., *Bringing Europe and Third Countries Closer Together through Renewable Energies—WP3 North Africa Case Study* (Project Report 2015. Intelligent Energy Europe Contract No.: IEE/11/845/SI2.616378), accessed May 15, 2017. http://better-project.net/sites/default/files/D3.5.%20Final%20Report%20on%20the%20EU-North%20Africa%20case%20study_0.pdf.

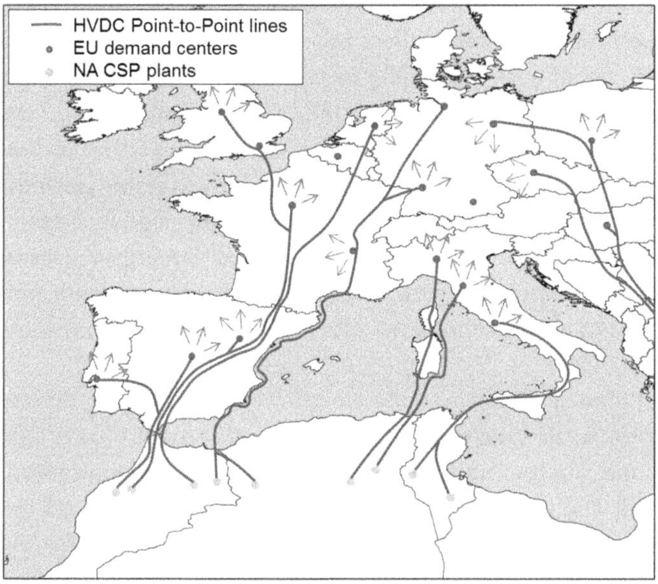

MAP 9.1 Sketch of point-to-point HVDC power lines interconnecting potential CSP plants in North Africa with demand centers in Europe.

High Voltage Direct Current (HVDC) links. A study published in 2006 showed the concept and potential impact of such venture.[4]

A point-to-point infrastructure of this type can be quantified precisely in technical and economic terms, with a specific site and size of the CSP plant in MENA, a specific center of demand into which the power is fed and distributed within the European electricity grid, and a specific HVDC power line connecting both points. The original task was to find a business case for such an infrastructure including provider region, off-taker region and transit corridors for solar electricity from CSP with solar thermal energy storage that would have a high added value as it can be produced and delivered just as required by the European demand centers.

Unfortunately, except for a first sketch in Chapter 5 of the TRANS-CSP study report,[5] the illustration of the original point-to-point interconnection concept

4 Franz Trieb et al., *Trans-Mediterranean Interconnection for Concentrating Solar Power (TRANS-CSP)* (Stuttgart, German Aerospace Center (DLR), 2006), accessed May 15, 2017. www.dlr.de/tt/trans-csp.

5 Trieb et al., *Trans-Mediterranean Interconnection for Concentrating Solar Power (TRANS-CSP)*, 129–136.

was not published in the early years, because the long-term vision of interconnecting all such power lines with all kinds of renewable energies in EUMENA was considered to be much more appealing to policymakers and the public. In fact, that vision lead to the creation of the DESERTEC Foundation and the DESERTEC Industrial Initiative in 2009, but at the same time it also put that initiative on a wrong track, as a potential, concrete, and quantifiable business case for a first point-to-point interconnection was then concealed by a sexier but less tangible vision of a futuristic infrastructure called "Supergrid". Due to that change of wording and point of view, solar electricity imports were suddenly (mis)understood as a long-term issue rather than a short-term business opportunity.[6]

While the general idea of importing solar electricity from MENA to Europe was received well overall, the long-term supergrid vision and the idea of first concrete point-to-point interconnections for dispatchable solar power were increasingly confused. The confusion became even greater when a third option, the interconnection of national electricity grids between neighboring countries, was also mixed up with the other two concepts.

The third option, interconnecting neighboring national electricity grids, has several advantages for both sides such as sharing load and power plant capacity that may lead to compensation effects and cost savings. This measure can be considered business as usual, but it does not necessarily lead to a business case for solar electricity exports from MENA to Europe. In fact, the Maghreb Region in North Africa has in the meantime been connected to and synchronised with the European electricity grid via Spain and Morocco,[7] but due to strongly growing power demand in North Africa and persistent overcapacity in Spain, electricity flows from North to South rather than in the opposite direction. The idea to increase renewable power production in North Africa until surpluses flow northwards is not resilient and does not provide a business case. Occasional electricity surplus arriving in Europe would have a relatively low value or might even be considered a disturbance and would not generate significant income for North Africa.

---

6 Florian Zickfeld et al., *DESERT Power 2050—Perspectives on a Sustainable Power System for EUMENA* (Munich: Dii, 2012), 71–72, accessed May 15, 2017. http://dii-desertenergy.org/publications/.

7 ENTSO-E—European Network of Transfer System Operators for Electricity, *Regional Investment Plan 2015—Continental South West Region* (Brussels: ENTSO-E, 2015), 18–19, accessed May 15, 2017. https://www.entsoe.eu/Documents/TYNDP%20documents/TYNDP%202016/rgips/Regional%20Investment%20Plan%202015%20-%20RG%20CSW%20-%20Final.pdf.

Regarding the supergrid, an open question was related to who would be in charge of building and operating supergrid interconnections to facilitate long-distance solar electricity trade between MENA and EU. In 2012 the DESERTEC Industrial Initiative (later called Dii GmbH) published its vision of an interconnected electricity grid in EUMENA in a comprehensive report, proposing a scenario for EUMENA with a share of about ninety per cent renewable electricity by 2050. The scenario was mainly based on variable generation from wind power and photovoltaics, and a supergrid was needed to distribute this large fluctuating power capacity anywhere in EUMENA. This virtual grid infrastructure was supposed to balance out fluctuations of supply and demand.[8]

The supergrid was modelled assuming a node in the middle of each country that was interconnected to other nodes by virtual HVDC power lines with sufficient transfer capacity. It was assumed that all electricity would freely flow through that supergrid in order to be produced and consumed anywhere in all connected countries.

As an example of the tremendous effort such an enterprise would require, the study calculated a necessary net transfer capacity (NTC) of the Spanish electricity grid of 170 gigawatt. This would be about 55 times higher than the NTC of the present grid infrastructure in Spain of about three gigawatt. Similar expansions of grid infrastructure would also be required in other countries. It was particularly interesting to see the evolution of the supergrid vision in the minds of its propagators: in the course of the Dii analysis, the supergrid had become a simplistic, completely virtual infrastructure with one virtual node at the geographic center of each country and single lines with extremely high transmission capacity interconnecting each node to its neighbors, which represents not only a much higher degree of abstraction, but also a severe change of paradigm when compared to the original concept illustrated in map 9.1.

In 2012, a doctoral thesis by Stetter revealed that introducing point-to-point CSP-HVDC interconnections according to the original TRANS-CSP study would leave existing EUMENA grid infrastructure undisturbed in spite of achieving a similarly high renewable energy share.[9] Providing only fifteen to twenty per cent dispatchable power from North African CSP plants to complement and compensate variable input from wind power and PV in Europe, and making use of flexible renewable sources in Europe such as biomass, hydro-,

8 Zickfeld et al., *DESERT Power 2050*, 71–72.

9 Daniel Stetter, "Enhancement of the REMix Energy Model—Global Renewable Energy Potentials Optimized Power Plant Siting and Scenario Validation." PhD diss., University Stuttgart and DLR, 2012), accessed May 15, 2017. https://elib.uni-stuttgart.de/handle/11682/6872?locale=en.

and geothermal power, the necessary effort for installing grid extensions and electricity storage facilities was reduced significantly. Unfortunately, the publication of the Dii supergrid vision in 2012 started a destructive public and political discussion on solar electricity imports. As a consequence, Dii GmbH abandoned this idea officially by the end of 2014. While this again created a lot of attention, the work of Stetter that shows the positive impact of potential CSP-HVDC point-to-point interconnections was only published within the scientific community and widely ignored by policy and the public.

In 2012, the European Agency for Competitiveness and Innovation (EACI) commissioned a study titled "Bringing Europe and Third Countries Closer Together through Renewable Energies—BETTER" in order to identify major barriers for renewable energy cooperation with neighbouring regions and establish how they could be overcome. In the case of North Africa, the answer was quite simple: the study recommends turning away from the abstraction of a virtual supergrid infrastructure and instead concentrating on a concrete business case of a first point-to-point HVDC line interconnecting a flexible concentrating solar power plant in North Africa with a European center of demand, as described in the TRANS-CSP study from 2006.[10]

The role of such infrastructure in providing access to remote sources of energy, the challenge related to its intercontinental size, and the required strategies for its realisation are analogous to the role, challenges, and success strategies related to conventional oil- and gas pipelines. Therefore, we call this infrastructure 'Solar Electricity Pipeline (SEP)'.

The North Africa Case Study of the BETTER Project concluded that the original concept of point-to-point interconnections should be assessed in the context of detailed feasibility studies, but has not yet been. Also, as an example, a first rough picture of this type of infrastructure interconnecting potential CSP plants in Morocco with points of high demand in Germany was presented (map 9.2),[11] the layout of the CSP plants being such that they could actually substitute the last nuclear power stations scheduled to be shut down in Germany in 2022. However, the study also concludes that the years lost to the supergrid diversion will make it difficult, if not impossible, to finalise such infrastructure in due time before that date.

---

10 Trieb et al, *Bringing Europe and Third Countries Closer Together through Renewable Energies*, 160–172.

11 Denis Hess, "Fernübertragung regelbarer Solarenergie von Nordafrika nach Mitteleuropa," (diploma diss., University of Stuttgart, 2013), accessed May 15, 2017. http://elib.dlr.de/83385/.

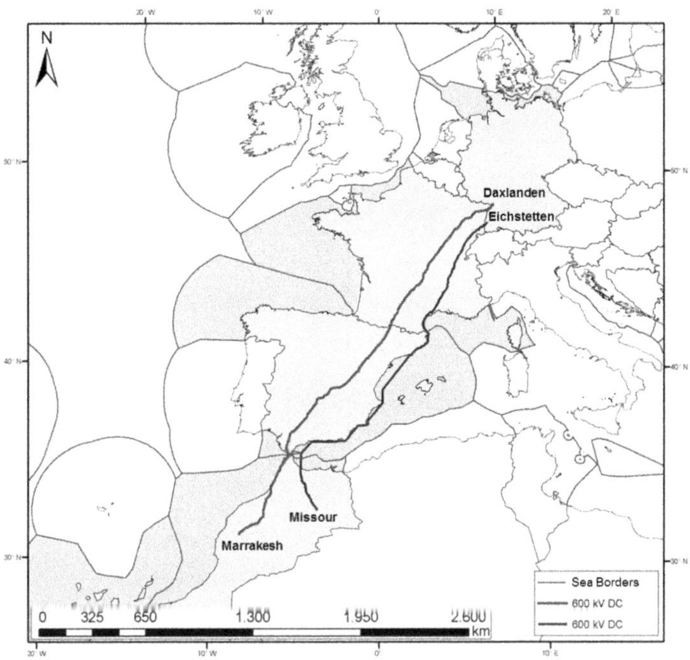

MAP 9.2 Example of two potential alternative HVDC transfer corridors from Marrakesh and Missour in Morocco to Daxlanden and Eichstetten in Germany identified by demand and supply side assessment and geographical analysis of the region.

3 Technical Background of Solar Electricity Transfers

The technical background of potential CSP-HVDC Solar Electricity Pipelines is illustrated in figure 9.1. The elements required for such infrastructure are all commercially proven technologies: a solar power station including a concentrating solar thermal collector field, high-temperature heat storage and steam cycle power block, transformers and rectifiers that allow for the conversion from low voltage to high voltage and from alternating current (AC) to direct current (DC) and vice-versa, and High-Voltage-Direct-Current (HVDC) overhead-, underground- or sea cables for efficient long-distance electricity transfer.[12] For security reasons, the maximum allowed outage capacity of single production

12 ABB, "HVDC," web page on HVDC applications, accessed May 15, 2017. http://new.abb.com/systems/hvdc.

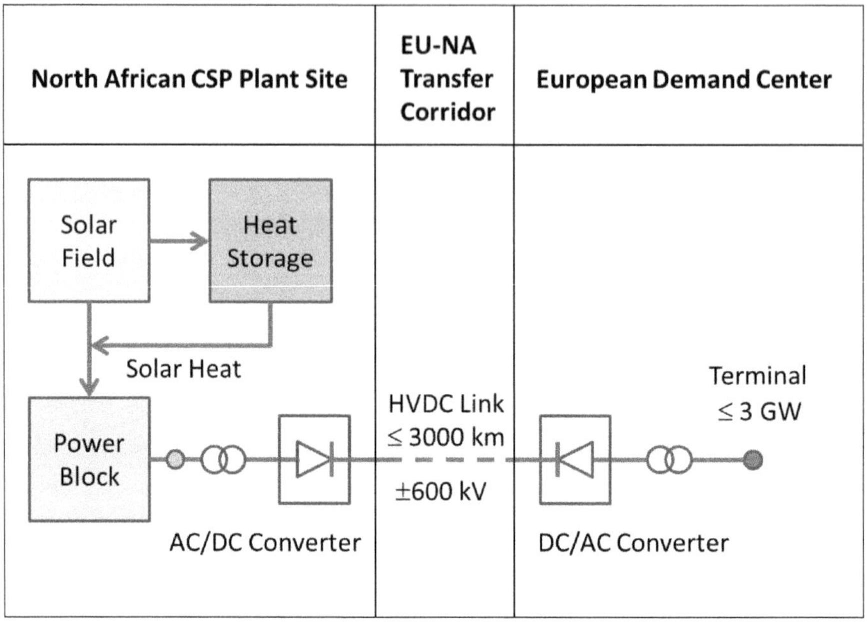

FIGURE 9.1 Components of a CSP-HVDC Solar Electricity Pipeline (SEP).

units in Europe is restricted to three gigawatt,[13] which also accounts for such interconnections.

The added value of solar electricity imports at the terminal stems from the flexibility of a CSP power plant that is capable of providing firm and flexible capacity just as needed by the load, no matter whether the sun is shining or not. The heat needed for steam and power production comes either directly from the solar field, from the heat storage, or—as a last resort—from a secondary fuel used for steam generation which may be natural gas, fuel oil, coal or—preferably—biomass. In this way, a solar electricity pipeline will provide firm capacity and power on demand just like a conventional power plant, independent from sunshine, but will consume much less or even no fossil fuel to do so.[14]

13 ENTSO-E—European Network of Transmission System Operators for Electricity, "Continental Europe Operation Handbook," accessed May 15, 2017. https://www.entsoe.eu/publications/system-operations-reports/operation-handbook/.

14 Franz Trieb, Tobias Fichter, and Massimo Moser, "Concentrating Solar Power in a Sustainable Future Electricity Mix," *Sustainability Science* 9.1 (2014): 47–60. DOI 10.1007/s11625-013-0229-1.

3.1 *Concentrating Solar Thermal Power Plants*

In contrast to wind and photovoltaic power plants, a CSP plant is capable of producing electricity on demand, not only when the wind blows or when the sun shines. This is due to the fact that a CSP plant uses a conventional steam cycle for power generation. A conventional power plant uses coal or natural gas to generate pressurised high temperature steam that drives a turbine to generate electricity. Instead of coal or gas, a CSP plant uses heat from a solar collector field for the same purpose. The size of the solar field can be extended to produce extra heat that can be stored for night-time operation or for times with little sunshine, thus allowing for steam generation and power production at any time, not only during the sunny hours of a day. In principle, a CSP plant can be operated just like a conventional power plant, but if correctly designed and placed on a good sunny site, it will not consume any or very little fossil fuel.[15]

This explains the high value of this type of power plant: it provides dispatchable power and firm capacity that can be freely adapted to the need of consumers, and at the same time it makes use of a very large and renewable source of energy. This makes CSP comparable to hydropower, biomass or geothermal power, though with much greater technical and economic potential than those resources.[16] The availability of CSP depends very much on the specific site where such plants are installed: solar energy can only be concentrated on clear days to produce high temperature heat, and the heat storage capacity is limited to about 24 hours of full load operation for technical and economic reasons. This means that the number of clear days per year directly defines the availability of solar power from a CSP plant.

Another impact on availability is related to the position of the sun in the sky: at high latitudes like in Europe, the sun is very low in the sky in winter which leads to rather unfavorable angles of incidence and lowers the efficiency of the concentrating solar collectors, while in lower latitudes like in MENA, the geometrical position of the sun towards the collectors is more favorable

---

15 Trieb, Fichter, and Moser, "Concentrating Solar Power in a Sustainable Future Electricity Mix."

16 Franz Trieb, et al., *Characterization of Solar Electricity Import Corridors from MENA to Europe—Export Potentials, Infrastructures and Costs* (Stuttgart: DLR, 2009), 79–80. http://www.dlr.de/tt/Portaldata/41/Resources/dokumente/institut/system/publications/Trieb_Presentation_SolarPaces_Berlin_09-2009-02.pdf; and Technical Note TN2.3 for REACCESS project, funded by European Commission 7th Framework Programme. www.dlr.de/tt/csp-resources.

throughout the year. This and the higher number of sunny days lead to a much better availability of power from CSP plants in MENA than in Europe (fig. 9.2).[17]

In principle it is possible to extend the technical components of such a plant, e.g. in order to avoid consuming fossil fuel for co-firing by using biomass for additional steam generation, or by using PV during the day to cover the internal parasitic electricity demand for collector tracking, pumping, and control. It is also possible to form virtual power stations that operate like one single unit including other renewable sources.

Concentrating solar power plants have been in operation worldwide since the 1980s. Recently, a large plant of this type with 160 megawatt rated power capacity has been commissioned in Morocco (fig. 9.3).[18] The global cumulative CSP capacity was about five gigawatt (5,000 megawatt) in 2015, which still is little compared to 230 gigawatt photovoltaic and 430 gigawatt wind power.[19] Different CSP technologies like parabolic trough, linear fresnel or central receiver collectors are available. Molten salt, concrete, or rocks are used for heat storage, and steam turbine (Rankine) or gas turbine (Brayton) power cycles are used for power generation.[20] Up to now, most CSP capacity has been installed in the US and Spain. Lately, China has set up an ambitious plan to install five gigawatts of additional CSP power capacity until 2020.[21]

3.2 *High Voltage Direct Current Transmission Links*

In technical terms, the installation of HVDC power lines over thousands of kilometres is challenging, but it is a proven technology. The option of including a dispatchable, but remote source of renewable energy in the national power plant portfolio by transferring electricity over long distances from a remote power plant to a local centre of demand has a fifty-year-long history on all continents except in Europe. Examples are the HVDC interconnections of the Inga and Itaipu hydropower plants in Central Africa and Brazil respectively, the HVDC lines connecting geothermal power stations on one island with centers

17 Trieb et al., *Trans-Mediterranean Interconnection for Concentrating Solar Power (TRANS-CSP)*, 55–56; and Trieb, Franz., Tobias Fichter, and Massimo Moser. "Concentrating Solar Power in a Sustainable Future Electricity Mix."

18 © MASEN.

19 REN21—Renewable Energy Policy Network of the 21st Century, "Renewables 2016 Global Status Report" (Paris: REN21, 2016), 19, accessed May 15, 2017. http://www.ren21.net/REN21Activities/GlobalStatusReport.aspx.

20 ESTELA—European Solar Thermal Electricity Association, "What is Solar Thermal Electricity?," accessed May 15, 2017. http://www.estelasolar.org/ste-means-eu-energy-security/the-ste-technologies/.

21 CSP Focus, "China Officially Targets 5GW of CSP Allocation by 2020," accessed May 15, 2017. http://www.cspfocus.cn/en/market/detail_193.htm.

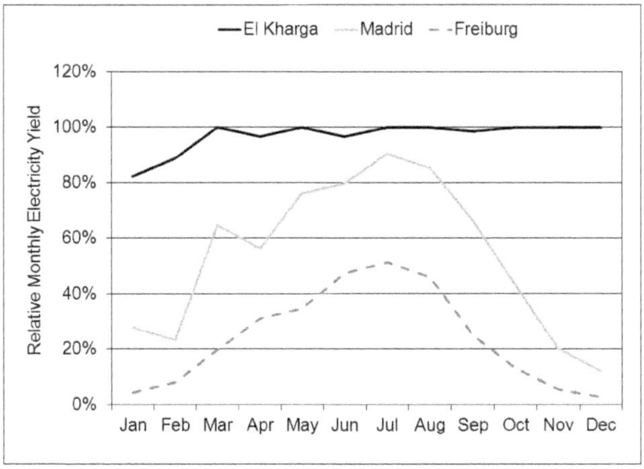

FIGURE 9.2 Modelled solar power availability of identical CSP plants with large solar field and heat storage at different sites in Egypt (El Kharga), Spain (Madrid) or Germany (Freiburg). The availability of the plant in El Kharga, Egypt is equivalent to over 7,500 full load operating hours per year, similar to the typical utilisation of nuclear plants, but at lower cost than those.

FIGURE 9.3 Steam cycle power station and solar heat storage in the center of the concentrating solar thermal collector field of NOOR 1, a CSP plant with 160 megawatt power capacity at Ouarzazate, Morocco commissioned in early 2016.

of demand on other islands in the Philippines, or the HVDC links from hydropower dams with a total capacity of almost forty gigawatt to centers of demand in China.[22]

While HVDC links in China have up to seven gigawatt capacity each, an HVDC interconnection in Europe would be limited by the maximum outage capacity a single unit is allowed to have. This capacity is limited to about three gigawatt because this is the maximum that can be compensated by the European interconnected system in case of failure.[23]

In order to provide high flexibility for power supply, HVDC interconnections in Europe would be based on Voltage Source Converter (HVDC-VSC) technology that allows for black start and for the generation of reactive power and is highly flexible for part load operation or stand-by. While maximum voltage in some HVDC links in China reaches up to 1,000 kilovolt, European HVDC-VSC links would be limited to about 600 kilovolt operating voltage.[24] Moreover, the HVDC links would use bipolar interconnections with a positive and a negative pole and a third, redundant line that helps to keep up fifty per cent of capacity in case one of the main lines breaks. HVDC links can be composed of different sections of overhead lines, underground cables or sea cables depending on local requirements such as e.g. bridging large bodies of water or environmentally sensitive areas like bird migration routes.

3.3 CSP-HVDC Overall System Performance

On average over the year, 15–17 per cent of solar energy reaching a concentrating solar collector field can be transformed into net electricity output through a CSP plant. While typical AC/DC or DC/AC converter stations lose about one per cent of power input each, typical losses of the HVDC overhead lines with an operative voltage of 600 kilovolt are in the order of three to four percent per 1,000 kilometre transfer distance. Including all losses, the overall annual efficiency of CSP-HVDC imports from Morocco to Germany over a 2,600-kilometre distance in terms of net electricity received at the European center of demand versus solar energy input to the North African plant would typically be on the order of 14. 1 per cent. This value is comparable to the overall efficiency of present grid-connected PV systems. However, the electricity output of CSP-HVDC links can be adapted to demand and provide dispatchable power just as

22 ABB, "HVDC," web page on HVDC applications, accessed May 15, 2017. http://new.abb.com/systems/hvdc.

23 ENTSO-E—European Network of Transmission System Operators for Electricity, "Continental Europe Operation Handbook."

24 Denis Hess, "Fernübertragung regelbarer Solarenergie von Nordafrika nach Mitteleuropa."

required by consumers and for grid stabilisation, which is a strong added value of this type of power generation. A techno-economic analysis of potential CSP-HVDC interconnections from MENA to Europe has been published by Trieb et al. in 2012.[25]

Typical technical and economic characteristics have been obtained from modelling a potential solar electricity pipeline from Morocco to Germany. The model results shown in Table 9.1 are based on first estimates of the technical and economic performance of such a structure.[26]

TABLE 9.1 Estimates of some techno-economic performance indicators of a solar electricity pipeline project

Power capacity:	ca. 3 GW (30 CSP plants á 100 MW)
Electricity produced:	ca. 800 TWh (20,000 GWh/year × 40 years)
Solar collector field area:	ca. 300 km²
Area of oasis incl. solar collector field	ca. 600–900 km²
Length of HVDC link:	ca. 2,600 km
Transmission losses of HVDC link:	ca. 10%
Planning and construction time:	ca. 10 years
Water consumption of the CSP plant:	ca. 10 million m³/year
Production of sea water desalination plant:	ca. 150 million m³/year
Electricity consumption of desalination plant:	ca. 100 MW, 600 GWh/year
Length of water pipeline from sea shore:	ca. 300 km
Direct employment:	ca. 5,000 persons
Operation time:	ca. 40 years
Electricity cost CIF Germany:	ca. 0.12 €/kWh (excl. taxes)
Investment incl. HVDC transfer and water:	ca. 30 billion €
Investment Desalination plant:	ca. 2 billion €
Annual turnover:	ca. 2.4 billion €/year
Compensation payment for land use:	ca. 50 million €/year

25 Franz Trieb et al., "Solar Electricity Imports from the Middle East and North Africa to Europe," *Energy Policy* 42 (2012): 341–353.

26 Hess, "Fernübertragung regelbarer Solarenergie von Nordafrika nach Mitteleuropa,"; Trieb et al, *Bringing Europe and Third Countries Closer Together through Renewable Energies*, 192–218.

4 Economic Performance and Market Perspectives

The BETTER project provides a first rough estimate of the economic performance of such a structure with a 1.5 gigawatt net capacity and roughly ten billion kWh annual electricity production equivalent to about 6,500 full load operating hours—including a CSP plant with thermal energy storage, an HVDC link across 2,600 kilometres from Morocco to Germany according to fig. 9.1, and the necessary infrastructure for seawater desalination and transport to cover the plant's freshwater demand, revealing a CAPEX of about 15 billion euros.[27] The specific investment for such infrastructure of around 10,000€ per kilowatt would be lower than e.g. that for the nuclear power plant Hinkley Point C planned recently in the UK with equivalent capacity and power output.[28] With a weighted average cost of capital (WACC) of about five per cent per year and an OPEX of 0.045€/kWh including compensation payments for the land used by the total infrastructure in all countries involved, a total average electricity cost of around 0.12€/kWh for first projects around 2025 and 0.07 to 0.11€/kWh for later ones could be achieved.[29] This cost level looks rather high compared e.g. to the cost of electricity from wind power or coal, but this is natural, as CSP-HVDC imports represent a higher quality, being at the same time dispatchable as power from a coal plant but also renewable and emission-free like wind power. As this quality is hard to find in Europe, we have a promising business case for solar electricity imports here (tab. 9.2).[30]

From the European off-taker point of view, the question is not so much whether domestic or imported wind or photovolataic power would be cheaper or more expensive than CSP imports. The question is whether there is an added value for CSP imports in Europe, just like there obviously is an added value for fuel oil and natural gas imports because these are highly dispatchable, valuable forms of energy. On the other hand, from the MENA point of view, there is no other form of energy except oil and gas with such added value that may find an interested off-taker when exported to Europe.

Large-scale international investments suffer from considerable risk perception, especially when we are talking about regions like North Africa and

27 Ibid; Trieb, et al., "Solar Electricity Imports from the Middle East and North Africa to Europe,".

28 European Commission, "State Aid: Commission Concludes Modified UK Measures for Hinkley Point Nuclear Power Plant are Compatible with EU Rules," accessed May 15, 2017. http://europa.eu/rapid/press-release_IP-14-1093_en.htm.

29 Trieb et al., "Solar Electricity Imports from the Middle East and North Africa to Europe".

30 Trieb et al, *Bringing Europe and Third Countries Closer Together through Renewable Energies*, 160–172.

TABLE 9.2 Quality and availability of renewable energy and fossil fuel resources in Europe and North Africa and opportunities for a potential business case for exports from North Africa to Europe

Source	Type	Quality	North Africa	Europe	Business Case
Biomass	renewable	dispatchable	scarce	available	no
Hydropower	renewable	limited dispatchable	scarce	available	no
Geothermal	renewable	limited dispatchable	scarce	scarce	no
CSP	**renewable**	**dispatchable**	**plentiful**	**scarce**	**yes**
PV	renewable	intermittent	plentiful	plentiful	no
Wind Power	renewable	intermittent	plentiful	plentiful	no
Coal / Lignite	fossil	limited dispatchable	not available	available	no
Natural Gas	**fossil**	**dispatchable**	**available**	**scarce**	**yes**
Oil Products	**fossil**	**dispatchable**	**available**	**scarce**	**yes**
Nuclear	fossil	limited dispatchable	not available	available	no

the Middle East. High risk always means high interest rates for banks and private equity investors, which translates into high capital cost for such projects. Especially renewable energy projects—that are primarily based on capital investment—risk becoming unprofitable if the expected capital cost is high.

Mitigating the risk of such projects is therefore a very effective measure to reduce the cost for consumers without the need for subsidies like soft loans or grants. One of the best measures is a power purchase agreement between credible partners that can be backed by international guarantees.[31] Making solar electricity pipelines very good (AAA) investment opportunities will certainly trigger real world projects in the short and medium term. Unfortunately, due to DESERTEC's failure and the related misconceptions, political support for such enterprises is currently at a historical minimum, as though all possible support was already spent on the DESERTEC initiative. A revival of support can only

31 Oliver Waissbein, Yannik Glemarec, Hande Bayractar and Tobias S. Schmidt, *De-risking Renewable Energy Investment—A Framework to Support Policymakers in Selecting Public Instruments to Promote Renewable Energy Investments in Developing Countries* (New York, United Nations Development Programme (UNDP), 2013), accessed May 15, 2017. www.undp.org/drei.

be expected if the original concept of concrete point-to-point solar electricity pipelines is finally understood and pursued by serious entrepreneurs.

5 Political Framework

The possibility of producing renewable energy in the Middle East and North Africa region (MENA) and exporting it to the European Union (EU) has received a lot of interest among scientists, industry leaders, and policymakers alike. However, despite the promising outcomes of such cooperation, numerous and various barriers remain in place preventing this concept from becoming reality. Among other pre-requisites such as having sufficient financial resources to undertake investment in such power plants, designing a sustainable energy system for the Euro-Mediterranean region requires strong political will as well as international collaboration and coordination among various actors. To date, various initiatives, policies, and regulations have been implemented on both sides of the Mediterranean to promote the large-scale deployment of renewable energy and also to allow a cross-border renewable electricity exchange between North Africa and Europe. Success, however, has been limited.[32]

One of the instruments aimed at fostering such cooperation was introduced by the Directive 29/2009/EC of the European Commission, also known as the Renewable Energy Directive—RED.[33] The RED is part of the EU Climate and Energy package that came into force in June 2009 and sets mandatory national renewable energy targets for EU member states (MS) in order to achieve a twenty per cent renewable energy share of the EU's gross final energy consumption by 2020. In order to grant member states more flexibility as well as to reach the overall EU target in a cost-effective manner, the RED allows member states to partially fulfill their national targets in other countries with large renewable energy potential and low generation costs. Such cooperation can take place within Europe (Articles 6, 7 and 11) but also with third countries (Article 9), in that latter case called the 4th cooperation mechanism.

---

32 Johann Lilliestam et al., "Understanding the Absence of Renewable Electricity Imports to the European Union," *International Journal of Energy Sector Management* 10.3 (2016): 291–311, accessed May 15, 2017. DOI 10.1108/IJESM-10-2014-0002.

33 European Commission, "Directive 2009/28/EC" of the European Parliament and of the Council of 23 April 2009 on the promotion of the use of energy from renewable sources and amending and subsequently repealing Directives 2001/77/EC and 2003/30/EC, L 140/16 particularly Art.9 related to Joint projects between member states and third countries, accessed May 15, 2017. https://ec.europa.eu/energy/en/topics/renewable-energy/renewable-energy-directive.

In order to implement Article 9, projects must meet various conditions. Among others, (i) the project must have been completed after 2009, (ii) electricity must be physically transferred from plants installed in the origin non-EU countries and consumed in the destination EU countries, (iii) the amount of electricity exported must be tied to the allocated interconnection capacity by all responsible transmission system operators involved, (iv) electricity produced should not have received support from a third country other than investment aid granted to the power plant and, finally, (v) the amount of electricity nominated for export and the production of electricity from renewable energy sources are referring to the same period of time.

The idea of solar electricity transfers found its way into the concept of the Mediterranean Solar Plan (MSP) of the Union for the Mediterranean and got strong forward momentum with the DESERTEC Foundation and the DESERTEC Industrial Initiative (later called Dii GmbH) in 2009.[34] Those were followed by several other initiatives like Transgreen and later MedGrid in 2010.[35]

On the regulatory side, some North African countries also took important steps in that direction such as through the Moroccan Renewable Energy Law of 2009 that includes the option of renewable electricity exports to Europe, also considering the construction of the necessary transmission lines physically separate and independent from the national electricity supply and grid infrastructure.[36] Similarly, in 2011, Algeria launched its Renewable Energy and

34 Matthias Ruchser, *Desertec or the Mediterranean Solar Plan—Whose Sun is Shining Brighter?* (Bonn: Deutsches Institut für Entwicklungspolitik (DIE), 2010), https://www.die-gdi.de/uploads/media/Column_Ruchser.02.08.2010.pdf; Union for the Mediterranean (UfM), Secretariat of the Union for the Mediterranean, "The UfM: An Action-Driven Organisation with a Common Ambition," accessed May 15, 2017. http://ufmsecretariat.org/wp-content/uploads/2017/10/UfM-Roadmap-for-action-2017.pdf; DESERTEC Foundation, "Archive of Press Releases Since 2008," accessed May 15, 2017. http://www.desertec.org/press; Dii 2017, "Great Idea: Power from the Desert," accessed May 15, 2017. http://dii-desertenergy.org/.

35 Stijn Cole et al., "A European Supergrid: Present State and Future Challenges" (17th Power Systems Computation Conference, Stockholm, August 22–26, 2011), accessed May 15, 2017. https://www.pscc-central.org/uploads/tx_ethpublications/sp2.pdf; Observatoire Mediterraneén de l'Energie (OME), "Mediterranean Energy Perspectives 2011," in *L'Energie*, ed. O,M.d. (Paris: O.M.d., 2011), accessed May 15, 2017. http://www.ome.org/publications/mep-2011/; European Commission, "Commission Welcomes Desertec and Medgrid Cooperation on Solar Energy in North Africa and the Middle East," last modified February 19, 2018, accessed May 15, 2017. http://europa.eu/rapid/press-release_IP-11-1448_en.htm.

36 Ministère de l'Energie, des Mines, de l'Eau et de l'Environnement (MEMEE), *Loi 13–09 relative aux énergies renouvelables* (Rabat: MEMEE, 2010), accessed May 15. 2017.http://www.mem.gov.ma/SiteAssets/PdfDocumentation/LoiEnergiesRenouvelables.pdf.

Efficiency Program that includes ten gigawatt of renewable power capacity scheduled for export to Europe.[37]

Summarising, the great renewable potential in the North African region, potential business opportunities, a favourable regulatory environment, and the interest of governments on both sides of the Mediterranean led to great expectations of the attractiveness and feasibility of exporting renewable electricity to the European continent. However, despite great expectations that were created around Article 9 of the RED, not a single project has seen the light as of today and this remains unlikely to happen in the short term. There are various factors that can partially explain this fact such as the current lack of demand for cross-border RES imports from third countries within the EU, the lack of grid infrastructure including high voltage transmission lines across the Mediterranean Sea, and the regulatory and legislative gaps related to the harmonisation of the EU and Southern Mediterranean energy markets.

While the abovementioned hurdles could have been overcome in principle, the long-term vision of a future EUMENA Supergrid, as explained before, avoided the need to make a concrete short-term business case that would have tapped the added value of dispatchable solar power for the benefit of both sides of the Mediterranean.

6 Pilot Study Required for a First Solar Electricity Pipeline

Since 2009, several initiatives proposed to establish 'pilot' projects to demonstrate the 4th cooperation mechanism on a smaller scale. However, just as in the case of natural gas pipeline projects, a smaller scale demonstration prior to a real project is impossible in the case of a solar electricity pipeline project because the related effort would be almost the same as that required for a full size project, but would not be justified by the smaller electricity output. This and the erroneous concept that formed the basis of those considerations up to now has hindered the realisation of concrete projects under the 4th cooperation mechanism although, as has been shown in the BETTER report, flexible renewable energy imports will most probably be needed in Europe shortly after 2020 and planning should be started as soon as possible in order to open up such options.[38] Therefore the European Commission should consider releasing

37 Ministère de l'Énergie et des Mines (MEM), "Renewable Energy and Energy Efficiency Program," accessed May 15, 2017. www.mem-algeria.org/francais/uploads/enr/Programme_ENR_et_efficacite_energetique_en.pdf.

38 Trieb et al, *Bringing Europe and Third Countries Closer Together through Renewable Energies*; Franz Trieb et al., "Rescuing the Concept of Solar Electricity Transfers from

TABLE 9.3 Topics of a tentative CSP-HVDC pilot study

GEOGRAPHICAL ASSESSMENT
- Demand side analysis of potential off taker regions in EU for dispatchable solar electricity imports.
- CSP production site screening of potential supplier regions in NA.
- Geographic analysis of potential HVDC corridors in selected transit regions.

TECHNICAL AND FINANCIAL ASSESSMENT
- Technical design and performance study related to the prior demand side analysis.
- Economic and financial engineering study including de-risking opportunities, compensation payments and citizen participation.
- Assessment of the necessary logistics and time schedule.

IMPACT ASSESSMENT
- Environmental impact study.
- Socio-economic impact study.
- Social acceptance and citizen involvement study including a concept for a prior public visualization of the infrastructure and for citizen dialogue and participation.
- Analysis of the replication potential of such projects.

POLITICAL AND REGULATORY FRAMEWORK ASSESSMENT
- Analysis of existing and missing political framework requirements including the role of authorities at different levels.
- Analysis of existing and missing regulatory framework requirements including the role and identification of permitting authorities.

a call for a CSP-HVDC Pilot Study that may contain, among others, the topics listed in table 9.3 to be assessed in reasonable detail.

Such a study could be commissioned directly by the European Commission or included in the HORIZON 2020 Call.[39] It would help the commission to evaluate the potential and assess the importance of such infrastructure measures for the EU's future energy infrastructure.

North Africa to Europe," *International Journal of Energy Sector Management* 10.3 (2016): 448–473, accessed May 15, 2017. www.emeraldinsight.com/1750-6220.htm.

39 European Commission, "HORIZON 2020—The EU Framework Programme for Research and Innovation—Call for projects 2014–2020," accessed May 15, 2017. http://ec.europa.eu/programmes/horizon2020/en.

In addition, a recently published "Joint Declaration on the Establishment of a Roadmap for Sustainable Electricity Trade between Morocco and the European Internal Energy Market" between the Federal Republic of Germany, the French Republic, the Kingdom of Spain, the Portuguese Republic and the Kingdom of Morocco has opened a possible avenue for such activities after the COP 22 in Marrakech, Morocco.[40]

However, HORIZON 2020 and the "Roadmap for Sustainable Electricity Trade (SET)" do not include the concept of solar electricity pipelines in their portfolio of possible options.

7 Social Context and Acceptance

In the public and political discourse about DESERTEC after 2009, doubts were raised about the security of supply related to power from MENA, the vulnerability of such infrastructure, to the environmental protection of desert areas, water consumption in the desert, and the colonial-style exploitation of natural resources in developing countries that eventually would not profit from such ventures.[41] Although there are satisfactory answers to these doubts, the populist spread of distrust against a solar electricity partnership between MENA and Europe was much more effective than any of those answers.

A very popular argument refers to neo-colonialism, stating that solar energy from the Sahara belongs to Africa, not to Europe. This is very true, but the quality and value of solar energy is very different to e.g. that of commodities like natural gas or mineral oil. Exploiting the latter for export will certainly deprive Africa of the opportunity to use them for its own development as the resources will be gone after being exported. Exploiting solar energy for export, on the other hand, is certainly a way for Africa to tap that wealth. In other words: the only way to deprive (North) Africa of its wealth of solar energy is not to make use of it for export. This is a fundamental difference between solar energy and limited commodities such as natural gas or crude oil. Using solar

40 European Commission, "Joint Declaration on the Establishment of a Roadmap for Sustainable Electricity Trade between Morocco and the European Internal Energy Market between The Federal Republic of Germany, The French Republic, The Kingdom of Spain, The Portuguese Republic, and The Kingdom of Morocco," executed at Marrakech during COP22 on the 17th November 2016, accessed May 15, 2017. https://ec.europa.eu/energy/sites/ener/files/documents/2016_11_13_set_roadmap_joint_declaration-vf.pdf.

41 Euractive, "First Steps to Bring Saharan Solar to Europe," last modified February 22, 2010, accessed May 15, 2017. http://www.euractiv.com/section/energy/news/first-steps-to-bring-saharan-solar-to-europe/.

energy for export will leave more than enough potential for domestically used solar power plants[42] that are already being installed in many places.[43] Creating added value and selling dispatchable solar power to Europe will create additional income for the North African countries. This income can then help e.g. to finance their own solar energy supply, by no means depriving them of the option to do so, but on the contrary providing finance and intensifying transfers of solar industry and economic know-how.

Another very popular argument against EUMENA solar energy partnerships postulates a "dangerous" dependence of Europe on "insecure" North African states. This argument is particularly interesting in view of massive European natural gas imports from Russia, a country with an OECD country risk rating of 4 which is involved in armed conflicts in Ukraine and Syria, while Morocco, a major candidate for solar exports to Europe, has a rating of 3 and, needless to say, no armed conflict with European or Mediterranean countries.[44] Even if both countries (Morocco and Russia) had the same risk rating, adding another energy supplier to the European portfolio would, by simple logic, always reduce, but never increase European energy supply risk. Further answers to the risk myth have been given by Lilliestam and Ellenbeck.[45]

Regarding the security of grid stability, the European Network of Transfer System Operators for Electricity (ENTSO-E) has a rule called n-1 Criterion which states that no single unit can be larger than the maximum capacity that can be compensated within a TSO's control area in case of an emergency outage.[46] This rule is applied to power plants and all other equipment and also applies to potential future CSP-HVDC solar electricity pipelines. The approximate size of this maximum capacity limit is about three gigawatts and depends on the position of the respective device's interconnection within the grid. This limits the size of a single solar electricity pipeline to a net import capacity of about three gigawatts.

---

42 Trieb et al., *Characterization of Solar Electricity Import Corridors from MENA to Europe—Export Potentials, Infrastructures and Costs.*

43 Moroccan Agency for Solar Energy (MASEN), *Endless Power for Progress* (Rabat, MASEN, 2016), accessed May 15, 2017. http://www.masen.ma/media/uploads/documents/MASEN_Brochure_instit_Anglais_finale_FPeQ8Wd.pdf.

44 Organisation for Economic Co-Operation and Development (OECD), "OECD Country Risk Classification," last modified December 21, 2017, accessed May 21, 2018. http://www.oecd.org/tad/xcred/crc.htm.

45 Johann Lilliestam and Saskia Ellenbeck, "Energy Security and Renewable Electricity Trade—Will Desertec Make Europe Vulnerable to the 'Energy Weapon'?," *Energy Policy* 39.6 (2011): 3380–3391.

46 ENTSO-E—European Network of Transmission System Operators for Electricity, "Continental Europe Operation Handbook."

A common argument against operating steam turbines in desert regions is the lack of water for cooling the condenser of the power plants. This is a very good argument, as there is usually little water available in the desert, and if so, there are better and more important uses for it than evaporating it for cooling a power cycle.

However, many desert regions actually have steam cycle power plants running, so the question is, how can they do so? The answer again is very simple: they use air-cooled condensers.[47] Air cooled condensers use air to cool the engine, just like most automobiles do. Some plants use fans to move air through the condensing heat exchangers, some, like the Heller cooling towers, use natural convection in a large chimney. Both concepts work well in many arid regions and have been in use world-wide for many years in places where water for evaporation is scarce.

One more point of criticism is that air-cooling will reduce the efficiency of power plants compared to evaporation cooling. This is true, but only critical for plants that burn fossil fuel for power generation, because in that case more fuel will be needed per kilowatt-hour of electricity produced. When using air-cooling in a concentrating solar power plant, its design efficiency will certainly be lowest on hot days—usually days with a lot of sunshine and excess heat from the solar collectors—but highest on cold days in winter (and at night), when solar energy is scarce. So for solar steam cycle power plants, air-cooled condensers will reduce surplus and curtailment of solar heat rather than reduce overall efficiency, the impact of dry-cooling being just a question of proper design and operation.[48]

Nevertheless, it is true that large scale CSP plants in the middle of the desert will require a lot of water. As an example, a three-gigawatt plant producing about 20,000 gigawatt-hours of electricity per year will consume about ten million cubic meters of fresh water per year for operation and maintenance, cleaning of the concentrating solar collectors, and drinking water for the staff. As this water should not be taken from local sources, each plant must reserve about one to two per cent of its energy output for seawater desalination and for pumping about 100–150 million cubic meters per year from the sea shore to the plant site in the desert (tab 9.1). This way, each solar electricity pipeline will export solar energy from the desert to populated places in Europe and at

47 Nathan Bracken et al., *Concentrating Solar Power and Water Issues in the U.S. Southwest, Technical Report, NREL/TP-6A50-61376* (Denver: The Joint Institute for Strategic Energy Analysis, 2015.

48 Massimo Moser, "Techno-economic Analysis of Enhanced Dry Cooling for CSP," *Energy Procedia* 49 (2014): 1177–1186, accessed June 12, 2020. http://www.sciencedirect.com/science/article/pii/S1876610214005815.

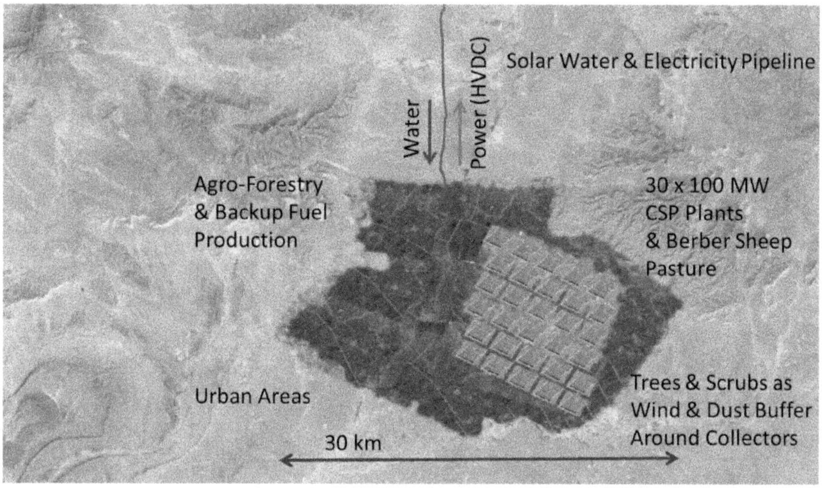

FIGURE 9.4 Potential oasis formed at the southern end of a three gigawatt solar electricity pipeline from MENA to Europe including agro-forestry and related communal services. Artist view created with Google Earth.

the same time import freshwater to the plant site in the desert in order to feed the needs of the plant and the surrounding communities. Seawater desalination can not only be combined with concentrating solar power plants built for export, but also with those serving the national electricity supply.[49]

An important aspect of high quality of living and working in the desert—we are talking about approximately 5,000 direct employees in a three-gigawatt CSP plant plus indirect services needed—is shade from trees and all kinds of plants that will need water for irrigation. A side-effect of the agro-forestry that can be started when importing sufficient additional amounts of freshwater is the local availability of food, labour, and manifold opportunities for local income. As an example, drip water from cleaning the collectors, together with shade from the collectors, will enhance the growth of grass and small vegetation all over the solar collector field. That growth must be controlled because it presents a fire hazard. A herd of sheep may effectively keep the grass and plants short enough and at the same time use the solar collector field as pasture, providing labour and income for locals.

49 Massimo Moser, Franz Trieb, and Tobias Fichter, "Potential of Concentrating Solar Power Plants for the Combined Production of Water and Electricity in MENA Countries," *Journal of Sustainable Development of Energy, Water and Environment Systems* 1.2 (2013): 122–140, accessed May 15, 2017. DOI: http://dx.doi.org/10.13044/j.sdewes.2013.01.0009.

Scrubs and trees separating the collector fields of each power plant unit will reduce the wind and dust load on the collectors and provide additional biomass for backup steam generation for the few occasions when there is not enough solar energy available over several days in the year, at the same time creating additional local labour opportunities. Depending on the local annual rainfall levels that will complement freshwater from seawater desalination, an oasis measuring about six hundred to nine hundred square kilometres can be created around each solar electricity pipeline project (fig. 9.4). This is particularly interesting for reactivating former agricultural areas that have been lost to desertification.

Host countries may (and should) participate in the financing of such infrastructure depending on their readiness to invest and their interest in sharing the investment. In addition to and independent from that, they will receive regular compensation payments from each project that will be quantified in accordance with the amount of land used for the power plants and for the power transmission lines. For a three-gigawatt unit, this payment would be on the order of fifty million euro per year for the total lifetime of forty years scheduled for such plants.[50]

On such a basis, agricultural, urban, and eventually also touristic areas could be developed in the vicinity of such plants, and strong economic zones could emerge in many desert regions throughout the Sahara. Another aspect is education and training, opening fascinating opportunities in very innovative as well as traditional fields ranging from energy and water engineering to desert agro-forestry and urban planning that will provide useful knowledge for students not only from the host countries, but from around the world.

Recently, the German Federal Ministry for Economic Cooperation and Development (BMZ) published the idea for an initiative called "Marshall Plan with Africa". Although the German Section of the Club of Rome and the Global Marshall Plan Initiative proposed including the concept of solar electricity exports in that initiative[51] it was unfortunately not taken into consideration in the paper released by the ministry in January 2017.[52]

---

50 Denis Hess, "Fernübertragung regelbarer Solarenergie von Nordafrika nach Mitteleuropa."

51 Bert Beyers et al., *Migration, Nachhaltigkeit und ein Marshall Plan mit Afrika—Denkschrift für die Bundesregierung* (Ulm: Club of Rome and Senat der Wirtschaft, 2017), accessed May 2017. http://www.faw-neu-ulm.de/portfolio-item/denkschrift-bundesregierung.

52 Bundesministerium für wirtschaftliche Zusammenarbeit und Entwicklung (BMZ), *Afrika und Europa—Neue Partnerschaft für Entwicklung, Frieden und Zukunft—Eckpunkte für einen Marshallplan mit Afrika* (Berlin: BMZ, 2017), accessed March 25, 2017. http://www.bmz.de/de/laender_regionen/marshallplan_mit_afrika/einleitung/index.jsp.

8 Conclusions

In spite of the failure of DESERTEC, the concept of solar electricity exports from MENA to Europe has not lost any relevance for sustainable development in both regions. On the contrary, affordable solar electricity that can be delivered on demand is key to the European "Energiewende" as it provides the same quality as conventional power plants without producing any emissions. This high added value for Europe is the basis for concrete business cases, power purchase agreements, and real world projects in the short-, medium- and long-term future.

The added value of such infrastructure for MENA is not less as it is not limited to the monetary value of such exports. The creation of new economic zones in the desert is crucial for a growing population in a severely limited natural environment. The availability of energy leads to the availability of water, food, labour, and income at the solar production sites, creating multiple opportunities for jobs, training, tourism, and living. Synergies of technical innovation and traditional agro-forestry will create new and fascinating fields of employment in an environment that otherwise would be lost to the desert and left unused.

The major challenge is to overcome the misconceptions related to DESERTEC's failure. As long as politicians and the public do not recognise the role and the potential benefits of point-to-point interconnections for dispatchable solar power, the idea of EUMENA solar electricity imports will remain on hold.

CHAPTER 10

Connecting Religion and the Environment in Islamic Education: Raising Awareness, Exploring Perspectives, and Defining Best Practice

Ursula Kowanda-Yassin

Abstract

The following chapter will investigate the connection between two educational topics: environmental protection and Islamic education. What are the reasons for addressing religion and ecology together though other important issues no doubt exist within each respective field? When did the discourse on Islamic ethics and the environment begin at an intellectual level? How have predominantly Muslim communities reacted to environmental challenges? Such questions will be discussed and followed up with suggestions around Islamic ecological education here. This chapter outlines possible approaches to environmental education for Muslim students, seeking ways of encouraging students to adopt an environmentally friendly lifestyle. A number of role-models will be discussed, including activists who have already incorporated religion into their ecological projects, highlighting the Islamic perspective on ecology within traditional Islamic sources. There will also be food for thought on possible educational approaches and goals for environmental improvement.

1 Introduction

Ecology is, without a doubt, one of the primary issues that will concern future generations. With this in mind, developing appropriate strategies and learning goals for students and incorporating ecology into religious education is highly recommended. Religion is part of a believer's identity and the belief in God can act as an amplifier for an awareness of the treatment of our natural environment. For this reason, it makes sense to examine environmentally friendly principles.

The following chapter will investigate the combination of two educational topics: the protection of the environment, and Islamic education. The chapter is divided into four main thematic sections. Part one shows representative publications by researchers on Islam and the environment. Part two and three

 | DOI:10.1163/9789004444973_012

contain the primary focus of this article: part two highlights the key reasons why looking at Islam and ecology together makes sense. It describes ecological imperatives and outlines how environmental awareness, social initiatives, and the development of identity can be fostered. Part three presents opportunities that can be leveraged to put an ecological lifestyle into practice, e.g. through experiential learning or by making theoretical facts comprehensible. Part four describes approaches to Islamic ecology from a theological, philosophical, social, and spiritual perspective.

2 Environmental Issues within Islamic Studies—A Short Introduction

The increasing worldwide concern over sustainability and the many efforts towards long-term environmental improvement have, in recent decades, urged Muslims to start searching for answers within their own religion.[1] Principles for the care of creation are defined and highlighted in it, as recent research on Islam and ecology has shown. Publications by Mawil Izzi Dien,[2] Harfiyah Abdel-Haleem,[3] Othman Llewellyn,[4] Fazlun Khalid,[5] Agwan A.R.[6] and others focus on theological research and the quest for Islamic principles on Islam and ecology. Richard C. Foltz has published various books on world views, animals, and environmentalism.[7] Ibrahim Özdemir studied Islamic ethics with a specific focus on the environment and has written several articles on the subject, including an article about the situation in Turkey.[8] Anne-Marieke Schwencke has studied Muslim ecological awareness.[9] Books on popular science such

1 Ibrahim Özdemir, "Turkey," in *Environmentalism Environmentalism in the Muslim World*, ed. Richard C. Foltz (New York: Nova Science Publishers, Inc., 2005), 29–32.

2 Mawil Izzi Dien, *The Environmental Dimensions of Islam* (Trobridge, Wilshire: Redwood Books, 2000).

3 Harfiyah Abdel-Haleem, *Islam and the Environment* (London: Ta-Ha-Publishers Ltd., 1998).

4 Mohamed Al Sayyed Al Glenid et al., *Environmental Protection in Islam* (Cambridge: World Conservation Union, 1994).

5 Fazlun Khalid and Joanne O'Brien, *Islam and Ecology* (London: World Wide Fund for Nature, 1992).

6 A.R. Agwan, ed., *Islam and the Environment* (New Delhi: Institute of Objective Studies, 1997).

7 Richard C. Foltz, ed., *Environmentalism in the Muslim World* (New York: Nova Science Publishers, Inc., 2005).

8 Özdemir, "Turkey," 17–36.

9 Anne-Marieke Schwencke, *Globalized Eco-Islam. A Survey of Global Islamic Environmentalism* (Leiden: Leiden Institute for Religious Studies (LIRS), Leiden University, 2012).

as *Green Deen. What Islam Teaches about Protecting the Planet*[10] written by Ibrahim Abdulmatin, which suggests practical solutions, or *Öko-Dschihad Öko-Dschihad. Der grüne Islam—Beginn einer globalen Umweltbewegung*[11] by Ursula Kowanda-Yassin, which explores global Muslim environmental commitment have been published.

Muslim-majority countries are currently paying greater attention to climate change, overpopulation, and resource scarcity, although the seriousness of the global ecological situation is often only partly acknowledged.[12] The Turkish ministry of the environment, for example, only has a small budget and environmentalists have suggested that the reason for this is that the state "doesn't consider environmental protection a high priority."[13]

At educational institutions such as schools and universities, ecological topics are of growing interest to students and teachers of Islamic pedagogy. Students are scrutinising topics such as animal welfare, Islamic ethics, and ecological education for their MA theses.[14] Asma El-Maaroufi-Ulzheimer, a German researcher at the University of Münster, is investigating animal ethics and Islam. She is also an activist and runs her own blog "Greenukum—Your Green" which she publishes with her husband under the pen name 'Asma El'.[15]

A study by Mark Bryant and Sophie Gilliat-Ray investigates the question: "Are British Muslims 'green'?"[16] and Bouchrah Chrebah examines the environmental awareness and behaviour of German and Turkish students.[17]

---

10 Ibrahim Abdul-Matin, *Green Deen. What Islam Teaches about Protecting the Planet* (Leicestershire: Kube Publishing, 2012).

11 Ursula Kowanda-Yassin, *Öko-Dschihad. Der grüne Islam—Beginn einer globalen Umweltbewegung* (Salzburg, Wien: Residenz-Verlag, 2018).

12 Foltz, *Environmentalism*. This publication includes articles on the countries Iran (Richard C. Foltz), Turkey (Ibrahim Özdemir), Egypt (Safei-Eldin A.Hamed), Pakistan (Ali Raza Rizvi) and Nigeria (Ali Ahmad) and on activists voices (Fazlun M. Khaled, Ayman Ahwal).

13 Özdemir, "Turkey," 22–23.

14 Here are some examples: Kübra Mentes (Germany) compares the two German organisations, Hima e.V. and NourEnergy, their activities and their motivation. Zarife Yavuz Bilir (Austria) wrote her thesis on the subject of "Supporting Muslim Pupils to Develop Environmental Awareness," and Fatima El-Meeyuf (Great Britain) wrote her thesis on "The Non-Engagement of Young Muslims in the Eco-Movement."

15 Kowanda-Yassin, *Öko-Dschihad*, 144–146.

16 Mark Bryant and Sophie Gilliat-Ray, "Are British Muslims 'green'? An Overwiew of Enviromental Activism among Muslims in Britain," *Journal for the Study of Religion, Nature and Culture* 5.3. (2013): 284–306.

17 Bouchrah Chrebah, *Umweltbewusstsein und Umweltverhalten—Ein Vergleich von deutschen und syrischen Studierenden* (PhD.diss., Oldenburg: Fakultät für Mathematik und Naturwissenschaften der Carl von Ossietzky Universität Oldenburg, 2009).

The beginning of this increased interest in Islam and ecology can be traced back to a discussion on religion, anthropocentrism, and the role of religion in the exploitation of the earth which was started by Lynn White Junior at the end of the 1960s.[18] In ecological discourse, religions are sometimes held responsible for man's frequently ruthless exploitation of natural resources.[19] In ecological discourse, religions are sometimes held responsible for man's frequently ruthless handling of natural resources.[20] Qur'ānic Verses such as verse 65 of the *Sūrat al-Ḥāǧǧ* [The Pilgrimage] express that mankind has nature at its disposal: *'A-lam tara 'anna Allāha saḫḫaralakum mā fī-l-'arḍi* [...] [Have you not seen how God has subjected to you all that is in the earth].[21]

One of the results of this discussion was that the issue of Islam and anthropocentrism gained traction. In recent decades, a discussion has begun amongst religious scholars on how man's role on earth is to be understood. In contemporary Islamic interpretations, man is seen primarily as the custodian of creation, and as a thinking individual is responsible for treating everything in nature with care.[22]

3 Why Combine Religious Education with Environmental Issues? Ecological Challenges, Muslim Awareness, Religious Theories, and Pedagogical Approaches

3.1 *Ecological Necessities*

Predominantly Muslim societies are facing serious ecological problems. Iran has "the greatest degree of biodiversity in Southwestern Asia",[23] but "virtually

18 Lynn jr. White, "Die historischen Ursachen unserer ökologischen Krise," in *Gefährdete Zukunft. Prognosen angolamerikanischer Wissenschaftler*, ed. Michael Lohmann (München, Wien: Hanser, 1970).

19 Seyyed H. Nasr, *Man and Nature. The Spiritual Crisis in Modern Man* (ABC International Group Inc., 1997).

20 Ursula Kowanda-Yassin, *Mensch und Naturverständnis in den Grundlagen des sunnitischen Islam. Ein Beitrag zum aktuellen Umweltdiskurs* (Würzburg: Ergon-Verlag, 2011), 67–80; White, "Die historischen Ursachen," 24–25.

21 Qur'ān, 22:65.

22 Fazlun M. Khalid, "Applying Islamic Environmental Ethics," in *Environmentalism in the Muslim World*, ed. Richard C. Foltz (New York: Nova Science Publishers, Inc., 2005), 104; Ursula Kowanda-Yassin, *Mensch und Naturverständnis*, 80–104.

23 Richard C. Foltz, "Iran," in *Environmentalism in the Muslim World*, ed. Richard C. Foltz (New York: Nova Science Publishers, Inc., 2005), 4.

all species are endangered"[24] and Iran's environmental issues are among the most critical in the world.[25] Further problems in Iran include the increasing levels of air pollution caused by millions of cars—many of which are more than twenty years old—as well as factories and oil refineries. Every year, 4,000 people die of pollution-related lung cancer, and many more are still suffering from the aftermath of a drought which lasted from 1999 to 2001 and in which which fifty villages had to be evacuated.[26] In Turkey, air, water, soil, and noise pollution are major environmental problems, and solid waste landfills pose a risk to local inhabitants due to the threat of methane gas explosions.[27] As well as facing such environmental problems, Egypt will also have to deal with the degradation of agricultural land due to salination, pollution of the water of the Nile, desertification and the scarcity and poor quality of available water.[28]

Several steps are being taken in predominantly Muslim countries like Turkey, Egypt, Iran, and Pakistan to cope with these challenges, and environmental education is beginning to emerge in diverse forms. Morocco can be held up as a good example for its ecological efforts. In the last fifty years, various national strategies have been adopted, including developing solar energy, sustainable tourism, craftsmanship, academic research, a national environmental plan, reforestation, and the celebration of Earth Day, to name but a few. Morocco has also invested heavily in the improvement of education and has received support for this:

> [...] this programme has received financial and logistic support from the Hassan II Fund for economic and social development, and has also mobilised support from a range of stakeholders, including ministries, public offices, locally elected councils, industry and NGOs.[29]

At an academic level, departments for environmental studies, research, and professional training programmes are being set up at universities, in schools,

---

24 N. Patrick Peritore, "Iran: From Revolution to Ecological Collapse," in *Third World Environmentalism: Case Studies from the Global South* (Gainsville: University Press Florida, 1999), 209, cited in Richard C. Foltz, "Iran," 3.

25 Ibid., 3.

26 Ibid., 3.

27 Özdemir, "Turkey," 18–19.

28 A. Hamed Safei-Eldin, "Egypt," in *Environmentalism in the Muslim World*, ed. Richard C. Foltz (New York: Nova Science Publishers, Inc., 2005), 41.

29 Ahmed Legrouri, and Khalid Sendide, "Education for Sustainable Development in Morocco," in *National Journeys: Towards Education for Sustainable Development*, ed. UNESCO (Paris: UNESCO, 2013), 34–35.

and in civil society.[30] Despite rising environmental awareness in Egypt and environmental concerns articulated amongst Egyptians:[31]

> this trend has not significantly modified peoples' attitudes, daily habits, or willingness to take initiatives in confronting environmental problems or to participate in conservation programs.[32]

Safei-Eldin describes:

> When some municipalities made the effort of cleaning streets and public open spaces, or planting trees, people continued dumping their garbage on sidewalks or vandalizing parks.[33]

Noted writers have held the government to account for this behaviour, noting the lack of role models and the state's mismanagement of natural resources.[34]

However, the activities of NGOs are sometimes viewed with distrust, especially when undertaken by religious groups: "The secular establishment in Turkey has always been suspicious of any form of religiosity and religion-based activity."[35]

Bouchrah Chrebah's findings illustrate that there are differences in the ecological awareness of German and Syrian students, but both groups show great engagement with environmental topics.[36]

On the other hand, the research conducted by Gülay Kýzýlocak and Martina Sauer showed that Turkish migrants are conscious of the need for ecological awareness, but are often unable to understand complex interconnections: "Environmental awareness of Turkish migrants is ambivalent. There is a positive stance for first steps towards environmental communication but a lack of knowledge regarding complex issues."[37]

---

30 Özdemir, "Turkey," 23–24; Safei-Eldin, "Egypt," 50–53.

31 Safei-Eldin, "Egypt," 42.

32 Ibid., 42–43.

33 Ibid., 42–43.

34 Ibid., 43.

35 Özdemir, "Turkey," 30.

36 Chrebah, *Umweltbewusstsein und Umweltverhalten*, 76–109.

37 Gülay Kýzýlocak and Martina Sauer, *Environmental Awareness and Behaviour of Turkish Migrants in Germany. Research on the Current Situation and Determination of the Possibility of their Inclusion* (Duisburg-Essen: UNESCO-Verbindungsstelle für Umwelterziehung: 2002), 16–19.

3.2 *Islamic Awareness of Ecology: Taking Care of Nature—A Religious Islamic Duty?*

Practising Muslims are generally convinced that their responsibility for the environment also means being responsible for future generations who will otherwise have to cope with the destruction of nature.[38] A comparative study on the attitudes of Christian, Muslim, and secular students towards "environmental issues and carbon capture and storage (CCS) technologies" showed that the Muslims interviewed referred to the Qur'ān and Sunna when talking about ecological responsibility.[39] They explained that they were motivated by the fact that they will be held responsible for their deeds on the day of Resurrection: "God is omniscient and all-seeing, [...] therefore every deed done by man will be known to the Creator."[40]

Muslim activists who were interviewed for the publication *Öko-Dschihad*[41] argued that they found reasons for ecological action in their religion and this gave them the conviction that they were responsible individuals wanting to treat God's creation with care. They admitted in the interviews that they had changed their lifestyle and started projects after learning about Islam and the environment. For example, Fazlun Khaled first decided to study Islamic law (Arab. *fiqh*) and then discovered that Islam provides ecological principles of consequence. He founded the first Muslim environmental organisation, the Islamic Foundation for Ecology and Environmental Science (IFEES).[42]

Furthermore, the Muslims who were interviewed answered that according to the Qur'ān, nature was created by God and given in trusteeship (Arab. *amānah*)[43] to humanity. In this context, Muslim authors[44] and participants of the study[45] refer to the terms of stewardship (Arab. *ḫalīfah*.[46] They argue that this means acting wisely and being mindful of the future because all actions have consequences and our treatment of the environment is no exception. Therefore, religion appears to provide important guidelines on how to act in

---

38 From the Christian perspective on ethics and nature, see also Anton Neumann-Lechner and Florian Preißer, *Umweltpädagogik in Erwachsenenbildung und Gemeinwesenarbeit*, Band 6 (Bad Heilbrunn: Verlag Julius Klinkhardt, 1991), 32–35, 44–47.

39 Aimie L.B Hope and Christopher R. Jones, "The Impact of Religious Faith on Attitudes to Environmental Issues and Carbon Capture and Storage (CCS) Technologies: A Mixed Methods Study," *Technology in Society* 38 (2014): 48–59.

40 Hope and Jones, "The Impact of Religious Faith on Attitudes," 52–54.

41 Kowanda-Yassin, *Öko-Dschihad*.

42 Interview in Kowanda-Yassin, *Öko-Dschihad*, 45–47.

43 See also Part 4.1. of this chapter.

44 E.g. Mawil Izzi Dien, *The Environmental Dimensions of Islam*, 74–77.

45 Hope and Jones, "The Impact of Religious Faith on attitudes," 55–57.

46 For the term *ḫalīfah* see also Part 4.1. of this chapter.

an environmentally friendly manner and is an additional factor in sensitising people to environmental awareness. Mawil Izzi Dien argues that nature itself must be protected, not only because humanity needs it to survive, but because it has its own intrinsic value as God's creation.[47] The Turkish Sufi poet Yunus Emre (born in 1240) described this as respecting nature as God's creation and thus showing respect to the Creator himself.[48]

Sekem, an organisation promoting sustainable development, was started in 1977 when its founder began to plant organic cotton in an inhospitable part of southern Egypt together with the people that lived there, resulting in the regeneration of the region. The fundamental principle of Sekem is that all major religions believe in the principle of man as a trustee. Sekem promotes the idea that man must take a holistic approach in order to effectively activate his resources. This is needed in order to build a sustainable future.[49]

An ethical approach to ecology is essential in forming an awareness of the value of nature.[50] The "conservation of Creation" is understood as a significant topic for religious education.[51] Matthias Bahr and Stephan Leimgruber recommend observing several points when discussing this topic with pupils in Christian Religious Education:[52]

(1) acknowledging pupils' anxieties around the future, as caused by negative predictions;
(2) discussing how to deal with nature in a responsible way;
(3) encouraging people to act creatively and independently;
(4) understanding nature as God's creation by examining religious texts;[53]
(5) connecting ethical and spiritual concepts within one's own lifestyle.

From a pedagogical perspective, this also means "picking up the pupils where they are", as described by the German neurobiologist Gerald Hüther.[54] This means finding out where their interest lies, where they stand in their

---

47 Mawil Izzi Dien, *The Environmental Dimensions of Islam*, 2000.
48 Annemarie Schimmel, *Yunus Emre. Ausgewählte Gedichte* (Köln: Önel-Verlag, 1991), 10.
49 Sekem, "Ecology," accessed November 1, 2019. https://www.sekem.com/en/ecology/.
50 Neumann-Lechner, and Preißer, *Umweltpädagogik*, 29–31.
51 Matthias Bahr and Stephan Leimgruber, "Verantwortung für die Eine Welt," in *Religionsdidaktik. Ein Leitfaden für Studium, Ausbildung und Beruf*, ed. Georg Hilger, Stephan Leimgruber, Hans G. Ziebertz (München: Kösel-Verlag, 2010), 482–483.
52 Bahr and Leimgruber, Verantwortung für die Eine Welt, 482.
53 Bahr and Leimgruber believe that thinking about divine texts and being aware of the value of creation is essential for ecological and religious pedagogy.
54 Deutschlandfunk, "Den Schatz der frühen Kindheit bewahren," accessed June 28, 2018. https://www.deutschlandfunkkultur.de/den-schatz-der-fruehen-kindheit-bewahren.954.de.html?dram:article_id=144099.

development, and designing appropriate pedagogical concepts, methods, and content with this in mind.

3.3 *Individual Responsibility and Participation*

The environmental movement has always argued that every action counts and that everyone can bring about change.[55] The notion that "individual action makes no difference as long as others continue to live their life in excess" was never considered an acceptable argument. It is not enough to know about environmental change and devastation, it is taking action that is crucial in order to shift things and implement change. Thus, action must be an intrinsic part of religious environmental education.

In environmental education, social participation is a basic principle, because "communities, stakeholders, groups, or societies learn how to innovate and to adapt in response to changing social and environmental conditions."[56]

The Egyptian organisation Sekem holds events[57] in Cairo in order to support the development of individual skills and to encourage participantsto become activewithin their own community in creative and socially relevant ways. Numerous workshops and discussions are held at these events, as well as creative activities. These provide an opportunity for exchange between educators in order to facilitate a sustainable and socially just future.

Peter Posch states that environmental school projects give pupils positive impressions:

> [...] that they are capable of achievements that are appreciated by society; that information and a readiness to stand up for one's interests can indeed lead to changes and; that argumentation skills, knowledge and courage are prerequisites of human progress in a democracy.[58]

Thus, pupils learn to take on responsibility through these projects. Getting regional organisations and communities on board is essential for the Association for Forests, Development & Conservation (AFDC) in Lebanon, which believes

---

55 Many advices are given by Greenpeace, e.g.: Greenpeace, "Jeder Einzelne kann etwas für den Klimaschutz tun," accessed January 1, 2019. https://www.greenpeace-muenchen.de/images/stories/Gruppen/Energie/2016%2008%20spartipps.pdf.

56 Arjen E.J. Wals, ed., *Social Learning towards a Sustainable World. Principles, Perspectives, and Praxis* (Wageningen: Academic Publishers 2007), 140.

57 Sekem, "Events," accessed November 1, 2019. https://www.sekem.com/en/events/.

58 Peter Posch, "The Study: 'Environment and School Initiatives': Phase One," in *Evaluating Innovation in Environmental Education* (Paris: OECD documents, 1994), 24.

that otherwise, change will not have a lasting impact.[59] A youth project has already planted 10,000 trees. Young people have made contacts and have acquired skills that will enable them to continue their involvement in this field—which will be useful for their future careers.[60]

All the while, it must be remembered that when situations seem threatening or hopeless, this can cause distress.[61] When contexts appear too complex and too difficult to understand, it is hard to find solutions.[62] To lessen the likelihood of students becoming disheartened, Bahr and Leimgruber suggest teaching religious environmentalism in a solution-oriented way.[63] Certain competencies and skills, such as knowledge, problem-solving, the ability to plan and to implement ideas, the aptitude for teamwork and planning are required.[64] Further characteristics of self-reliant activity are the ability to self-organise, build one's experience, and develop an experimental attitude.[65] A good example is that of the "DeenTravellers", two Dutch Muslims who biked to Mekka and collected rubbish in the areas of the pilgrimage. Their intention was to highlight the consequences of littering the environment while drawing attention to the fact that the pilgrimage has now become an environmental challenge and Muslims should think about environmentally friendly ways of performing it.[66] The entire project was documented, with the young men sharing films and pictures over social media so that their followers could vicariously participate in the adventure.[67]

NGOs and activists who base their work on Islamic ethics tend to design and organise activities in a pedagogically sound manner, for greater impact.[68]

---

59 AFDC, "Association for Forests, Development & Conservation," accessed November 1, 2019. http://www.afdc.org.lb/.

60 AFDC, "Earn and Learn," accessed November 1, 2019. http://www.afdc.org.lb/content/earn-and-learn.

61 Bahr and Leimgruber, *Verantwortung für die Eine Welt*, 482.

62 Ernst-D. Lantermann, "Von den Schwierigkeiten umweltschützendes Handeln auszubilden," in *Methoden der Umweltbildungsforschung*, ed. Dietmar Bolscho and Gerd Michelsen (Opladen: Leske und Budrich, 1999), 124–131.

63 Bahr, and Leimgruber, *Verantwortung für die Eine Welt*, 472–484.

64 Gärnter and Hellberg-Rode, "Umweltbildung und Gestaltungs-kompetenz," 20.

65 Hans G. Ziebertz, and Ulrich Riegel, "Eigenverantwortliches Lernen im Religionsunterricht: Projekt und Freiarbeit," in *Religionsdidaktik. Ein Leitfaden für Studium, Ausbildung und Beruf*, ed. Georg Hilger, Stephan Leimgruber, and Hans G. Ziebertz (München: Kösel-Verlag, 2010), 550–551.

66 Kowanda-Yassin, *Öko-Dschihad*, 86–88.

67 Facebook, "Deen Travellers," accessed January 09, 2019. https://www.facebook.com/deentravellers/videos/vb.387084731679586/469915303396528/?type=2&theater.

68 Kowanda-Yassin, *Öko-Dschihad*, 47, 52–54, 55–67, 70, 75–80, 86–89, 91–95, 102–104, 105–111, 125–149, 154–158.

Muslim activities and projects function as examples of how individual action can make a difference. Some NGOs, such as the Muslim Action for Development and Environment (MADE),[69] encourage action by organising fun activities like the "Tour de Salah",[70] where the challenge is to cycle through London alongside other participants, taking breaks to pray at mosques on the way. Participants learn that cycling is an enjoyable alternative means of transport and benefit from taking part in a group activity at the same time. With this initiative, MADE also seeks to show the public that environmental protection is an important issue for British Muslims.[71]

Muslim activists combine social initiatives with religious advocacy based on the Qur'ān to show the importance of individual deeds, highlighting that in order to receive divine support, one must act. The British organisation MADE calls individuals to action with invitations like: "Join us and invest in a positive future for our youth" and by citing Qur'ānic verses, such as verse 11 of *Sūrat ar-Raʿd* [The Thunder]: "*inna-llāha lā yuġaiyiru mā bi-qawmin ḥattā yuġaiyirū mā bi- ʾanfusihim* [Allah does not change the condition of people until they change what is in themselves]".[72]

In mosques or schools which provide classes for Religious Education, it is stressed that, according to prophetic traditions, it is within everyone's capability to do helpful deeds.[73] In Islamic education, this is also taught to pupils as a religious duty.[74] Within this context, prophetic traditions, like the tradition of planting a fruit tree, are cited. It is said that whoever plants a fruit tree will be rewarded because the tree provides shadow and fruit for humans as well as animals and offers a habitat for insects and birds.[75] Learners are also reminded that the result is not in human hands because it is all God's doing. What counts is one's intention, the purpose of one's action. Actions are judged by

69 MADE, "Muslim Action for Development & Environment," accessed January 01, 2019. www.made.ngo.

70 MADE, "Tour de Salah," accessed August 22, 2018. https://tourdesalah.org.uk/.

71 According to an interview conducted by Kowanda-Yassin in 2017 with Fatima El- Meeyuf for the publication *Öko-Dschihad*.

72 MADE, "Muslim Action for Development & Environment"; Qur'ān 13:11.

73 According to interviews conducted by Kowanda-Yassin in 2017 for the publication *Öko-Dschihad*.

74 E.g. Amena Shakir, Said Topalovic, *Islamstunde 7. Wie werde ich stark? Religionsbuch für die Sekundarstufe I*, (Wien: Oldenburg-Verlag, 2015), 76–70.

75 Ḥadīṯ, Ṣaḥīḥ al-Buḫārī, Kapitel 36/Ḥadīṯ No. 2320: "*mā min muslimin yaġrisu ġarsan ʾaw yazraʿu zarʿan fa-yaʾkulu minhu ṭayrun ʾaw insānun ʾaw bahīmatun ʾillā kāna lahu bihī ṣadaqa*", Islamische Datenbank, „Ṣaḥīḥ al-Buḫārī", accessed April 8, 2019, http://islamische-datenbank.de/sahih-al-buchari?chapterno=36&hadithid=1&action=display.

the intention and not the results.[76] With regard to the importance of taking action, teachers cite Quranic verses which stress that man should always try to change things for the better. Religious scholars and teachers argue that Islam does not only mean believing in God, but also requires action appropriate to Islamic knowledge. Muslims are asked to strive to improve any negative situation, either through deeds, words, or through their heart.[77] Many verses in the Qur'ān speak of people as "those who believe and do righteous deeds" (Arab. *inna llaḏīna āmanū wa-ʿamil ū ṣ-ṣāliḥāt*)[78] and regard actions as proof of one's belief in God. Environmental education is only effective when accompanied by actions.

3.4 *Strengthening a Child's Sense of Identity*

Awakening an awareness of the environment in children can have a powerful and positive influence which must not be underestimated, especially when it comes to developing a child's sense of self. Feeling responsible and able to make a difference can strengthen a child's self-confidence and pave the way towards personal initiative and individual responsibility.[79] From a teachers' point of view, ecological education is also seen as an important tool to help young people find their own identity.[80] This is why committed teachers carry out environmental project work at school. Posch describes four target dimensions as examples of school projects:

> [...] to experience the environment as a sphere of personal experience; to examine the environment as a subject of interdisciplinary learning and research; to shape the environment as a sphere of socially important action; to accept the environment as a challenge to initiative, independence, and responsible action.[81]

Projects were conducted with these objectives in mind and the outcomes are described by Posch as follows:

---

76 Regarding to the ḥadīṯ, "*ʾinnamā-l-aʿmālu bi-n-nīyāt* [Indeed all actions are based on the intentions (...)]". Muḥammad Aḥmad Rassoul, ed., *Ḥadīt für Schüler. An-Nawawī's Vierzig Ḥadīṯe mit Kommentar*. Aus dem Arabischen von ʿ Abdullah aṣ-Ṣāmit Frank Bubenheim, 32, 38,38 (Köln: IB Islamische Bibiliothek, 1987).

77 Rassoul, (ed.), *Ḥadīt für Schüler*, 214–216.

78 Qur'ān, 31:8, 19:96.

79 Posch, "The Study", 21–26.

80 Neumann-Lechner, and Preißer, *Umweltpädagogik*, 47–49.

81 Posch, "The Study", 21–26.

> […] pupils experience several things: 1. That they are capable of achievements that are appreciated by society; 2. That information and a readiness to stand up for one's interests can indeed lead to change and 3. argumentation skills, knowledge, and courage are prerequisites of human progress in a democracy.[82]

Engaging in topics that concern each individual within society gives a child the feeling of being an important part of that society. It also helps the child to engage with and define his or her own opinions.[83] The Sekem organisation in Egypt runs various educational events. Sekem sees the basis for developing a strong identity as follows:

> The people need to be guided to their individual inspirational sources, whether these are sciences, philosophy, religion, arts or beauty. This can only be achieved through freedom in cultural life through the establishment of connections with other local cultures through a global network. Therefore, free education and spiritual development of all human beings must be the highest priority. Through a holistic approach every individual is able to unfold his full potential and skills and enables himself to improve his own life and contribute to the development of the community and country."[84]

The organisation "Muslim Action for Development and Environment" (MADE) in Great Britain strives to empower youths and encourages them to take part in ecological projects, an approach which has been highly successful. MADE claims to have reached 4,000 young people and trained 660 students to initiate environmental campaigns with their projects.[85]

---

82 Ibid., 24.

83 Georg Hilger, Ulrich Kropač, and Stephan Leimgruber, "Der sozialisationsbegleitende oder therapeutische Religionsunterricht," in *Religionsdidaktik. Ein Leitfaden für Studium, Ausbildung und Beruf*, ed. Georg Hilger, Stephan Leimgruber, and Hans G. Ziebertz, 55–58 (München: Kösel-Verlag, 2010); Matthias Bahr, "Handlungsorientiertes Lernen," in *Religionsdidaktik. Ein Leitfaden für Studium, Ausbildung und Beruf*, ed. Georg Hilger, Stephan Leimgruber, and Hans G. Ziebertz, 542–548 (München: Kösel-Verlag, 2010); Hans-Georg Ziebertz, "Wozu religiöses Lernen?—Religionsunterricht als Hilfe zur Identitätsbildung," in: *Religionsdidaktik. Ein Leitfaden für Studium, Ausbildung und Beruf*, ed. Georg Hilger, Stephan Leimgruber, and Hans G. Ziebertz, 142–151 (München: Kösel-Verlag, 2010), 142.

84 Sekem, "Index," accessed November 1, 2019. https://www.sekem.com/en/index/.

85 Kowanda-Yassin, *Öko-Dschihad*, 67.

Self-identity plays a key role in Religious Education too.[86] As Ziebertz explains:

> Pupils are not to be seen as recipients of EDUCATIONAL CONTENTS but ARE INSTEAD the centre of all learning and teaching efforts. [...] Contents are presented by the teacher as a basis for investigation, acceptance, rejection or transformation.[87]

For Ziebertz the question "Who am I?" is ever-present in plural societies, as there are so many different concepts of life and adolescents in particular struggle to accept themselves as they are. Religious Education discusses existential human topics and can give guidance on how to cope with daily life.[88]

3.5 *Ecology Connects People*

There are similarities between Islam and the other Abrahamic religions in their ethical tenets for protecting the environment and understanding the Earth as God's creation.[89] They share the image of God as the creator of all living beings. Thus, ecology becomes an interesting topic for interreligious discussion.[90] ARC[91] in Great Britain and Pilgrim,[92] in Austria, are good examples of organisations that are already working for the environment on an interreligious basis.

Through this shared understanding, a common goal is found, and this can encourage people to become connected through ecological action and strive to achieve this goal together. A representative of ARC explained that, as a result of their work, people holding opposing beliefs, or who are prejudiced against each other and would not normally agree debate, actually came together. Now

86 Ziebertz, "Wozu religiöses Lernen?," 142–153.

87 Ibid., 143.

88 Ibid., 154.

89 Elisabeth Birnbaum, "Wüste, Sintflut, Garten Eden. Ambivalente Naturerfahrung im Alten Testament," in *Gärten in der Wüste, Schöpfungsethik zwischen Wunsch und Wirklichkeit,* ed. Severin J. Lederhilger (Regensburg: Verlag Friedrich Pustet, 2018), 51, 57–59; Franz Gruber, "Der bedrohte Garten. Was uns der Schöpfungsglaube zu denken und zu tun gibt," in *Gärten in der Wüste, Schöpfungsethik zwischen Wunsch und Wirklichkeit,* ed. Severin J. Lederhilger (Regensburg: Verlag Friedrich Pustet, 2018), 106–110; Matthias Bahr and Stephan Leimgruber, "Verantwortung für Eine Welt," in *Religionsdidaktik*, ed. Hilger, Leimgruberand Ziebertz (München: Kösel-Verlag, 2010), 472–484.

90 Ibid., 472–481.

91 Alliance of Religions and Conservation, "About ARC," accessed August 13, 2018. http://www.arcworld.org/.

92 Pilgrim, "Pilgrim aktuell," accessed August 13, 2018. http://pilgrim.at/aktuell.html.

they take part in discussions organised by the ARC on local ecological projects in their search for solutions.[93]

ARC offers a secular platform and supports communities in seeking religious narratives that are based on ecological principles.[94] Through this cooperation, many prejudices are eradicated; people get to know each other at a personal level as human beings and can explain their religious beliefs to one another.[95] A wide range of conferences[96] are organised by various institutions to discuss religious ways of thinking and theological materials on ecology. However, it is not only collaboration between people of different religions; academic institutions and professors are now working closer together to look at questions of sustainability and development. In order for this exchange to happen, Sekem organises a multi-day event in Cairo:[97] People from all over the world and from different educational backgrounds share ideas on creative pedagogical practices to unfold individual potential holistically and thus help build a sustainable future. Arab countries are also working together on a wider basis in the Arab Forum for Environment and Development (AFED). One of the goals of the AFED is:

> Working with educational institutions, universities, and research centres towards the development of environmental education programmes at all levels, both for promoting constructive environmental behaviour and attitudes, and building up capacities in the domains of environmental sciences and management.[98]

A draft for a university course promoting an awareness of environment and sustainability has been developed by the AFED and is currently being revised with the help of professors from 25 different universities. Many universities have already shown interest in the course.[99]

93 This was explained in an interview in 2017 for the publication Kowanda-Yassin, *Öko-Dschihad*, 71.

94 "About ARC." See also part 3.3.3 of this chapter.

95 Stephan Leimgruber and Hans-Georg Ziebertz, "Interkulturelles und interreligiöses Lernen," in *Religionsdidaktik. Ein Leitfaden für Studium, Ausbildung und Beruf*, ed. Stephan Georg Hilger, Leimgruber, Hans G. Ziebertz (München: Kösel-Verlag, 2010), 467–469.

96 E.g.in 2015 in New York (USA), 2016 in Köln and Stuttgart (D), 2017 in Kremsmünster (A).

97 Sekem, "Events," accessed November 1, 2019. https://www.sekem.com/en/events/.

98 Arab Forum for Environment and Development, "AFED Course of Action," accessed November 1, 2019. http://www.afedonline.org/en/inner.aspx?menuID=6.

99 Arab Forum for Environment and Development, "AFED Announces Report and Conference on Environmental Education," accessed November 1, 2019. http://www.afedonline.org/en/inner.aspx?contentID=1370.

4 Putting an Ecological Lifestyle into Practice—A Pedagogical Approach

In the following section, various ways of teaching people how to practise an ecologically friendly lifestyle will be presented. A collection of pedagogical ideas and methods which are already present in the work carried out by educators and NGOs will also be examined. "The Islamic Foundation for Ecology and Environmental Sciences" (IFEES) is an example of a non-governmental organisation advocating ecology and environmental studies which is highly appreciated by Muslim communities. NGOs can support future developments on a professional basis and can serve as role models of how to move and motivate people to deal with environmental issues in a conscious manner. Adults should also be made aware of the importance of conserving resources such as water and electricity. Therefore, seminars are held for ministry staff as well as engineers from government agencies by the NGO Kuwait Water Association, with the ultimate goal to train all government employees in Kuwait.[100]

4.1 *Experiential Learning and Empathy*

Teachers need to be provided with teaching materials; and in order to constantly improve these materials, it is necessary to look at science-based evidence. The Arab Forum for Environment and Development (AFED) carried out a study[101] on teaching environmental issues at 57 universities in 22 Arab nations which focused, among other things, on their curricula. The results of this study will be presented and discussed at a conference in Beirut in November 2019. The report takes a comprehensive look at how environmental issues are taught in Arab countries and presents recommendations for further development as well as a chapter on innovations which was put together by the research team under the leadership of Dr Ahmed Gaber, professor at the University of Cairo. In addition to this, the AFED has developed a programme for environmental education which provides teachers with material from which to teach. The "AFED Environmental Education Handbook"[102] addresses 12 topics suitable for 12- to 18-year-old pupils for teacher training. It presents concepts for activities and projects and gives information about air pollution and other

100 Kuwait Water Association, "Training Programs," accessed November 1, 2019. https://www.kwa.org.kw/copy-of-proposed-projects.

101 Arab Forum for Environment and Development, "AFED Announces Report and Conference on Environmental Education."

102 Arab Forum for Environment and Development, "AFED Environmental Education Program," accessed November 1, 2019. http://www.afedonline.org/en/inner.aspx?contentID=180.

environmental challenges. The Handbook also provides texts as reading material, exam questions, and videos on general environmental themes.

4.1.1 Experiential Learning[103]

An effective way to raise awareness is by giving pupils the opportunity to come into close contact with the subject of their study.[104] Walking through the woods, smelling the green grass, being surrounded by natural materials, walking on the ground barefoot, and listening to birds singing and insects humming all give a firm impression of the beauty of nature.[105] Experiencing nature with all senses is revitalising. It is also an opportunity to learn more about oneself. Learning is imprinted on the learner's mind and in his or her feelings.[106] To that end, special nature learning paths have been developed and ideas for teaching these are discussed in workshops[107] and specially designed seminars.[108] With regard to environmental education, gardening in particular is an effective way of showing the results of carefully handling plants by sharing in the fruits of the effort.[109] By experiencing nature with all senses, learners feel part of nature and the wish to conserve its beauty arises naturally. During this process, they also come to notice the detrimental impact of pollution and the resulting extinction of many animals.[110] A concrete example of this is seen in the series of lessons created by Neumann-Lechner and Preißer.[111]

The British Muslim NGO Wisdom in Nature (WIN), for instance, organises outings for people to experience nature with all their senses.[112] Mansoureh Shodjai, an Iranian teacher in child environmental education programmes, organises "nature concerts" where children listen to the sounds of birds singing

103 Posch, "The Study", 23.

104 Many examples are given by Herbert Österreicher, *Natur- und Umweltpädagogik für sozialpädagogische Berufe* (Köln: Bildungsverlag EINS, 2014), and concrete examples are described on the pages 115–148. See also Max Mittendorfer and Michael Nader, "Der Schulgarten als Ort lebensbezogenen Lernens," in *Ernte und Aussaat, Spiritualität und Nachhaltigkeit—Überlegungen und Handlungsimpulse*, vol. 5, ed. Gabriele Hösch-Schagar, Birgit Karre, Erhard Mayerhofer (Wien, Berlin: LIT-Verlag, 2011), 123–130.

105 Neumann-Lechner, Preißer, *Umweltpädagogik*, 61–63.

106 Österreicher, *Natur- und Umweltpädagogik*,11–53.

107 Agrar- und Umweltpädagogik, "Fort- und Weiterbildung," accessed June 8, 2018. http://www.agrarumweltpaedagogik.ac.at.

108 Bundesministerium Nachhaltigkeit und Tourismus, "Natur- und Artenschutz," accessed June 8, 2018. https://www.bmnt.gv.at/umwelt/natur-artenschutz/wdavf_2018.html.

109 Österreicher, *Natur- und Umweltpädagogik*, 185–219.

110 Neumann-Lechner, and Preißer, *Umweltpädagogik,* 24.

111 Ibid., 107–126.

112 "Wisdom in Nature," accessed June 28, 2018. http://wisdominnature.org/.

and the wind blowing in the trees.[113] Ibrahim Abdul Matin, author of Green Deen, depicts a similar experience when spending time in the mountains praying in the open air.[114] Sufis make their experience of nature and religion visible in their poems. They feel that getting closer to nature is also a way of getting closer to their creator.[115] Jelaludin Rumi (born in 1207 in Balkh, Afghanistan) describes nature in his poem *The Grasses*, as follows:

> The same wind that uproots the trees
> makes the grasses shine
> The lordly wind loves the weakness
> and the lowliness of grasses. [...]
>
> There is no reality but God,
> says the [...] sheikh,
> who is an ocean for all beings.[116]

Within the context of religious environmental education, this can be taken as a reminder to go outdoors into nature and stay close to the earth, since creation is understood as a proof of God's existence.[117] From a spiritual perspective, this is the moment when learners experience nature with all their senses and this might shape their consciousness of nature.

4.1.2 Experiencing the Plights of Others as a Factor in Building Greater Empathy

Social responsibility and justice are matters which play a vital role in environmentalism. When people experience the challenges of others,[118] like when taking on new roles and activities in daily life, this can spark a more empathetic,

113 Foltz, "Iran," 12.

114 Abdul-Matin, *Green Deen*, 1–3.

115 Annemarie Schimmel, *Die Zeichen Gottes. Die religiöse Welt des Islams* (München: C. H. Beck, 1995), 27, 31.

116 Coleman Barks, trans., *The Essential Rumi. New Expanded Edition* (New York: Harper One, 2004).

117 In *Sūrat al-Mulk* [The Sovereignty] (Qur'ān, 67/3–4), the believer is asked to look around and to discover the wonder of God's perfect creation: "*allaḏī ḫalaqa saba'a samāwātin ṭibāqan mā tarā fī-ḫalq ir-raḥmāni min tafwāwutin fa-r-rǧ 'iu l-baṣara hal tarā min futūrin* [It is He who has created skies in seven layers. You see no disharmony in the creation of the All-Merciful (...)]."

118 Friedemann Schulz von Thun, Kathrin Zach, and Karen Zoller, *Miteinander reden von A-Z. Lexikon der Kommunikationspsychologie* (Reinbek bei Hamburg: Rowohlt-Verlag GmbH, 2012), 58–59.

ethical way of thinking. MADE, for example, invites its members and interested persons to join the so-called "Water Challenge"[119] which takes place at three levels: At the easiest level, participants are asked to use as little water as possible while washing themselves for their ritual prayers. At the second level, they must try to take a shower with only one bucket of water at their disposal. And at the most difficult level, participants are asked to walk six kilometres with a bucket full of water balanced on their head, thus experiencing the daily difficulties which many women all over the world must face in order to procure clean water for their families. The prize for this challenge was chosen thoughtfully: a water well will be built in Zanzibar and dedicated to the winner. By trying out these practices, participants get a glimpse of the difficulties of other people's lives, developing a greater awareness of the plight of others. Furthermore, people's willingness to take action and give aid tends to increase and they feel a greater appreciation of their own situation. Muslim environmentalists and teachers remind members and learners that gratitude is an important feature of Islam. It is described in the Qur'ān,[120] where the concept of taking pity on both humans and animals plays a considerable role. MADE sees the developed perspective of participants after the "Water Challenge" as another step in creating environmental consciousness. Thinking about one's own deeds leads to self-reflection, as required in the Qur'ān.[121] Further to this, the psychological and pedagogical perspective confirms that reflection helps to develop a conscious and empathetic personality.[122] Yet there are also difficulties to be considered: Sympathising with others and feeling moved by their plight does not always empower people to act.[123]

119 MADE, "The Ultimate Water Challenge," accessed June 6, 2018. https://www.made.ngo/the-ultimate-water-challenge-leader-board/.

120 Qur'ān, 46:15.

121 Qur'ān, 16:4, 2:31; Schulz von Thun, *Miteinander reden*, 178.

122 Ibid., 178; Helmut Gärnter and Gesine Hellberg-Rode,"Umweltbildung und Gestaltungskompetenz für nachhaltige Entwicklung," in: *Umweltbildung und nachhaltige Entwicklung*, ed. Helmut Gärtner and Gesine Hellberg-Rode, 20 (Hohengehren, Schneider-Verlag, 2001).

123 Ernst-D. Lantermann, "Von den Schwierigkeiten, umweltschützendes Handeln auszubilden," in *Methoden der Umweltbildungsforschung*, ed. Dietmar Bolscho and Gerd Michelsen (Opladen: Leske und Budrich, 1999), 121–124.

4.2 *Increasing Theoretical Knowledge*

4.2.1 Teaching Facts and Encouraging People in Their Own Independent Quest for Information

In religious Islamic education the importance of life-long learning is underlined,[124] referring to the following *ḥadīṯ*: "*ṭalabu l-ʿilmi farīḍatun ʿalā kulli muslimin* [Seeking knowledge is an obligation placed on every Muslim]."[125] In the Qurʾān the believer is reminded: "*wa-qul rabbi zidnī ʿilman* [And say: My Lord, increase my knowledge]."[126] Thus the constant search for knowledge is understood to be part of religious practise. Having an inquisitive mind is an effective way of learning and therefore also an important tool in ecological education.[127]

One's independent quest for information as a consumer is essential in raising environmental awareness. In Turkey "new environmental studies courses have been implemented at different levels of the educational system",[128] but:

> […] the education of the public and the raising of environmental awareness should not be left to formal educational institutions alone. NGOs, the mass media and local authorities can also play a role in increasing public awareness about the environmental challenge and working towards an environmentally friendly life.[129]

Gülay Kýzýlocak and Martina Sauer argue that there is an "[…] urgent need to take on measures to increase awareness and pass on explanations with the

124 Said Topalovic et al., *Islamstunde 7. Wie werde ich stark? Religionsbuch für die Sekundarstufe I*, ed. Amena Shakir (Wien: Oldenburg-Verlag, 2015), 54–57; Hacer Geçgel et al., *Islamstunde 5. In Freundschaft leben. Religionsbuch für die Sekundarstufe I*, ed. Amena Shakir (Wien: Oldenburg-Verlag, 2015), 76 and; Aida Tuhcic et al., *Islamstunde 6. Die Schöpfung bewahren, Religionsbuch für die Sekundarstufe I*, ed. Amena Shakir (Wien: Oldenburg-Verlag, 2015), 127–129, ed. Amena Shakir (Wien: Oldenburg-Verlag, 2015).

125 Ḥadīṯ related by Ibn Māǧa, al-Muqaddima, 17. Ditib, "Streben nach Wissen ist Pflicht für jede Muslimin und jeden Muslim," accessed April 9, 2019, http://www.ditib.de/detail_predigt1.php?id=215&lang=de.

126 Qurʾān, 20:114.

127 Österreicher, *Natur- und Umweltpädagogik*, 11–53.

128 Özedmir, "Turkey," 23.

129 Ibid., 24.

consideration of making use of specific information methods"[130] and "[...] then identifying patterns of reasoning."[131]

It makes sense for individuals to not be content with the information that is being given to consumers by multi-national companies and politicians. Scientific knowledge, studies on climate change, species extinction, and the development of new technologies concern each individual within society and research results should be accessible for the mainstream public so that individual and non-governmental organisations may build on this information.[132] By presenting research results and solutions concisely and accessibly, researchers can support NGOs and individuals as they learn of complex interconnections and improve their knowledge of environmentalism.[133] The Qur'ān, too, requires that Muslims develop their own independent thinking and reflect on facts.[134]

4.2.2 Making Theoretical Facts Comprehensible

A topic has little relevance if presented in an abstract manner. Showing pictures and examples that are connected to the reality of the learner helps to develop their personal interest and investment in the subject.

Awareness can be raised more effectively by making the format in which information is presented practical and accessible. A honey bee flies the circumference of the globe three times in order to collect enough nectar to produce one kilo of honey.[135] Together with the fact that bees are mentioned in the Qur'an, where honey is called medicine for humans,[136] this data point reinforces the value of bees for Muslim believers. Likewise, when we take the example of an apple: every conventional apple is sprayed on average 31 times with insecticides before being sold to the consumer. By contrast, there are no pesticides to be found in an organic apple.[137] According to the Qur'ān, God

130 Kýzýlocak and Sauer, *Environmental Awareness and Behaviour*, 16.

131 Ibid., 16.

132 Özedmir, "Turkey," 20.

133 For example the nine suggestions given by researchers to enable climate change: Zeit Online. "Klimawandel. Wenn Klimaforscher die Welt regieren würden," accessed December 21, 2018. https://www.zeit.de/wissen/umwelt/2018-12/klimawandel-erderwaermung-co2-steuer-fleischkonsum-umweltschutz?

134 Qur'ān, 16:12, 16:17.

135 Propolis, "Wie weit fliegen Bienen für Honig," accessed June 3, 2018. https://propolis-honig.de/wie-weit-fliegen-bienen-fuer-honig/.

136 Qur'ān, 16:69.

137 Wirtschaftswoche, "Greenpeace-Test. 90 Prozent der deutschen Äpfel mit Pestiziden belastet," accessed June 3, 2018. https://www.wiwo.de/technologie/umwelt/greenpeace

advised the people of Israel to eat of the good food that God provides.[138] From an ecological perspective, this could also be understood as a call on mankind to eat natural, organic food. Another example of making environmental topics more accessible is in giving dramatic demonstrations, an approach which was adopted by the Moroccan activist Mamoun Ghallab. In his presentation about plastic production, he makes the excessive use of plastic bottles visible by bringing the actual rubbish on-stage. He shows his audience how many bottles one single person accumulates in one year (the bottles had been collected over one year), amounting to a huge pile of plastic measuring three metres by one metre and weighing twelve kilograms.[139]

4.2.3 Passing on Information within One's Community

The most effective way to reach people, make them aware of ecology, and help them change their habits is to have them address their own family, their friends, and those who are actively involved in their community. With this approach, individuals are more likely to be heard and can also gain access to other members.[140] In this way, emerging ecological information such as new approaches to waste reduction or the installation of solar cells can be passed on efficiently. The "Green Hand" organisation in Lebanon trains young people to be able to work with their communities on issues which affect their region, including environmental issues. Through workshops and working groups, young people gain the necessary skills while also socialising. The goal of "Green Hand" is for young people to find their own solutions and initiate projects. Their projects include, for example, eco-tourism, organic.[141] Hima e.V., a German Muslim environmental NGO, sees stakeholders as important partners for their projects.[142] When there is no contact to provide a closer connection to a community, it becomes harder to be heard. This was the reason why the team at "NourEnergy",[143] a German Muslim organisation that specialises in installing solar cells, decided to first find trustworthy contacts within

-test-90-prozent-der-deutschen-aepfel-mit-pestiziden-belastet/12477848.html. For more examples see also Neumann-Lechner, and Preißer, *Umweltpädagogik*, 28.

138 Qur'ān, 20:81.

139 Mamoun Ghallab, "Zero Zbel Xperience," accessed August 15, 2018. https://www.youtube.com/watch?v=K-BtrzYME2Y.

140 This was underlined by activists who were interviewed for the publication Kowanda-Yassin, *Öko-Dschihad*.

141 Green Hand, "Project Description," accessed November 1, 2019.http://www.greenhand.org.lb/projects/youth/project-description.aspx.

142 This has been described in an interview with Ursula Kowanda-Yassin for investigations for *Öko-Dschihad*.

143 NourEnergy, "NourEnergy," accessed August 15, 2018. http://www.nour-energy.com/.

educational institutions and mosques and successfully install their solar cells within these institutions.[144] Other projects like "Green up your Community" or "Green Mosque" are rapidly expanding with the help of individuals sharing ideas with their communities. Another way of gaining access to people is through religious authorities and sharing religious recommendations. For example, after giving a workshop on the importance of taking care of nature in Islam, Fazlun Khaled from IFEES experienced a high level of willingness amongst fishermen to change their habits.[145] A conference on Islam and ecology where religious authorities and scientists discussed the topic and presented a paper with the results was held in Istanbul in 2015. It lays out Islamic principles and concrete recommendations for both Muslim believers and international corporations.[146] The clear statement being made here by Muslim experts on Islam and ecology can be understood as a signal for worldwide communities to take on responsibility for nature and its exploitation.

4.3 *Overcoming Limitations and Finding Sustainable Solutions Through Teamwork*

New solutions can be found by leveraging teamwork, focused discussions, and examining role models. In contemporary religious instruction, pupils are encouraged to work together[147] referring to a *ḥadīṯ* saying that there is a special blessing in communal work.[148] In modern companies, teamwork and team building are considered as some of the most important skills to develop.[149] A community gives its members strength, along with the feeling of being part of something important. Moreover, in comparison to working alone, working in a group can facilitate reaching objectives. Supporting each other and experiencing success together can strengthen an individual's self-esteem and in the

144 This has been described in an interview with Ursula Kowanda-Yassin for investigations for *Öko-Dschihad*.

145 IFEES, "Zanzibar," accessed August 15, 2018. http://www.ifees.org.uk/zanzibar/.

146 IFEES, "Islamic Declaration on Global Climate Change," accessed August 15, 2018, http://www.ifees.org.uk/declaration/.

147 For example in the school books *Islamstunde*, vols 1–8, ed. Amena Shakir, where, in many units, pupils work together on topics in group work.

148 "*Kulū ǧamīʿan wa-lā tafarraqū fa-ʾinna l-barakata maʿa l-ǧamā ʿati* [Eat together and do not eat separately, for the blessing is in being together]." Related by ʾIbn Māǧah. English reference: Vol. 4, Book 29, Ḥadīṯ-No. 3286, Sunna.com, "Ibn Maagah," accessed April 9, 2019, https://sunnah.com/ibnmajah/29.

149 Manfred Gellert and Claus Nowak, *Teamarbeit, Teamentwicklung, Teamberatung: Ein Praxisbuch für die Arbeit in und mit Teams* (Limmer Verlag, 2010); and Friedrich Glasl, *Konfliktmanagement: Ein Handbuch für Führungskräfte, Beraterinnen und Berater* (Bern-Haupt-StuttgartVerlag Haupt, 2013).

case of failure or exhaustion, individuals within a community can support and encourage one another to keep going.[150]

Depending on the country and circumstances, challenges may differ, and individual solutions are required.[151] Transferring solutions from one area to another becomes futile when prevailing circumstances differ. Local individuals and organisations are experts in their given field. The members of ARC[152] look for solutions within their area, taking into consideration the mentality, level of education, economic limitations, issues, and traditions within their society which may damage the environment.[153]

When undertaking teamwork to overcome ecological challenges, a strong sense of achievement is important to keep motivation high and ensure sustainable activity. Each person needs to be aware thathis or her actions are meaningful. What is the reason for change, and what impact can be expected? And if there is no obvious effect, what should be changed to improve the results? From a pedagogical point of view, positive feedback is a necessary and effective tool, and so are the respect and esteem of others. Knowing the intrinsic value of one's own deeds and experiencing joy while doing them is helpful and keeps motivation high. When a subject is looked at only from a negative perspective, this can cause fear and distress[154] and leads to a feeling of helplessness that might block one's creativity.[155] Another important point is appreciation, which is seen as a precious facet of Islam and is recommended to Muslim believers. The importance of joy in keeping motivation high was also stressed in prophetic traditions, where a smile is understood to be like giving a gift of charity (*Arab. ṣadaqa*).[156] NGOs practise this and show appreciation for accomplished work by awarding prizes to their members.[157]

---

150 Österreicher, *Natur- und Umweltpädagogik*, 425–430.

151 Foltz, *Environmentalism*.

152 ARC, accessed August 15, 2018. http://www.arcworld.org/.

153 See also part 2.5 of this chapter.

154 Hans Zeier, "Umwelt, Angst und Stress," in *Umweltpädagogik. Ansätze, Analysen, Ausblicke*, ed. Felix von Cube Volker and Storch (Heidelberg: Edition Schindele, 1988), 79–92.

155 Michael Rosenberger, "'Wenn ich wüsste, dass morgen die Welt untergeht ...'. Christliche Schöpfungsspiritualität zwischen Angst und Hoffnung," 39–49, in: *Gärten in der Wüste, Schöpfungsethik zwischen Wunsch und Wirklichkeit*, ed. Severin J. Lederhilger (Regensburg: Verlag Friedrich Pustet, 2018), 44–45.

156 Related by Muslim, Ḥadīṯ-No. 121, Buch 1, Kapitel 13, Nr.121, Islamische Datenbank, "Riyāḍu ṣ-Ṣāliḥīn," accessed April 9, 2019. http://islamische-datenbank.de/riyadhu-s-salihin?action=anzeigen&hadithno=121.

157 E.g. "The Ultimate Water Challenge": As described in the foregoing text, MADE is donating a water well built in Zanzibar as the prize to be won in a water challenge.

The Organisation Friends of the Environment Centre in Doha, Qatar, specialises in bird conservation and works with various activities and events such as annual school-oriented competitions on environmental aspects. The NGO seeks to create sustainable change in habits by implementing a combination of responsibility and joyful experiences:

> Our belief consists of preserving the environment, to use and enjoy it without harming its natural resources, lands, and seas. We work sincerely to do our best in preserving the beautiful environment—to keep it clean and safe for our future generations.[158]

Finally, it is important to bring to light the work that has already been done. This helps demonstrate good results and achievements, gives participants a feeling of success, and provides motivation to continue the work. It also gives the people who were not involved in these projects greater insight, enabling further discussion and reflection on the topic.[159]

5 Various Approaches to Islamic Ecology and Learning Goals

Depending on the context and focus, there are various ways of learning about Islam and Nature. It is therefore helpful to consider a wide range of approaches to Islam and ecology, especially when it comes to teaching. In the following chapter, a few approaches will be described.

5.1 *Principles of Islam and Ecology—A Theological Approach*

A characteristic approach to Islamic ecological education is pointing out the religious principles supporting environmentalism.[160] These are similar to common environmental principles.[161] The learning goals can either be practical, giving examples from daily life, or they can instigate a reflection on humans and nature through philosophical discussion. This practical approach to Islam and ecology fits in with the subjects children study in school such as biology and is relevant to topics they hear about in their daily life such as

158 Friends of the Environment Center, accessed November 1, 2019. https://arab.org/directory/friends-of-the-environment-center/ and http://fec.qa/.

159 How to present such achievements is described in Hilbert Meyer, *Unterrichtsmethoden. II: Praxisband*, 14th. edition (Berlin: Cornelsen Verlag Scriptor GmbH & Co.KG, 2011), 167–180.

160 See also part 1 of this chapter.

161 See also part 2.5 of this chapter.

environmental pollution. Thus, Muslim pupils can see the link between these school subjects and universal principles of environmentalism in their religion.

Applying principles like cleanliness,[162] economising, and simplicity[163] in one's life, especially concerning food, clothing, and sanitation, is supported in Qur'ānic verses and prophetic traditions. The "Muslim Declaration on Nature" of 1986 published in Assisi emphasises several salient terms:[164] *ḫalīfah* is interpreted as the trustee: humans as responsible for God's creation. The term *āḫira*, the hereafter, refers to the Islamic belief that humans will be questioned on their deeds on the day of judgement and judged accordingly. *Tawḥīd*, the belief in the Oneness of God, is understood as meaning that all creation is one and therefore all creations are related to one another. The *amānah* is understood as a trust that has to be returned in its original good condition and the term *ʿubudīya*" describes humans as God's servants who are obliged to live according to divine and natural laws.

By teaching these principles, Muslim pupils learn on the one hand to develop a deeper understanding of their religion, and on the other hand, to understand the different possibilities when it comes to interpreting their religion for the good of ecology, underpinned by a strong theological foundation.[165]

5.2 *The Islamic Perspective of Humanity's Nature—A Philosophical Approach*

The nature of humanity and his behaviour in the environment is a topic which lends itself well to further reflection and discussions. When broaching a specific topic that affects learnings in daily life, workshops can stimulate fruitful discussions and promote thoughtfulness, prompting participants to think about the broader field.[166] In the following chapter, human characteristics as described by Abdoljavad Falaturi and Udo Tworuschka will be examined alongside their religious sources.[167] Falaturi describes humanity from the

---

162 *Aṭ-ṭuhūru šarṭu l-ʾimāni* [Cleanliness is part of Faith],"in *Ḥadīṯ für Schüler*, ed. Rassoul, 167–169.

163 Qur'ān, 2:11.

164 IFEES, "The Islamic Declaration on Global Climate Change," accessed August 15, 2018. http://www.ifees.org.uk/declaration/. Those terms were also described in other publications, for Example: Izzi Dien, *The Environmental Dimensions of Islam*.

165 In Tuhcic Aida (sic.) et al., *Islamstunde 6*, ed. Amena Shakir, the Islamic principles are tought for children, 72–85.

166 On nutrition see e.g. Österreicher, *Natur- und Umweltpädagogik*, 310–337 and; Willi Lindner, "Nachhaltige Ernährung—Wie durchbrechen wir die gläserne Decke," in *Ernte und Aussaat*, ed. Gabriele Hösch-Schagar, Birgit Karre, and Erhard Mayerhofer, 111–115 (Münster: LIT, 2011).

167 Abdoljavad Falaturi and Udo Tworuschka, *Der Islam im Unterricht. Beiträge zur interkulturellen Erziehung in Europa. Grundwissen über den Islam*, vol. 3 (Braunschweig:

Islamic perspective as creatures with the highest intelligence, created with an intellect and the ability to choose how to live their life. Humanity is created "in the best of stature"[168] in their physical and mental capacities and is a creature of dignity.[169] Believing in God is in their nature,[170] but they themselves decide to follow the path of faith or to neglect it. Their duty as *ḫalīfah*, a trustees on earth, is a test that shows whether they will act carefully with creation or live a selfish life.[171] If they reflect on their own habits and actions, they can fulfil their character to the best, but if they remain selfish and seek power over the rest of creation, they can be the most evil creatures.[172] Humanity's insatiable nature is described in *Sūrat at-Takāthur* [Competition], where they are reminded that no one can take possessions with them to the grave, meaning that materialistic acquisitions will not count on the day of judgement, but only good actions and behaviour.[173] With their intellect, humans can develop special abilities and they are required to use their minds.[174] At the same time, they must not forget to be grateful[175] for the mercy they receive from God. Humans are responsible for their actions, and they will be questioned on the day of judgement to justify them with no one else to take on this responsibility.[176] Social ethics such as helping others[177] and being merciful with all of creation are emphasised in Islamic sources.[178]

The most crucial topic when it comes to Islam and ecology is the discussion about anthropocentrism.[179] Is humanity the most important creature because of their intellect? And does this allow them to exploit resources as they are

---

Georg—Eckert-Institut für internationale Schulbuchforschung, 1996), 17–19; Kowanda-Yassin, *Mensch und Naturverständnis*. 53, 54, 90–104.

168 Qur'ān, 95:4.

169 See also Tariq Ramadan, *Der Islam. Gott-Mensch-Welt. Begegnung mit dem Islam* (Berlin: Greenpalace, 2002), 83.

170 Qur'ān, 7:172, with a commentary of Muhammad Asad, *Die Botschaft des Korans. Übersetzung und Kommentar* (Düsseldorf: Patmos, 2011), 301.

171 See also Abdul Hamid Abu Sulayman, *The Qur'anic Worldview. A Springboard For Cultural Reform* (London/Washington: The International Institute of Islamic Thought, 2011), 76–84.

172 Qur'ān, 103:2–3, 91:7–10; Falaturi and Tworuschka, *Der Islam im Unterricht.*

173 Qur'ān, 102:1–2.

174 Qur'ān, 16:4, 2:31.

175 Qur'ān, 46:15.

176 Qur'ān, 74:38.

177 Qur'ān, 2:177.

178 Ṣaḥīḥ Muslim, Ḥadīṯ No. 4282/Kapitel 43, Islamische Datenbank, "Ḥadīṯ No. 4282," accessed April 8, 2019. http://islamische-datenbank.de/sahih-muslim?action=viewhadith-&chapterno=43&min=20&show=10: "*man lā yarḥam lā yurḥam.* [He who is not merciful will not get mercy]." and see also Falaturi, *Der Islam im Unterricht*, 17.

179 Kowanda-Yassin, *Mensch und Naturverständnis*, 23–90.

doing? What is the duty of humanity on earth? And what are the responsibilities pupils can take on? By presenting this picture of human characteristics, a lively discussion on contemporary ecological challenges, their causes, and the influence of humanity can be started and pupils may reflect on their attitude towards a responsible lifestyle.[180]

5.3 *Social Ethics and Ecology—Thinking about One's Own Habits*

In contemporary Islamic Religious Education, the importance of social justice, mercy with all creatures, and helpfulness is pointed out and reflected on in religious sources.[181] The school books *Islamstunde* for Islamic education in state schools in Austria stress social ethics, underpinning these with *ḥadīṯ*s and Qur'ānic verses.[182] Pupils are taught that believing in God is not enough, that real faith is demonstrated by the way they practise their religion and by interacting sensitively with their social and natural environment. In the following paragraphs, the social ethics which are stressed in Islamic instruction will be discussed and religious sources will be quoted. In the Qur'ān, faith and good deeds are very often named together.[183] What one wishes for oneself applies also to one's brothers and sisters,[184] including neighbours, who are to be treated in the best manner:[185] Muslim believers are reminded that having power over other creatures is a great responsibility and must not be exploited to further one's own interests.[186]

Muslims are enjoined to stand up against injustice and to rectify grievances[187] as far as this lies within their ability.[188] They are obliged to give their best and try to do beneficial deeds for society.[189] All these areas remind learners that they are a part of society and need to take responsibility for their actions. They can decide to be a valuable member of society and help others in order

180 See also part 1. of this chapter.

181 For example Qur'ān, 49:11–12; 2:177.

182 Said Topalovic et al., *Islamstunde 8. Die Anderen respektieren. Religionsbuch für die Sekundarstufe I*, 61–79, and *Islamstunde 7*, 61–65, 78–81 both edited by Amena Shakir.

183 Qur'ān, 31:8, 19:96.

184 Rassoul, ed., *Ḥadīṯ für Schüler*, 127.

185 Ibid. 135.

186 Muhammad Asad, *Die Prinzipien von Staat und Regierung im Islam* (Mössingen: Edition Bukhara, 2011).

187 Rassoul, ed., *Ḥadīṯ für Schüler*, 214–216. "*man ra'ā minkum munkaran fa-l-yuġaiyirhu bi-yadihi fa-in lam yastaṭiʿ fa-bi-lisānihi fa-in –lam yastaṭiʿ fa-bi-qalbihi ḏālika aḍʿafu l-imāni* [Whosoever of you sees evil, let him change it with his hand; and if he is not able to do so, then (let him change it) with his tongue; and if he is not able to do so, then with his heart—and that is the weakest of faith]."

188 Asad, *Die Prinzipien*, 53–63.

189 Rassoul, ed., *Ḥadīṯ für Schüler*, 146, 167–173, 204–206.

to create a peaceful, socially just society. They are called to put into practise their religious conscience and to consider other people's plight and how they can help to alleviate them.[190] By getting to know these religious traditions and discussing the topic, pupils can reflect on religious content, become sensitised to an environmentally friendly approach in Islam, and bear witness in their own life.

5.4 *A Spiritual Approach—The Wonder of Nature in the Qur'ān*[191]

A spiritual approach to religious education is to talk about God's signs in nature. In his poem, Yunus Emre describes the production of honey by bees as an act of praise because they "bless the prophet a thousand times when they enter the bee hive".[192] This spiritual approach is often associated with Sufism[193] as Sufi poets took a mystical view of nature and religion. Sufi poets refer to the Qur'ānic verses which point to the value of nature and the effects of the experience of nature upon the human soul.[194] For Yunus Emre, the basis of the Qur'ān is a literal description of the wonders in nature.[195]

The organisation "Wisdom in Nature", which embraces mysticism, connects "Permaculture & Sufism" by offering workshops. One such workshop, "A Return to the Natural Patterning" was offered in 2014 starting with a quotation by Jalaludin Rumi: "The speech of water, the speech of earth, and the speech of mud are apprehended by the sense of them that have hearts".[196]

When pupils get to know their religion and its spirituality, they can find their own individual ways in which to practise their belief and come closer to God.[197] They have the chance to philosophise and reflect on their own thoughts and feelings. Charles le Gai Eaton compares the verses of the Qur'ān to nature as divine signs for humankind: both nature and the Qur'ān have the same creator and the same Arabic name: *āya*.[198]

---

190 Rassoul, ed., *Ḥadīṯ für Schüler*, 146, 167, 184, 217.

191 As afore-mentioned in this article many verses in the Qur'an point out the wonder of God's creation and its effect on mankind. Kowanda-Yassin, *Mensch und Naturverständnis*, 67–68.

192 Schimmel, *Yunus Emre*, 86.

193 On the praise of God by all creation see e.g. Navid Kermani, *Gott ist schön. Das ästhetische Erleben des Koran* (München: C.H. Beck, 2011), 420.

194 Qur'ān, 45:3–4. See also part 4.4 of this chapter.

195 Qur'ān, 45/3–4, 67/3–4.

196 Wisdom in Nature, "Permaculture and Sufism. A Return to the Natural Patterning," accessed November 01, 2018. http://wisdominnature.org/category/sufism/.

197 Bahr, and Leimgruber, *Verantwortung für die Eine Welt*, 482–484.

198 Eaton, *Der Islam und die Bestimmung des Menschen*, 379, 422.

5.5 *Animal Welfare*

This issue deserves wider treatment than it can be given within the scope of this article.[199] Therefore, it will only be presented here as a short excerpt of Islamic principles. The importance of treating animals with care has been handed down in many *ḥadīṯ*s. According to prophetic traditions, showing mercy to animals is an obligation and mistreating them is forbidden.[200] In a *ḥadīṯ*, a camel complains about its owner failing to treat it well.[201] In Qur'ānic verses, animals are regarded as creatures that praise their creator at all times[202] and are understood as useful for humans to satisfy their own daily needs like nourishment, warmth or safety.[203]

This topic in particular highlights the ambivalence within societies and the need for discussion in Muslim communities. In daily life, situations contrary to the above recommendations can be found in many countries, especially when it comes to the treatment of animals. There is therefore an increasing number of ecologically conscious Muslims who have become vegetarian and given up the consumption of meat. Over the past few decades, we have seen a discussion on vegetarianism and Islam in various contexts.[204]

Combining environmental and religious education is an idea which is already being practised by teachers and NGOs. Islamic environmentalism is becoming more widespread, underpinned by religious sources and traditions. Interest in the topic is growing, and researchers are studying theological sources on Islam and ecology as well as Muslim awareness of the environment.

Notable practical achievements can be seen in the work of NGOs and the numerous projects which have been brought to completion. In the long term, implementing a change in habits and increasing awareness of ecologically sensitive alternatives becomes more complex, and it may well be a long process until these are internalised. As studies have shown, religion can act as an amplifier in the process. It is thus fitting to incorporate this topic in Religious Education and to develop interdisciplinary ecological education in connection with religion, paving the way to a more enlightened, cleaner future.

---

199 For further discussion on Animal Welfare see also Kowanda-Yassin, *Mensch und Naturverständnis*, 165–194.

200 Isabel Schatzschneider, *Animal Welfare in Ḥalāl Market Standards. European Conceptions Coined in Islamic Terms*, Bonner Islamwissenschaftliche Hefte, Heft 42 (Berlin: EB, 2018) 16–17.

201 Schatzschneider, *Animal Welfare*, 17.

202 Qur'ān, 57:1.

203 Qur'ān, 36:71–73.

204 Richard Foltz, *Animals in the Islamic Tradition and Muslim Cultures* (Oxford: Oneworld Publications, 2006); Basri Masri, *Islamic Concern for Animals* (Petersfield: The Athene Trust, 1987); Kowanda-Yassin, *Mensch und Naturverständnis*, 188–192.

Bibliography

ABB. "HVDC." Web page on HVDC applications. Accessed May 15, 2017. http://new.abb.com/systems/hvdc.

Abbaspour, Madjid and Alieh Sabetraftar. "Review of Cycles and Indices of Drought and their Effect on Water Resources, Ecological, Biological, Agricultural, Social and Economical Issues in Iran." *International Journal of Environmental Studies* 62.6 (2005).

Abdel-Haleem, Harfiyah. *Islam and the Environment.* London: Ta-Ha-Publishers Ltd., 1998.

Abdul-Matin, Ibrahim. *Green Deen. What Islam Teaches about Protecting the Planet.* Leicestershire: Kube Publishing, 2012.

Abu Sulayman, Abdul Hamid. *The Qur'anic Worldview. A Springboard For Cultural Reform.* London/Washington: The International Institute of Islamic Thought, 2011.

Adams, Robert McC and Hans J. Nissen. *The Uruk Countryside.* Chicago, London: The University of Chicago Press, 1972.

AFDC. "Association for Forests, Development & Conservation." Accessed November 1, 2019. http://www.afdc.org.lb/.

AFDC. "Earn and Learn." Accessed November 1, 2019. http://www.afdc.org.lb/content/earn-and-learn.

AghaKouchak, Amir, Hamidreza Norouzi, Kaveh Madani, Ali Mirchi. "Aral Sea Syndrome Desiccates Lake Urmia: Call for Action." *Journal of Great Lakes Research* 41.1 (2015): 307–311.

Agrar- und Umweltpädagogik. "Fort- und Weiterbildung." Accessed June 8, 2018. http://www.agrarumweltpaedagogik.ac.at.

Agwan, A.R. *Islam and the Environment.* New Delhi: Institute of Objective Studies, 1997.

Ahmad, Khaliq and Kazeem Ogunsola. "An Empirical Assessment of Islamic Leadership Principles." *International Journal of Commerce and Management* 21.3 (2011): 292.

al-Akhbar. "ISIS Captures Iraqi Dam, Floods Areas." April 11, 2014. https://english.al-akhbar.com/node/19385.

Alipour, Habib and Hossein Ghasemi Tangal Olya. "Sustainable Planning Model toward Reviving Lake Urmia." *International Journal of Water Resources Development* 31.4 (2015): 519–539.

al-Karaji, Muhammad. *Inbaṭ al-miyāh al-ḫafīya* [The Extraction of Hidden Waters]. Tehran: Bonyad Farhang Iran Publication, 1966.

al-Khashab, Wafiq H. *The Water Budget of the Tigris and Euphrates Basin.* Chicago: University of Chicago Press, 1958.

Allam, Hossam and Simon Inauen. *E-Waste Management Practices in the Arab Region.* Cairo: CEDARE, 2009. Accessed October 19, 2017. http://citeseerx.ist.psu.edu/viewdoc/download?doi=10.1.1.625.2567&rep=rep1&type=pdf.

Allan, Tony (J.A.). *The Middle East Water Question. Hydroploitics and the Global Economy.* I.B. Tauris, London, New York, 2002.

Alliance of Religions and Conservation. "About ARC." Accessed August 13, 2018. http://www.arcworld.org/.

al Sayyed al Glenid, Mohamed, Abou Bakr Ahmed Ba Kader, Abdul Latif Tawfik El Shirazy Al Sabbagh, Mouel Yousef Samarrai Izzidien, and Othman Abd-ar-Rahman Llewellyn. *Environmental Protection in Islam.* Cambridge: World Conservation Union, 1994.

al Tamimi, Samar A. Abid. "Reshaping Higher Education in the Gulf States: Study Abroad Trends and Student Experiences." *Gulf Affairs* (2017): 10–13.

Amiet, Pierre. *La glyptique mésopotamienne archaïque.* Paris: Centre Nationale de la Recherche Scientifique, 1980.

Amiet, Pierre. "Susa in the Ancient Near East." In *The Royal City of Susa: Ancient Near Eastern Treasures in the Louvre,* edited by Prudence O. Harper, Joan Aruz, and Françoise Tallon. New York: Harry N. Abrams, 1992.

Andrae, Anders and Tomas Edler. "On Global Electricity Usage of Communication Technology: Trends to 2030." *Challenges* 6.1 (2015): 117–157.

Anthonioz, Stéphanie. *L'eau, enjeux politiques et théologiques, de Sumer à la Bible.* Leiden: Brill, 2009.

Anthonioz, Stéphanie. "Water(s) of Abundance in the Ancient Near East and in Hebrew Bible Texts: A Sign of Kingship." In *Thinking of Water in Judah,* edited by Ehud Ben Zvi and Christoph Levin. Berlin: de Gruyter, 2014.

Anvari, Arezoo, and Mohammad Valaie. "Evaluation of the Effects of Water Level Decline of Urmia Lake in Sustainable Rural Development: Case Study: Central Marhamat Abad Rural District, Miandoab County." *Bulletin of Environment, Pharmacology and Life Sciences* 4.7 (2015): 65–71.

Aqrawi, Adnan A.M. "Stratigraphic Signatures of Climatic Change during the Holocene Evolution of the Tigris—Euphrates Delta, Lower Mesopotamia." *Global and Planetary Change* 28 (2001): 267–283. doi:10.1016/S0921-8181(00)00078-3.

Arab Forum for Environment and Development. "AFED Announces Report and Conference on Environmental Education." Accessed November 1, 2019. http://www.afedonline.org/en/inner.aspx?contentID=1370.

Arab Forum for Environment and Development. "AFED Course of Action." Accessed November 1, 2019. http://www.afedonline.org/en/inner.aspx?menuID=6.

Arab Forum for Environment and Development. "AFED Environmental Education Program." Accessed November 1, 2019. http://www.afedonline.org/en/inner.aspx?contentID=180.

Armstrong, Karen. *A Short History of Myth*. Edinburgh: Canongate, 2005.

Armstrong, Karen. *The Great Transformation—The Beginning of Our Religious Traditions*. New York: Alfred A. Knopf, 2006.

Arz, Helge W., Frank Lamy, Jürgen Pätzold, Peter J. Müller, and Maarten Prins. "Mediterranean Moisture Source for an Early-Holocene Humid Period in the Northern Red Sea." *Science* 300 (2003): 118–121. doi:10.1126/science.1080325.

Arz, Helge W., Frank Lamy, and Jürgen Pätzold. "A Pronounced Dry Event Recorded around 4.2 ka in Brine Sediments from the Northern Red Sea." *Quaternary Research* 66 (2006): 432–441. doi:10.1016/j.yqres.2006.05.006.

Asad, Muhammad. *Die Botschaft des Korans. Übersetzung und Kommentar*. Düsseldorf: Patmos, 2011.

Asad, Muhammad. *Die Prinzipien von Staat und Regierung im Islam*. Mössingen: Edition Bukhara, 2011.

Asem, Alirez, Amin Eimanifar, Morteza Djamali and Patricia De los Rios. "Biodiversity of the Hypersaline Urmia Lake National Park (NW Iran)" *Diversity* 6 (2014): 102–132.

Asem, Alireza, Amin Eimanifar, and Michale Wink. "Update of 'Biodiversity of the Hypersaline Urmia Lake National Park (NW Iran)'." *Diversity* 8.6 (2016).

Assmann, Jan. *Das kulturelle Gedächtnis. Schrift. Erziehung und politische Identität in frühen Hochkulturen*. München: C.H. Beck, 1992.

Assmann, Jan. *Achsenzeit. Eine Archäologie der Moderne*. München: C.H. Beck, 2018.

Attinger, Pascal. *Eléments de linguistique sumérienne: la construction de du*$_{11}$*/e/di „dire"*. Fribourg, Suisse: Ed. Universitaires, 1993.

Aṭ-ṭuhūru šarṭu l-ʹimāni [Cleanliness is part of Faith]." In *Ḥadīṯ für Schüler*, edited by Rassoul, 167–169.

Ayboğa, Ercan and Akgün Ilhan. "Iran's Dam Policy and the Case of the Lake Urmia." Last modified July 31, 2012, accessed May 18, 2018. http://www.academia.edu/9613644.

Baalousha, Husam M. and Omar K.M. Ouda. "Domestic Water Demand Challenges in Qatar." *Arabian Journal of Geosciences* 10 (2017).

Bagg, Ariel M. *Assyrische Wasserbauten. Landwirtschaftliche Wasserbauten im Kernland Assyriens zwischen der 2. Hälfte des 2. und der 1. Hälfte des 1. Jahrtausends v. Chr.* Mainz: Verlag Philipp von Zabern, 2000.

Bahadori, Mehdi N. "Passive Cooling Systems in Iranian Architecture." *Scientific American* 238. 2 (1978): 144–154.

Bahadori, Mehdi N. and A. Dehgani Sanij. *Wind Towers—Architecture, Climate and Sustainability*. Basel: Springer International Publishing, 2014.

Bahr, Matthias. "Handlungsorientiertes Lernen." In *Religionsdidaktik. Ein Leitfaden für Studium, Ausbildung und Beruf*, edited by Georg Hilger, Stephan Leimgruber, and Hans G. Ziebertz, 542–548. München: Kösel-Verlag, 2010.

Bahr, Matthias and Stephan Leimgruber. "Verantwortung für die Eine Welt." In *Religionsdidaktik. Ein Leitfaden für Studium, Ausbildung und Beruf*, edited by Georg Hilger, Stephan Leimgruber, Hans G. Ziebertz. München: Kösel-Verlag, 2010.

Bahri, Media Z. Interview given to Rizki Nugraha and Ayu Purwaningsih/Deutsche Welle on June 14, 2018. Last accessed January 28, 2019. http://www.dw.com/en/indonesia-combines-islam-with-environmental-activism/a-44220277.

Baldé, Kees, Vanessa Forti, Vanessa Gray, Ruediger Kuehr, Paul Stegmann. *The Global E-waste Monitor*. Bonn, Geneva, Vienna: United Nations University (UNU), International Telecommunication Union (ITU), and International Solid Waste Association (ISWA), 2017.

Barks, Coleman trans. *The Essential Rumi. New Expanded Edition*. New York: Harper One, 2004.

Bar-Matthews, Miryam, Avner Ayalon, Mabs Gilmour, Alan Matthews, and Chris J. Hawkesworth. "Sea-land Oxygen Isotopic Relationships from Planktonic Foraminifera and Speleothems in the Eastern Mediterranean Region and their Implication for Paleorainfall during Interglacial Intervals." *Geochimica et Cosmochimica Acta* 67 (2003): 3181–3199. doi:10.1016/S0016-7037(02)01031-1.

Bar-Matthews, Miryam and Avner Ayalon, "Mid-Holocene Climate Variations Revealed by High-resolution Speleothem Records from Soreq Cave, Israel and their Correlation with Cultural Changes." *The Holocene* 21 (2011): 163–171. doi:10.1177/0959683610384165.

Basri Masri. *Islamic Concern for Animals*. Petersfield: The Athene Trust, 1987.

Bauchot, Roland, ed. *Schlangen: Evolution, Anatomie, Physiologie, Ökologie und Verbreitung, Verhalten, Bedrohung und Gefährdung, Haltung und Pflege*. Augsburg: Bechtermünz Verlag, 1994.

Baylson, Joshua C. *Territorial Allocation by Imperial Rivalry. The Human Legacy in the Near East*. Chicago: University of Chicago, Department of Geography, 1987.

Beaumont, Peter. "Qanat Systems in Iran." *Hydrological Science Journal* 16 (1971): 39–50.

Beaumont, Peter, Gerald H. Blake, and J. Malcolm Wagstaff. *The Middle East. A Geographical Study*. London, New York, Sydney, Toronto: John Wiley and Sons, 1979.

Beaumont, Peter and Michael E. Bonine. *Qanat, Kariz and Khattara: Traditional Water Systems in the Middle East and North Africa*. London: University of London, 1989.

Beazley, Elizabeth. "Some Vernacular Buildings of the Iranian Plateau." *Iran* XV (1977): 89–102.

Becker, Helmut, Margarete van Ess, and Jörg Fassbinder. "Uruk: Urbane Strukturen im Magnet- und Satellitenbild." In *Uruk. 5000 Jahre Megacity. Begleitband zur Ausstellung 'Uruk—5000 Jahre Megacity' im Pergamonmuseum. Hrsg. von der Curt-Engelhorn-Stiftung für die Reiss-Engelhorn-Museen Mannheim, dem Deutschen Archäologischen Institut—Orient-Abteilung, der Deutschen Orient-Gesellschaft e. V. und dem Vorderasiatischen Museum—Staatliche Museen zu Berlin*, edited by Nicole Crüsemann, Margarete van Ess, Markus Hilgert, and Beate Salje, 355–361. Petersberg: Imhof Verlag, 2013.

Benima, Tamarah. *Joodser dan dit krijgt u het niet. De levenskunst van de Joodse Beschaving*. Amsterdam: Prometheus, 2015.

Berkhout, Frans and Julia Hertin. *Impacts of Information and Communication Technologies on Environmental Sustainability: Speculations and Evidence. Report to the OECD*. Brighton: Science and Technology Policy Research Unit, University of Sussex, 2001.

Berry, Richard W., Gerald P. Brophy, and Adnan Naqash. "Mineralogy of the Suspended Sediment in the Tigris, Euphrates, and Shatt-al-Arab Rivers of Iraq, and the Recent History of the Mesopotamian Plain." *Journal of Sedimentary Research* 40 (1970): 131–139. doi:10.1306/74D71F05-2B21-11D7-8648000102C1865D.

Beyers, Bert. *Migration, Nachhaltigkeit und ein Marshall Plan mit Afrika—Denkschrift für die Bundesregierung*. Ulm: Club of Rome and Senat der Wirtschaft, 2017. Accessed June 12, 2020. http://www.faw-neu-ulm.de/portfolio-item/denkschrift-bundesregierung.

Bianca, Stefano. *Hofhaus und Paradiesgarten: Architektur und Lebensformen in der islamischen Welt*. Munich: C.H. Beck, 1991.

Birnbaum, Elisabeth. "Wüste, Sintflut, Garten Eden. Ambivalente Naturerfahrung im Alten Testament." In *Gärten in der Wüste, Schöpfungsethik zwischen Wunsch und Wirklichkeit*, edited by Severin J. Lederhilger. Regensburg: Verlag Friedrich Pustet, 2018.

Black, Jeremy and Anthony Green. *Gods, Demons and Symbols of Ancient Mesopotamia: An Illustrated Dictionary*. London: British Museum Press, 1992.

Bobek, Hans. "Die Hauptstufen der Gesellschafts- und Wirtschaftsentfaltung in geographischer Sicht." *Die Erde: Journal of the Geographical Society of Berlin* 90 (1959): 259–298.

Bobek, Hans. "The Main Stages in Socio-economic Evolution from a Geographical Point of View." In *Readings in Cultural Geography*, edited by P.L. Wagner, 218–247. Chicago: University of Chicago, Dept. of Geography, 1962.

Bobek, Hans. "Zum Konzept des Rentenkapitalismus." *Tijdschrift voor Economische en Sociale Geografie* 65 (1974): 73–78.

Bobek, Hans. "Rentenkapitalismus und Entwicklung in Iran." In *Interdisziplinäre Iran-Forschung. Beiträge aus Kulturgeographie, Ethnologie, Soziologie und Neuerer Geschichte*, 113–124. Wiesbaden: Dr. Ludwig Reichert, 1979.

Boehmer, Rainer N. *Uruk: Früheste Siegelabrollungen*. Mainz: Philipp von Zabern, 1999.

Bond, Gerald, William Showers, Maziet Cheseby, Rusty Lotti, Peter Almasi, Peter deMenocal, Paul Priore, Heidi Cullen, Irka Hajdas, and Georges Bonan. "A Pervasive Millennial-Scale Cycle in North Atlantic Holocene and Glacial Climates." *Science* 278 (1997): 1257–1266. doi:10.1126/science.278.5341.1257.

Bonine, Michael E. "The Morphogenesis of Iranian Cities." *Annals of the Association of American Geographers* 69 (1979): 208–224.

Bonine, Michael E. *Yazd and its Hinterland. A Central Place System of Dominance in the Central Iranian Plateau*. Marburg. Institut für Geographie der Universität Marburg, Selbstverlag, 1980.

Bonine, Michel E., Abbas Amanat, and Michael E. Gasper, ed. *Is there a Middle East? The Evolution of a Geopolitical Concept*. Redwood City: Stanford University Press 2011.

Bonine, Michael E., Eckart Ehlers, Th. Krafft, and G. Stöber, *The Middle Eastern City and Islamic Urbanism. An Annotated Bibliography of Western Literature*. Bonn: F. Dümmlers Verlag, 1994.

Bourke, Stephen. *The Middle East. The Cradle of Civilization*. London: Thames and Hudson, 2008.

Bracken, Nathan, Jordan Macknick and Angelica Tovar-Hastings, Paul Komor, Margot Gerritsen and Shweta Mehta. *Concentrating Solar Power and Water Issues in the U.S. Southwest, Technical Report, NREL/TP-6A50–61376*. Denver: The Joint Institute for Strategic Engergy Analysis, 2015.

Brand, Bernhard. *Les énergies renouvelables et la digitalisation des systèmes électriques: Tendances et défis pour les pays de la région MENA*. Working paper February 2017.

Brandes, Mark A. *Siegelabrollungen aus den archaischen Bauschichten in Uruk-Warka*. Wiesbaden: Steiner Verlag, 1979.

Brandimarte, Luigia, Giuliano Di Baldassarre, Guendalina Bruni, Paolo D'Odorico, and Alberto Montanari. "Relation between the North-Atlantic Oscillation and Hydroclimatic Conditions in Mediterranean Areas." *Water Resources Management* 25 (2011): 1269–1279. doi: 10.1007/s11269-010-9742-5.

Braun, Cornel. *Teheran, Marrakesch und Madrid. Ihre Wasserversorgung mit Hilfe von Qanaten. Eine stadtgeographische Konvergenz auf kulturhistorischer Grundlage*. Bonn: Ferd. Dümmlers, 1974.

Braun-Holzinger, Eva A. "Löwenadler." In *Reallexikon der Assyriologie und Vorderasiatischen Archäologie*, edited by Dietz O. Edzard. Berlin: Walter de Gruyter, 1987–1990.

Braun-Holzinger, Eva A. "Apotropaic Figures at Mesopotamian Temples in the Third and Second Millennia." In *Mesopotamian Magic: Textual, Historical, and Interpretative Perspectives*, edited by Tzvi Abusch and Karel van der Toorn. Groningen: Styx Publications, 1999.

Brice, William C. ed. *The Environmental History of the Near and Middle East*. London, New York, San Francisco: Academia Press, 1978.

Bridey, François. *L'iconographie du décor peint de la céramique de Suse I : Les coupes des collections du musée du Louvre et du musée d'Archéologie nationale de Saint-Germain-en-Laye*. Paris: RMN, 2011.

Brinkman, John A. "Settlement Surveys and Documentary Evidence: Regional Variation and Secular Trend in Mesopotamian Demography." *Journal of Near Eastern Studies* 43 (1984): 169–180. doi:10.1086/373078.

Brookes, John. *Gardens of Paradise. The History and Design of the Great Islamic Gardens*. London: G. Weidenfeld and Nicholson, 1987.

Brown, Leon C. *From Madina to Metropolis. Heritage and Change in the Near Eastern City*. Princeton: The Darwin Press, 1973.

Brückner, Helmut. "Uruk – A Geographic and Palaeo-Ecologic Perspective on a Famous Ancient City in Mesopotamia." *Geoöko* 24 (2003): 229–248.

Brückner, Helmut. "Wasserstraßen im Wüstensand." *Antike Welt* 44.3 (2013): 18–24.

Brückner, Helmut and Max Engel. "Noah's Flood—Probing an Ancient Narrative Using Geoscience." In *Palaeohydrology—Traces, Tracks and Trails of Extreme Events*, edited by Jürgen Herget and Alessandro Fontana, 135–151. Cham: SpringerNature, 2020. doi:10.1007/978-3-030-23315-0_7.

Bryant, Mark and Sophie Gilliat-Ray. "Are British Muslims 'green'? An Overwiew of Enviromental Activism among Muslims in Britain." *Journal for the Study of Religion, Nature and Culture* 5.3. (2013): 284–306.

Buber, Martin. *I and Thou*. New York: Charles Scribner's Sons, 1970.

Bundesministerium für wirtschaftliche Zusammenarbeit und Entwicklung (BMZ). *Afrika und Europa—Neue Partnerschaft für Entwicklung, Frieden und Zukunft—Eckpunkte für einen Marshallplan mit Afrika*. Berlin: BMZ, 2017.

Bundesministerium Nachhaltigkeit und Tourismus. "Natur- und Artenschutz." Accessed June 8, 2018. https://www.bmnt.gv.at/umwelt/natur-artenschutz/wdavf_2018.html.

Burney, Charles. *From Village to Empire. An Introduction to Near Eastern Archaeology*. Oxford: Phaidon Press Ltd., 1975.

Butzer, Karl W. *Early Hydraulic Civilization in Egypt. A Study in Cultural Ecology*. Chicago: University of Chicago Press, 1976.

Cancik-Kirschbaum, Eva. "'Menschen ohne König …': Zur Wahrnehmung des Königtums in sumerischen und akkadischen Texten." In *Das geistige Erfassen der Welt im Alten Orient: Sprache, Religion, Kultur und Gesellschaft*, edited by Claus Wilcke, 167–190 (Wiesbaden: Harrassowitz Verlag, 2007.

Carter, Elizabeth, Frank Hole, Zainab Bahrani, Agnès Spycket, and Joan Aruz. "Prehistoric Susa." In *The Royal City of Susa: Ancient Near Eastern Treasures in the Louvre*, edited by Prudence O. Harper, Joan Aruz, and Françoise Tallon, 25–46. New York: Harry N. Abrams, 1992.

Carter, Robert A. "The Social and Environmental Context of Neolithic Seafaring in the Persian Gulf." In *The Global Origins (and Development) of Seafaring*, edited by Atholl Anderson, James H. Barrett, and Katherine V. Boyle, 191–202. Cambridge: McDonald Institute of Archaeological Research, 2010.

Castells, Manuel. *The Rise of the Network Society: The Information Age: Economy, Society, and Culture*. Hoboken: John Wiley & Sons, Ltd., 2009.

Cavigneaux, Antoine. "La Pariade du Scorpion dans les Formules Magiques Sumériennes (Textes de Tell Haddad V)." *Acta Sumerologica* 17 (1995): 75–99.

Chen, Yi S. "Major Literary Traditions Involved in the Making of Mesopotamian Flood Traditions." In *Opening Heaven's Floodgates*, edited by Jason M. Silverman, 141–190. New York: Gorgias Press, 2013.

Chen, Yi S. *The Primeval Flood Catastrophe*. Oxford: Oxford University Press, 2013.

Chrebah, Bouchrah. *Umweltbewusstsein und Umweltverhalten—Ein Vergleich von deutschen und syrischen Studierenden*. PhD.diss., Oldenburg: Fakultät für Mathematik und Naturwissenschaften der Carl von Ossietzky Universität Oldenburg, 2009.

Clarke, Joanne Nick Brooks, Edward B. Banning, Miryam Bar-Matthews, Stuart Campbell, Lee Clare, Mauro Cremaschi, Savino di Lernia, Nick Drake, Marina Gallinaro, Sturt Manning, Kathleen Nicoll, Graham Philip, Steve Rosen, Ulf Dietrich Schoop, Mary Anne Tafuri, Bernhard Weninger, and Andrea Zerboni. "Climatic Changes and Social Transformations in the Near East and North Africa during the 'Long' 4th Millennium BC: A Comparative Study of Environmental and Archaeological Evidence." *Quaternary Science Reviews* 136 (2016): 96–121. doi:10.1016/j.quascirev.2015.10.003.

Claussen, Martin, Claudia Kubatzki, Victor Brovkin, Andrey Ganopolski, Philipp Hoelzmann, and Hans-Joachim Pachur. "Simulation of an Abrupt Change in Saharan Vegetation in the Mid-Holocene." *Geophysical Research Letters* 26 (1999): 2037–2040. doi:10.1029/1999GL900494.

Cohn, Ellen. "Growing an Environmental Ethic: The Conceptual Roots of Bal Tashchit." In *Compendium of Sources in Halacha and the Environment*, edited by Shenson, Ora, and Spetgant, Shai. Jerusalem: Canfei Nesharim, 2005.

Cole, Stijn, Til Kristian Vrana, Jean-Baptiste Curis, and Chen-Ching Liu. "A European Supergrid: Present State and Future Challenges." 17th Power Systems Computation Conference, Stockholm, August 22–26, 2011. Accessed May 15, 2017. https://www.pscc-central.org/uploads/tx_ethpublications/sp2.pdf.

Collon, Dominique. *First Impressions. Cylinder Seals in the Ancient Near East*. London: British Museum Publications, 1987.

Cooke, Gary A. "Reconstruction of the Holocene coastline of Mesopotamia." *Geoarchaeology* 2 (1987): 15–28. doi:10.1002/gea.3340020102.

Costello, Vincent F. *Keshan—A City and Region in Fars*. London, New York: Bowker Publications, 1976.

Crutzen, Paul J. "Geology of Mankind," *Nature* 415 (2002).

CSP Focus. "China Officially Targets 5GW of CSP Allocation by 2020." Accessed May 15, 2017. http://www.cspfocus.cn/en/market/detail_193.htm.

Cullen, Heidi M. Peter B. deMenocal, Sidney Hemming, G. Hemming, Frank H. Brown, Tom Guilderson, and Frank Sirocko. "Climate Change and the Collapse of the Akkadian Empire: Evidence from the Deep Sea." *Geology* 28 (2000): 379–382. doi:10.1130/0091-7613(2000)28<379:CCATCO>2.0.CO;2.

Cullen, Heidi M. and Peter B. deMenocal. "North Atlantic Influence on Tigris—Euphrates Streamflow." *International Journal of Climatology* 20 (2000): 853–863. doi:10.1002/1097-0088(20000630)20:8<853::AID-JOC497>3.0.CO;2-M.

Dalby, Simon and Zahra Moussavi. "Environmental Security, Geopolitics and the Case of Lake Urmia's Disappearance." *Global Change, Peace & Security* 29.1 (2017): 39–55.

Daley, Stephanie. *Myths from Mesopotamia. Creation, the Flood, Gilgamesh, and Others. With Introduction and Notes*. Oxford: Oxford University Press, 2008.

de Châtel, Francesca. "The Role of Drought and Climate Change in the Syrian Uprising: Untangling the Triggers of the Revolution." *Middle Eastern Studies*, 50 (2014): 521–535.

Deckers, Katleen and Simone Riehl. "Fluvial Environmental Contexts for Archaeological Sites in the Upper Khabur Basin (Northeastern Syria)." *Quaternary Research* 67 (2007): 337–348. doi:10.1016/j.yqres.2006.11.005.

Deconinck, Stefan. *Security as a Threat to Development: The Geopolitics of Water Scarcity in the Nile River Basin*. Royal High Institute of Defence, 2009.

de Gruchy, Michelle, Katleen Deckers, and Simone Riehl. "A Diachronic Reconstruction of the Northern Mesopotamian Landscape (4th to 2nd Millennia BCE) from three Separate sources of Evidence." *Journal of Archaeological Science: Reports* 8 (2016). doi:10.1016/j.jasrep.2016.05.047.

Dehgani-Sanij, Alireza and Ali Sayigh. *Cisterns: Sustainable Development, Architecture and Energy*. Aalborg, Delft, River Publishers, 2014.

de la Blache, P. Vidal. *Principes de géographie humaine*. Paris, 1922.

de la Blache, P. Vidal. "Les genres de vie dans la géographie humaine." *Annales de Géographie* 20 (1991).

Delju, Amir H. "Observed Climate Variability and Change in Urmia Lake Basin, Iran." *Theoretical and Applied Climatology* 111.1–2 (2013).

de Miroschedji, Pierre. "Le dieu élamite au serpent et aux eaux jaillissantes." *Iranica Antiqua* 16 (1981): 1–25.

de Planhol, Xavier. *Le Monde Islamique. Essai de Géographie Religieuse*. Paris: Presses Universitaires de France, 1957/1959.

de Planhol, Xavier. *Les Fondements Géographiques de l'Histoire de l'Islam*. Paris: Flammarion, 1968/1975.

de Planhol, Xavier. *Les Nations du Prophète. Manuel géographique de politique musulmane*. Paris: Fayard, 1993.

DESERTEC Foundation. *Clean Power form Deserts—The Desertec Concept for Energy, Water and Climate Security*. Bonn: Protext Verlag, 2009.

DESERTEC Foundation. "Archive of Press Releases Since 2008." Accessed May 15, 2017. http://www.desertec.org/press.

Deutschlandfunk. "Den Schatz der frühen Kindheit bewahren." Accessed June 28, 2018. https://www.deutschlandfunkkultur.de/den-schatz-der-fruehen-kindheit-bewahren.954.de.html?dram:article_id=144099.

Dinies, Michèle, Reinder Neef, Birgit Plessen, and Harald Kürschner. "When the Desert was Green: Grassland Expansion during the Early Holocene in Northwestern Arabia." *Quaternary International* 382 (2015): 293–302. doi:10.1016/j.quaint.2015.03.007.

Dinies, Michèle, Birgit Plessen, Reinder Neef, and Harald Kürschner. "Holocene Vegetation in Northwestern Arabia—Changing Natural Resources." In *Actualités des recherches archéologiques en Arabie*, edited by Julie Goy et al., 1–19. Paris: Routes de l'Orient, 2016.

Dittmann, Reinhard. "Seals, Sealings and Tablets: Thoughts on the Changing Pattern of Administrative Control from the Late-Uruk to the Proto-Elamite Period at Susa." In *Ǧamdat Naṣr, Period or Regional Style? Papers given at a symposium held in Tübingen*, edited by Uwe Finkbeiner and Wolfgang Röllig. Wiesbaden: Reichert Verlag, 1986.

Dowlatshahi, Hengameh. *Kashan—The Great Pearl*. Tehran: Gooya Art House, 2006.

Drechsler, Philipp. "The Arabian Peninsula." In *A Companion to the Archaeology of the Ancient Near East, vol. I*, edited by Daniel T. Potts, 485–500. Hoboken: Wiley-Blackwell, 2012. doi:10.1002/9781444360790.ch25.

Drucker, Peter F. *The Age of Discontinuity: Guidelines to Our Changing Society*. New Brunswick, London: Transaction Publishers, 2011.

Duff, Alistair S. David Craig, and David A. McNeill. "A Note on the Origins of the 'Information Society'." *Journal of Information Science* 22.2 (2016).

Duverdun, Gaston. *Marrakech. Des origines à 1912*, 2 vols. Rabat, Editions Techniques Nord-Africaines, 1959/1966.

Economic and Social Commission for Western Asia (ESCWA). *Impact of ICT on Arab Youth: Employment, Education and Social Change*. Beirut: United Nations Economic and Social Commission for Western Asia, 2013.

Economic and Social Commission for Western Asia (ESCWA). *Regional Profile of the Information Society in the Arab Region*. Beirut: United Nations Economic and Social Commission for Western Asia, 2016.

Economic and Social Commission for Western Asia (ESCWA). *Innovation Policy for Inclusive Sustainable Development in the Arab Region*. Beirut: United Nations Economic and Social Commission for Western Asia, 2017.

Economic and Social Commission for Western Asia (ESCWA). *The Innovation Landscape in Arab Countries: A Critical Analysis*. Beirut: United Nations Economic and Social Commission for Western Asia, 2017.

Economic and Social Commission for Western Asia (ESCWA). *Perspectives on the Digital Economy in the Arab Region*. Beirut: United Nations Economic and Social Commission for Western Asia, 2018.

Euractive. “First Steps to Bring Saharan Solar to Europe.” Last modified February 22, 2010. Accessed May 15, 2017. http://www.euractiv.com/section/energy/news/first-steps-to-bring-saharan-solar-to-europe/.

European Commission. “HORIZON 2020—The EU Framework Programme for Research and Innovation—Call for projects 2014–2020.” Accessed May 15, 2017. http://ec.europa.eu/programmes/horizon2020/en.

European Commission. “State Aid: Commission Concludes Modified UK Measures for Hinkley Point Nuclear Power Plant are Compatible with EU Rules.” Accessed May 15, 2017. http://europa.eu/rapid/press-release_IP-14-1093_en.htm.

European Commission. “Joint Declaration on the Establishment of a Roadmap for Sustainable Electricity Trade between Morocco and the European Internal Energy Market between The Federal Republic of Germany, The French Republic, The Kingdom of Spain, The Portuguese Republic, and The Kingdom of Morocco.” Executed at Marrakech during COP22 on the 17th November 2016. Accessed May 15, 2017. https://ec.europa.eu/energy/sites/ener/files/documents/2016_11_13_set_roadmap_joint_declaration-vf.pdf.

European Information Technology Observatory (EITO). “The Impact of ICT on Sustainable Development, 2002.” Accessed July 25, 2017. http://homepage.cs.latrobe.edu.au/sloke/greenIT/eito_forum_2002.pdf, 251.

Ehlers, Eckart. “The Dying Oases of Central Iran. Causes and Consequences.” In *Desertification in Extremly Arid Environment*, edited by Wolfgang Meckelein, 65–72. Stuttgart, Geographische Studien, 1980.

Ehlers, Eckart and Abbas Saidi. “Qanats and Pumped Wells—The Case of Assad’abad, Hamadan.” In *Qanat, Kariz and Khattara: Traditional Water Systems in the Middle East and North Africa*, edited by P. Beaumont, Michael Bonine, and Kaith McLachlan, 89–112. SOAS: The Middle East Centre, School of Oriental and African Studies. University of London, 1989.

Ehlers, Eckart, Abdoldjavad Falaturi, Günther Schweizer, Georg Stöber, and Gerd Winkelhane. *Der Islamische Orient. Grundlagen zur Länderkunde eines Kulturraumes*. Köln: Studien zum Islam in interkulturellen Wechselbeziehungen, 1990.

Ehlers, Eckart and Carl F. Gethmann, ed. *Environment across Cultures*. Berlin: Springer, 2003.

Ehlers, Eckart. *Das Anthropozän. Die Erde im Zeitalter des Menschen*. Darmstadt: Wissenschaftliche Buchgesellschaft, 2008.

Eimanifar, Amin and Feridon Mohebbi. “Urmia Lake (Northwest Iran): A Brief Review.” *Saline Systems* 3.5 (2007): 105–130. 10.1016/0031-0182(86)90120-3.

Eliade, Mircea. *Histoire des croyances et des idées religieuses* 4 vols. Paris: Payot, 1976–1983, 1992/1192.

El-Yacoubi, Hassan H. and J. El-Yacoubi. *Islam: The Language of Oneness: Tawhid, the Islamic Paradigm*. Colorado: Boulder, 1994.

Emadodin, Iraj and Hans Rudolf Bork. "Degradation of Soils as a Result of Long-term Human-induced Transformation of the Environment in Iran: An Overview." *Journal of Land Use Science* 7.2 (2012): 203–219.

Emeis, Kay-Christian, Ulrich Struck, Hans-Martin Schulz, Reinhild Rosenberg, Stefano Bernasconi, Helmut Erlenkeuser, Tatsuhiko Sakamoto, and Francisca, Martinez-Ruiz. "Temperature and Salinity Variations of Mediterranean Sea Surface Waters over the Last 16,000 Years from Records of Planktonic Stable Oxygen Isotopes and Alkenone Unsaturation Ratios." *Palaeogeography, Palaeoclimatology, Palaoecology* 158 (2000): 259–280. doi: 10.1016/S0031-0182(00)00053-5.

Emerging Earth Community. "Indonesia Combines Islam with Environmental Activism." Last accessed March 26, 2019. www.emergingearthcommunity.org/innerfeed?source_site=fore&pag_title=Indonesia+combines+Islam+with+environmental+activism&article_id=6&feed_type=news&news_url=news/item/indonesia-combines-islam-with-environmental-activism/.

Engel, Max, Helmut Brückner, Anna Pinta, Kai Wellbrock Andreas Ginau Peter Voss Matthias Grottker Nicole Klasen Peter Frenzel. "The Early Holocene Humid Period in NW Saudi Arabia—Evidence from Sediments, Microfossils and Palaeo-hydrological Modelling." *Quaternary International* 266 (2012): 131–141. doi:10.1016/j.quaint.2011.04.028.

Engel, Max and Helmut Brückner. "The South Qatar Survey Project (SQSP)—Preliminary Findings on Holocene Coastal Changes and Geoarchaeological Archives." *Zeitschrift für Orient-Archäologie* 7 (2014): 290–301.

Engel, Max and Helmut Brückner. "Late Quaternary Environments and Societies: Progress in Geoarchaeology." *Zeitschrift für Geomorphologie N.F* 58.2 (2014): 1–6. doi:10.1127/0372-8854/2014/S-00168.

Engel, Max, Albert Matter, Adrian G. Parker, Ash Parton, Michael D. Petraglia, Gareth Preston, and Frank Preusser. "Lakes or Wetlands? A Comment on 'The Middle Holocene Climatic Records from Arabia: Reassessing Lacustrine Environments, Shift of ITCZ in Arabian Sea, and Impacts of the Southwest Indian and African Monsoons' by Enzel et al.," *Global and Planetary Change* 148 (2017): 257–267. doi:10.1016/j.gloplacha.2016.11.001.

English, Paul W. *City and Village in Iran: Settlement and Economy in the Kirman Basin*. Madison, Milwaukee, London: The University of Wisconsin Press, 1966.

Englund, Robert K. "Texts from the Late Uruk Period." In *Mesopotamien: Späturuk-Zeit und Frühdynastische Zeit*, edited by Pascal Attinger and Markus Wäfler, 13–233. Freiburg, Göttingen: Universitätsverlag,Vandenhoeck and Ruprecht, 1998.

Englund, Robert K. "BU!" In *Why Should Someone Who Knows Something Conceal It? Cuneiform Studies in Honor of David I. Owen on His 70th Birthday*, edited by Alexandra Kleinerman and Jack M. Sasson. Bethesda: CDL Press, 2010.

ENTSO-E—European Network of Transfer System Operators for Electricity. *Regional Investment Plan 2015—Continental South West Region*. Brussels: ENTSO-E, 2015.

ENTSO-E—European Network of Transmission System Operators for Electricity. "Continental Europe Operation Handbook." Accessed May 15, 2017. https://www.entsoe.eu/publications/system-operations-reports/operation-handbook/.

Environment and Ecology. "The Green Guide for Muslims." Accessed March 26, 2019. http://environment-ecology.com/religion-and-ecology/76-the-green-guide-for-muslims.html.

Enzel, Yehouda, Yochanan Kushnir, and Jay Quade. "The Middle Holocene Climatic Records from Arabia: Reassessing Lacustrine Environments, Shift of ITCZ in Arabian Sea, and Impacts of the Southwest Indian and African Monsoons." *Global and Planetary Change* 129 (2015): 69–91. doi:10.1016/j.gloplacha.2015.03.004.

Erdmann, Lorenz and Lorenz M. Hilty. "Scenario Analysis: Exploring the Macroeconomic Impacts of Information and Communication Technologies on Greenhouse Gas Emissions." *Journal of Industrial Ecology* (2010): 826–843.

Esack, Farid. *On Being a Muslim*. Oxford: Oneworld Publications, 2002.

Escher, Anton and Eugen Wirth. "Die Medina von Fes. Geographische Beiträge zur Persistenz und Dynamik, Verfall und Erneuerung einer traditionellen islamischen Stadt." *Erlanger Geographische Arbeiten* 53 (1992).

ESTELA—European Solar Thermal Electricity Association. "What is Solar Thermal Electricity?" Accessed May 15, 2017.

Facebook. "Deen Travellers." accessed January 09, 2019. https://www.facebook.com/deentravellers/videos/vb.387084731679586/469915303396528/?type=2&theater.

Falaturi, Abdoljavad and Udo Tworuschka. *Der Islam im Unterricht. Beiträge zur interkulturellen Erziehung in Europa. Grundwissen über den Islam*, vol. 3. Braunschweig: Georg—Eckert-Institut für internationale Schulbuchforschung, 1996.

Farajzadeh, Jamileh, Ahmad Fakheri Fard, and Saeed Lotfi. "Modeling of Monthly Rainfall and Runoff of Urmia Lake Basin Using 'Feed-Forward Neural Network' and 'Time Series Analysis' Model." *Water Resources and Industry* 7–8 (2014): 38–48.

Farhangfar, Sarvenaz Mohammad Bannayan, Hamid Reza Khazaei, and Mohammad Mousavi-Baygi. "Vulnerability Assessment of Wheat and Maize Production Affected by Drought and Climate Change." *International Journal of Disaster Risk Reduction* 13 (2015).

Farr, Tom G., Paul A. Rosen, Edward Caro, Robert Crippen, Riley Duren, Scott Hensley, Michael Kobrick, Mimi Palle, Ernesto Rodriguez, Ladislav Roth, David Seal, Scott Shaffer, Joanne Shimada, Jeffrey Umland, Marian Werner, Michael Oskin, Douglas Burbank, Douglas Alsdorf. "The Shuttle Radar Topography Mission." *Reviews of Geophysics* 45 (2007): RG2004. doi:10.1029/2005RG000183.

Farshad, Amiraslani and Deirdre Dragovich. "Combating Desertification in Iran over the Last 50 Years: An Overview of Changing Approaches." *Journal of Environmental Management* 92 (2011): 1–13.

Feitelson, Eran and Amit Tubi. "A Main Driver or an Intermediate Variable? Climate Change, Water and Security in the Middle East." *Global Environmental Change* 44 (2017): 39–48.

Fick, Stephen E. and Robert J. Hijmans. "WorldClim 2: New 1-km Spatial Resolution Climate Surfaces for Global Land Areas." *International Journal of Climatology* (2017): 4302–4315. doi:10.1002/joc.5086.

Fink, Daniel B. "Judaism and Ecology: A Theology of Creation." Accessed June 11, 2018. fore.yale.edu/religion/judaism/.

Finkelstein, Jacob J. "Mesopotamia." *Journal of Near Eastern Studies* 21 (1962): 73–92. doi:10.1086/371676.

Flohr, Pascal, Dominik Fleitmann, Eduardo Zorita, and Aleksey Yu. Sadekov "Late Holocene Droughts in the Fertile Crescent Recorded in a Speleothem from Northern Iraq." *Geophysical Research Letters* 44 (2017): 1528–1536. doi:10.1002/2016GL071786.

Foltz, Richard. *Animals in the Islamic Tradition and Muslim Cultures*. Oxford: Oneworld Publications, 2006.

Foltz, Richard C. "Iran's Water Crisis: Cultural, Political, and Ethical Dimensions." *Journal of Agricultural and Environmental Ethics* 15 (2002): 357–359.

Foltz, Richard C. *Environmentalism in the Muslim World*. New York: Nova Science Publishers, Inc., 2005.

Foltz, Richard C. "Iran." In *Environmentalism in the Muslim World*, edited by Richard C. Foltz. New York: Nova Science Publishers, Inc., 2005.

Fragner, Bert G. "Der Rentenkapitalismus als islamwissenschaftliche Forschungsaufgabe." *Zeitschrift Deutsche Morgenländische Gesellschaft* (*ZDMG*) 5 (1983): 168–175.

Fragner, Bert G. *Rent-capitalism in the Light oft the Economic History of Early Iran. Proceedings of the Symposium held on the Occasion of the 50th Anniversary of the Wiener Institut für Völkerkunde*. Wien: Wiener Beiträge zur Ethnologie und Anthropologie 1 (1984): 205–227.

Frenken, Karen: *Irrigation in the Middle East Region in Figures. AQUASTAT Survey—2008*. Rome: Food and Agriculture Organization of the United Nations, 2009.

Gelenbe, Erol and Yves Caseau. "The Impact of Information Technology on Energy Consumption and Carbon Emissions." *Ubiquity* (2015): 1–15.

Gellert, Manfred and Claus Nowak. *Teamarbeit, Teamentwicklung, Teamberatung: Ein Praxisbuch für die Arbeit in und mit Teams*. Limmer Verlag, 2010.

George, Andrew R. *The Babylonian Gilgamesh Epic: Introduction, Critical Edition and Cuneiform Texts vol. 1*. Oxford: Oxford University Press, 2003.

Gerber, Christoph. "Die Schlange auf der Tonkugel." In *Beiträge zur Kulturgeschichte Vorderasiens: Festschrift für Rainer Michael Boehmer*, edited by Uwe Finkbeiner,

Reinhard Dittmann, and Harald Hauptmann, 175–176. Mainz: Philipp von Zabern, 1995.

Ghallab, Mamoun. "Zero Zbel Xperience." Accessed August 15, 2018. https://www.youtube.com/watch?v=K-BtrzYME2Y.

Gharatchedaghi, Cyrus. *Distribution of Land in Varamin: An Opening Phase of the Agrarian Reform in Iran*. Oplanden: Leske-Verlag, 1967.

Glacken, Clarence J. *Traces on the Rhodian Shore. Nature and Culture in Western Thought from Ancient Times to the End of the Eighteenth Century*. Berkeley, Los Angeles, London: University of California Press, 1967.

Glaser, Rüdiger. *Global Change. Das neue Gesicht der Erde*. Darmstadt: Primus Verlag, 2014.

Glasl, Friedrich. *Konfliktmanagement: Ein Handbuch für Führungskräfte, Beraterinnen und Berater*. Bern-Haupt-StuttgartVerlag Haupt, 2013.

Glassner, Jean-Jacques. *Écrire à Sumer: L'invention du cunéiforme*. Paris: Édition du Seuil, 2000.

Gleick, Peter. "Water, Drought, Climate Change, and Conflict in Syria." *Weather, Climate and Society* 6.3 (2014): 331–340.

Goblot, Henri. *Les Qanats. Une technique d'acquisition de l'eau*. Paris, La Haye, New York: Mouton Editeur, 1979.

Gohari, Alireza, Saeid Eslamian, Jahangir Abedi-Koupai, and Ali Reza Massah Bavani. "Climate Change Impacts on Crop Production in Iran's Zayandeh-Rud River Basin." *Science of the Total Environment* 442 (2013): 405–419.

Gong, Yushu. *Studien zur Bildung und Entwicklung der Keilschriftzeichen*. Hamburg: Verlag Dr. Kovač, 1993.

Gossart, Cédric. "Rebound Effects and ICT: A Review of the Literature." In *ICT Innovations for Sustainability. Advances in Intelligent Systems and Computing*, vol. 310, edited by Lorenz M. Hilty and Bernard Aebischer. Berlin: Springer, 2015.

Green Hand. "Project Description." Accessed November 1, 2019. http://www.greenhand.org.lb/projects/youth/project-description.aspx.

Green, Margret W. and Hans J. Nissen. *Zeichenliste der archaischen Texte aus Uruk*. Berlin: Gebr. Mann Verlag, 1987.

Greenpeace. "Jeder Einzelne kann etwas für den Klimaschutz tun." Accessed January 1, 2019. https://www.greenpeace-muenchen.de/images/stories/Gruppen/Energie/2016%2008%20spartipps.pdf.

Grey, David and Claudia W. Sadoff. "Sink or Swim? Water Security for Growth and Development." *Water Policy* 9 (2007): 545–571.

Gruber, Franz. "Der bedrohte Garten. Was uns der Schöpfungsglaube zu denken und zu tun gibt." In *Gärten in der Wüste, Schöpfungsethik zwischen Wunsch und Wirklichkeit*, edited by Severin J. Lederhilger. Regensburg: Verlag Friedrich Pustet, 2018.

Guagnin, Maria, Richard Jennings, Heidi Eager, Ash Parton, Christopher Stimpson, Christian Stepanek, Madlene Pfeiffer, Huw S. Groucutt, Nick A. Drake, Abdullah

Alsharekh, and Michael D. Petraglia. "Rock Art Imagery as Aproxy for Holocene Environmental Change: A View from Shuwaymis, NW Saudi Arabia." *The Holocene* 26 (2016): 1822–1834 . doi:10.1177/0959683616645949.

Hakim, Besim S. *Arabic-Islamic Cities. Building and Planning Principles*. London: KPI Limited, 1986.

Hamblin, Dora J. *The First Cities*. New York: The Life Books, 1973.

Hamdan, Mohamed A., Taro Asada, Fekri M. Hassan, Barry G. Warner, Ali Douabul, Majeed R. A. Al-Hilli, and Abdul-Ridah A. Alwan. "Vegetation Response to Reflooding in the Mesopotamian Wetlands, Southern Iraq." *Wetlands* 30 (2010): 177–188. doi:10.1007/s13157-010-0035-9.

Hancock, Rosemary. *Islamic Environmentalism: Activism in the United States and Great Britain*. New York: Routledge, 2018.

Hansen, Donald P. "Art of the Early City-states." In *Art of the First Cities: The Third Millennium BCE from the Mediterranean to the Indus*, edited by Joan Aruz and Ronald Wallenfels. New Haven: Yale University Press, 2003.

Hansen, Donald P. "Art of the Akkadian Dynasty." In *Art of the first Cities*, edited by Joan Aruz and Ronald Wallenfels. New Haven: Yale University Press, 2003.

Harrington, Cameron and Schuyler Null. "What Can Iraq's Fight over the Mosul Dam Tell Us about Water Security?" *NewSecurityBeat Wilson Center*, August 20, 2014. www.newsecuritybeat.org/2014/08/iraqs-fight-mosul-dam-water-security/.

Hartl, Martin. *Das Nadjafabadtal. Geographische Untersuchung einer Kanatlandschaft im Zagrosgebirge (Iran)*. Regensburg: Institut für Geographie an der Universität Regensburg, Selbstverlag, 1879.

Helck, Wolfgang. *Wirtschaftsgeschichte des Alten Ägypten im 3. und 2. Jahrtausend v. Chr*. Leiden, Köln: Brill, 1975.

Hellegers, Petra, Walter Immerzeel, and Peter Droogers. "Economic Concepts to Address Future Water Supply-Demand Imbalances in Iran, Morocco and Saudi Arabia." *Journal of Hydrology* 502 (2013): 62–67.

Henareh Khalyani, Azad, Audrey L. Mayer, and Emma S. Norman. "Water Flows toward Power: Socioecological Degradation of Lake Urmia, Iran." *Society & Natural Resources* 27.7 (2014): 759–767.

Heschel, Abraham J. *God in Search of Man: A Philosophy of Judaism*. New York: Farrar, Straus and Giroux, 1976.

Heschel, Abraham J. *Man Is Not Alone: A Philosophy of Religion*. New York: Farrar, Straus and Giroux, 1951.

Heschel, Abraham J. *The Sabbath*. New York: Farrar, Straus and Giroux, 1988.

Hess, Denis. "Fernübertragung regelbarer Solarenergie von Nordafrika nach Mitteleuropa." Diploma diss., University of Stuttgart, 2013.

Heyvaert, Vanessa M. A., and Cecile Baeteman. "Holocene sedimentary evolution and palaeocoastlines of the Lower Khuzestan plain (southwest Iran)." *Marine Geology* 242 (2007): 83–108. doi:10.1016/j.margeo.2007.01.008.

Heyvaert, Vanessa M.A., and Cecile Baeteman. "A Middle to Late Holocene Avulsion History of the Euphrates River: A Case Study from Tell ed-Dēr, Iraq, Lower Mesopotamia." *Quaternary Science Reviews* 27 (2008): 2401–2410. doi:10.1016/j.quascirev.2008.08.024.

Hilger, Georg. Ulrich Kropač, and Stephan Leimgruber. "Der sozialisationsbegleitende oder therapeutische Religionsunterricht." In *Religionsdidaktik. Ein Leitfaden für Studium, Ausbildung und Beruf*, edited by Georg Hilger, Stephan Leimgruber, and Hans G. Ziebertz, 55–58. München: Kösel-Verlag, 2010.

Hima. "Willkommen bei Hima." Accessed March 26, 2019. www.hima-umweltschutz.de/geschichte.

Hobhouse, Penelope. *Persische Gärten. Paradiese des Orients*. München: Knesebeck Verlag, 2005.

Hope, Aimie L.B. and Christopher R. Jones. "The Impact of Religious Faith on Attitudes to Environmental Issues and Carbon Capture and Storage (CCS) Technologies: A Mixed Methods Study." *Technology in Society* 38 (2014): 48–59.

Horner, Nathaniel C., Arman Shehabi, and Inês L. Azevedo. "Known Unknowns: Indirect Energy Effects of Information and Communication Technology." *Environmental Research Letters* 11.10 (2016). https://iopscience.iop.org/article/10.1088/1748–9326/11/10/103001.

Hourani, Albert H. and S.M. Stern. *The Islamic City. A Colloquium*. Oxford: University of Pennsylvania Press, 1970.

Huot, Jean-Louis. "Ubaidian Villages of Lower Mesopotamia." In *Upon these Foundations: The 'Ubaid Reconsidered, Proceedings from the 'Ubaid Symposium, Elisnore*, edited by Elizabeth F. Henrickson and Ingolf Thuesen, 19–42. Copenhagen: Museum Tusculanum Press, 1989.

Hummel, Ulrike. *Die grüne Welle*. Accessed January 29, 2019. http://de.qantara.de/inhalt/ islam-und-umweltschutz-die-gruene-welle.

Hutter, Manfred. *Religionen in der Umwelt des Alten Testaments I: Babylonier, Syrer, Perser*. Stuttgart: W. Kohlhammer, 1996.

Ifees Ecoislam. "Eco-Islam." Accessed March 26, 2019. http://www.ifees.org.uk/category/eco-islam/.

Ifees Ecoislam. "Islamic Declaration on Global Climate Change." Accessed April 14, 2019. http://www.ifees.org.uk.

Ifees EcoIslam. "Mosques as Green Project Exemplars." Accessed March 26, 2019. http://www.ifees.org.uk/mosques-as-green-project-exemplars/.

Ifees EcoIslam. "Zanzibar." Accessed August 15, 2018. http://www.ifees.org.uk/zanzibar/.

International Committee of the Red Cross, Geneva (ICRC). *Urban Services During Protracted Armed Conflict: A Call for a Better Approach to Assisting Affected People*. Geneva ICRC, 2015.

International Energy Agency. "Statistics." Accessed June 09, 2020. http://www.iea.org/statistics/statisticssearch/.

IRENA. *Renewable Energy in the Arab Region. Overview of Developments*. Abu Dhabi, International Renewable Energy Agency, 2016. Accessed July 26, 2017. http://www.irena.org/DocumentDownloads/Publications/IRENA_Arab_Region_Overview_2016.pdf. Also includes a comprehensive overview of the renewable energy targets of most Arab League member states.

International Telecommunication Union. "Smart Irrigation for Sustainable Agriculture in Abu Dhabi: Case Study—United Arab Emirates." Accessed July 28, 2017. https://itunews.itu.int/en/4853-Smart-irrigation-for-sustainable-agriculture-in-Abu-Dhabi.note.aspx.

International Telecommunication Union. *ICT as an Enabler for Smart Water Management ITU-T Technology Watch Report October 2010*. Geneva: ITU, 2011. Accessed March 26, 2018. https://www.itu.int/dms_pub/itu-t/oth/23/01/T23010000100003PDFE.pdf.

Islamische Datenbank. "Ḥadīṯ No. 4282." Accessed April 8, 2019. http://islamische-datenbank.de/sahih-muslim?action=viewhadith&chapterno=43&min=20&show=10.

Izzi Dien, Mawil. *The Environmental Dimensions of Islam*. Trobridge, Wilshire: Redwood Books, 2000.

Jacobsen, Thorkild. The Waters of Ur." *Iraq* 22 (1960): 174–185. doi:10.2307/4199683.

Jagersma, Abraham H. *A Descriptive Grammar of Sumerian*. Leiden: Internet Publication, 2010.

Javadi, Seyed M.R. *La maîtrise millenaire de la terre, de l'eau et du vent en Iran. Yasouj/Iran*. Razbar, 2016.

Jomehpour, Mahmoud. "Qanat Irrigation Systems as Important and Ingenious Agricultural Heritage: Case Study of the Qanats of Kashan, Iran." *International Journal of Environmental Studies* 66 (2009): 297–315.

Jotheri, Jaafar, Mark B. Allen, and Tony J. Wilkinson. "Holocene Avulsions of the Euphrates River in the Najaf Area of Western Mesopotamia: Impacts on Human Settlement Patterns." *Geoarchaeology* 31 (2016). doi:10.1002/gea.21548.

Kaplan, Edward and Samuel H. Dresner. *Abraham Joshua Heschel—Prophetic Witness*. New Haven and London: Yale University Press, 1998.

Karbassi, Abdolreza, Gholamreza Nabi Bidhendi, Amirhossein Pejman, Mehdi Esmaeili Bidhendi. "Environmental Impacts of Desalination on the Ecology of Lake Urmia." *Journal of Great Lakes Research* 36 (2010): 419–424. https://doi.org/10.1016/j.jglr.2010.06.004.

Kasperson, Jeanne X., Roger E. Kasperson, and B.L. Turner II, ed. *Regions at Risk. Comparisons of Threatened Environments*. Tokyo, New York, Paris: United Nations University Press: UNU Studies on Critical Environmental Risks, 1995.

Kay, Paul A. and Douglas L. Johnson. "Estimation of Tigris-Euphrates Streamflow from Regional Paleoenvironmental Proxy Data." *Climatic Change* 3 (1981): 251–263. doi: 10.1007/BF02423218.

Kennett, Douglas J. and James P. Kennett. "Influence of Holocene Marine Transgression and Climate Change on Cultural Evolution in Southern Mesopotamia." In *Climate Change and Cultural Dynamics: A Global Perspective on Mid-Holocene Transitions*, edited by David G. Anderson, Kirk A. Maasch, and Daniel H. Sandweiss, 229–264. San Diego: Academic Press, 2007. doi:10.1016/B978-012088390-5.50012-1.

Kermani, Navid. *Gott ist schön. Das ästhetische Erleben des Koran*. München: C.H. Beck, 2011.

Keshavarz, Marzieh, Ezatollah Karami, and Mansoor Zibaei. "Adaptation of Iranian Farmers to Climate Variability and Change." *Regional Environmental Change* 14 (2014).

Kessler, Rainer. *Staat und Gesellschaft im vorexilischen Juda. Vom 8. Jahrhundert bis zum Exil*. Leiden: Brill, 1992.

Khalid, Fazlun and Joanne O'Brien. *Islam and Ecology*. London: World Wide Fund for Nature, 1992.

Khalid, Fazlun M. "Islam and the Environment." In *Encyclopedia of Global Environmental Change*, edited by Ted Munn. Chichester: John Wiey & Sons, 2002.

Khalid, Fazlun M. "Applying Islamic Environmental Ethics." In *Environmentalism in the Muslim World*, edited by Richard C. Foltz. New York: Nova Science Publishers, Inc., 2005.

Khalid, Fazlun M. "Islam and the Environment—Ethics and Practice: An Assessment." *Religion Compass* 4.11 (2010) 707–716.

Khalid, Fazlun M. Interview given to Deutsche Welle by Franziska Badenschier on April 2, 2013. https://p.dw.com/p/188GU.

Khalighi, Shahram Sigaroodi, and Shiva Ebrahimi. "Effects of Land Use Change on Surface Water Regime (Case Study Orumieh Lake of Iran)." *Procedia Environmental Sciences* 2 (2010): 256–261.

Kibaroglu, Aysegül. *Natural Cooperation. Facing Water Challenges in the Middle East*. Washington, D.C.: Middle East Institute, 2016.

King, Marcus D., ed. *Water and Conflict in the Middle East*. London: Hurst, 2019.

Kinnier Wilson, James. "On the UD-ŠU-BALA at Ur Towards the End of the Third Millennium BC." *Iraq* 67 (2005): 47–60.

Koch, Heidemarie. *Frauen und Schlangen: Die geheimnisvolle Kultur der Elamer in Alt-Iran*. Mainz: Philipp von Zabern, 2007.

Kolb, Albert. "Die Geographie und die Kulturerdteile." In *Hermann von Wissmann-*Festschrift, edited by Leidlmair A. Tübingen: Geographisches Institut der Universität Tübingen, 1962.

Kortum, Gerhard. "Die Qanat-Bewässerung auf dem Hochland von Iran und ihre Ausbreitung über die Erde." In *Historische und zeitgenössische Probleme und Perspektiven in asiatischen und afrikanischen Gesellschaften*, edited by Hübner Ulrich and Antje Richter. Schenefeld: EB, 2004.

Kowanda-Yassin, Ursula. *Mensch und Naturverständnis im sunnitischen Islam: Ein Beitrag zum aktuellen Umweltdiskurs*. Würzburg: Ergon, 2011.

Kowanda-Yassin, Ursula. *Öko-Dschihad. Der grüne Islam—Beginn einer globalen Umweltbewegung*. Salzburg, Wien: Residenz-Verlag, 2018.

Krause, Rolf F. "Orient, Naher und Mittlerer Osten. Die Begriffe im Wandel der Zeit." *Geographische Rundschau* 45.1 (1993).

Kroeber, Alfred L. *The Nature of Culture*. Chicago: University of Chicago Press, 1952.

Kuzucuoglu, Catherine, Michel Fontugne, and Damase Mouralis. "Holocene Terraces in the Middle Euphrates Valley, between Halfeti and Karkemish (Gaziantep, Turkey)." *Quaternaire* 15 (2004): 195–206. doi:10.3406/quate.2004.1767.

Kuwait Water Association. "Training Programs." Accessed November 1, 2019. https://www.kwa.org.kw/copy-of-proposed-projects.

Kýzýlocak, Gülay and Martina Sauer. *Environmental Awareness and Behaviour of Turkish Migrants in Germany. Research on the Current Situation and Determination of the Possibility of their Inclusion*. Duisburg-Essen: UNESCO-Verbindungsstelle für Umwelterziehung, 2002.

Lambert, Wilfred G. "Dingir.šà.dib.ba Incantations." *Journal of Near Eastern Studies* 33 (1974): 267–322.

Lambton, Ann K.S. "The Origins, Diffusion and Functioning of the Qanat." In *Qanat, Kariz & Khattara. Traditional Water Systems in the Middle East and North Africa*, edited by Peter Beaumont, Michael Bonine, and Keith McLachan. London: University of London, 1998.

Lang, Bernhard. "The Social Organization of Peasant Poverty in Biblical Israel." *Journal for the Study of the Old Testament* 7 (1982).

Lange, Steffen and Tilman Santarius. *Smarte grüne Welt? Digitalisierung zwischen Überwachung, Konsum und Nachhaltigkeit*. München: oekom, 2018.

Lantermann, Ernst-D. "Von den Schwierigkeiten umweltschützendes Handeln auszubilden." In *Methoden der Umweltbildungsforschung*, edited by Dietmar Bolscho and Gerd Michelsen. Opladen: Leske und Budrich, 1999.

Lapidus, Ira M. *Middle Eastern Cities. A Symposium on Ancient, Islamic, and Contemporary Middle Eastern Urbanism*. Berkeley, Los Angeles: Univ. of California Press, 1969.

Larsen, Curtis E. "The Mesopotamian Delta Region: A Reconsideration of Lees and Falcon." *Journal of the American Oriental Society* 95 (1975): 43–57. doi:10.2307/599157.

Lawrence, Dan, Graham Philip, Hannah Hunt, Lisa Snape-Kennedy, and Tony J. Wilkinson. "Long Term Population, City Size and Climate Trends in the Fertile

Crescent: A First Approximation." *PLOS ONE* 11 (2016): e0157863. doi:10.1371/journal.pone.0157863.

Legrouri, Ahmed and Khalid Sendide. "Education for Sustainable Development in Morocco." In *National Journeys: Towards Education for Sustainable Development*, edited by UNESCO. Paris: UNESCO, 2013.

Leimgruber, Stephan and Hans-Georg Ziebertz. "Interkulturelles und interreligiöses Lernen." In *Religionsdidaktik. Ein Leitfaden für Studium, Ausbildung und Beruf*, edited by Stephan Georg Hilger, Leimgruber, Hans G. Ziebertz, 467–469. München: Kösel-Verlag, 2010.

Lembaga Pemuliaan Lingkungang Hidup Dan Sumber Daya Alam: Majelis Ulama Indonesia. "Wildlife Trafficking Fatwa." Accessed March 26, 2019. http://mui-lplhsda.org/wildlifetraffickingfatwa/.

Leng, Günter. "'Rentenkapitalismus' oder 'Feudalismmus'? Kritische Untersuchungen über einen (sozial)geographischen Begriff." *Geographische Zeitschrift* 62 (1974): 119–137.

Levi, Yehudah. "Ecological Problems: Living on Future Generations' Account." In *Compendium of Sources in Halacha and the Environment*, edited by Ora Shenson and Shai Spetgant. Jerusalem: Canfei Nesharim, 2005.

Levy, Natan. Harfiyah Haleem, and David Shreeve. *Sharing Eden: Green Teachings from Jews, Christians and Muslim*. Leicestershire: Kube Publishing, 2012.

Lilliestam, Johann and Saskia Ellenbeck. "Energy Security and Renewable Electricity Trade—Will Desertec Make Europe Vulnerable to the 'Energy Weapon'?" *Energy Policy* 39.6 (2011): 3380–3391.

Lilliestam, Johann, Saskia Ellenbeck, Charikleia Karakosta, N. Caldés. "Understanding the Absence of Renewable Electricity Imports to the European Union." *International Journal of Energy Sector Management* 10.3 (2016): 291–311.

Lindner, Willi. "Nachhaltige Ernährung—Wie durchbrechen wir die gläserne Decke." In *Ernte und Aussaat,* ed. Gabriele Hösch-Schagar, Birgit Karre, and Erhard Mayerhofer, 111–115. Münster: LIT, 2011.

Litt, Thomas, Christian Ohlwein, Frank H. Neumann, Andreas Hense, and Mordechai Stein. "Holocene Climate Variability in the Levant from the Dead Sea Pollen Record." *Quaternary Science Reviews* 49 (2012): 95–105. doi:10.1016/j.quascirev.2012.06.012.

Loretz, Oswald. "Die prophetische Kritik des Rentenkapitalismus." *Ugarit-Forschungen* 7 (1975): 271–278.

Löw, Ulrike. *Figürlich verzierte Metallgefäße aus Nord- und Nordwestiran*. Münster: Ugarit-Verlag, 1998.

Luic, Ljerka and Dejan Glumac. "The Role of ICT Technology in the Knowledge Society." Paper presented at the conferences "TELSIKS 2009–9th International Conference on Telecommunications in Modern Satellite, Cable, and Broadcasting Services," 2009.

MacKenzie, Debora. "Extremists in Iraq Now Control the Country's Rivers." *New Scientist*, June 12, 2014. www.newscientist.com/article/dn25722-extremists-in-iraq-now-control-the-countrys-rivers/.

Madani, Kaveh, Amir AghaKouchak, and Ali Mirchi. "Iran's Socio-economic Drought: Challenges of a Water-Bankrupt Nation." *Iranian Studies* 49.6 (2016): 997–1016.

MADE. "Muslim Action for Development & Environment." Accessed January 01, 2019. www.made.ngo.

MADE. "Tour de Salah." Accessed August 22, 2018. https://tourdesalah.org.uk/.

Magid, Shaul. "Nature, Exile, and Disability in R. Nahman of Bratslav's 'The Seven Beggars'." In *Judaism and Ecology: The Created World and Revealed World*, edited by Hava Tirosh-Samuelson. Cambridge, MA: Harvard University Press, 2002.

Maleksaeidi, Hamideh and Ezatollah Karami. "Social-Ecological Resilience and Sustainable Agriculture under Water Scarcity." *Agroecology and Sustainable Food Systems* 37.3 (2013): 262–290.

Manhart, Andreas, Markus Blepp, Corinna Fischer, Kathrin Graulich, Siddharth Prakash, Rasmus Priess, Tobias Schleicher, and Maria Tür. *Resource Efficiency in the ICT Sector. Final Report, November 2016*. Freiburg: Öko-Institut, 2016.

Mannan, Mehzabeen, Mohamed Alhaj, Abdel Nasser Mabrouk, and Sami G. Al-Ghamdi. "Examining the Life-cycle Environmental Impacts of Desalination: A Case Study in the State of Qatar," in *Desalination* 452 (2019): 238–246.

Marjani, Aliasghar and Mirmosadegh Jamali. "Role of Exchange Flow in Salt Water Balance of Urmia Lake." *Dynamics of* Atmospheres a*nd Oceans* 65 (2014): 1–16.

Marzouk, Evonne. "Lessons from Nature: What our *Tefilla* Teaches Us about Our World." In *Compendium of Sources in Halacha and the Environment*, edited by Shenson, Ora, and Spetgant, Shai. Jerusalem: Canfei Nesharim, 2005.

Maul, Stefan M. "Ringen um göttliches und menschliches Mass. Die Sintflut und ihre Bedeutung im Alten Orient." In *Schönheit und Mass. Beitäge der Eranos Tagungen 2005 und 2006*, edited by Erik Hornung und Andreas Schweizer. Basel: Schwabe, 2007.

Mayewski, Paul A., Eelco E. Rohling, J. Curt Stager, Wibjörn Karlén, Kirk A. Maasch, L. David Meeker, Eric A. Meyerson, Françoise Gasse, Shirley van Kreveld, Karin Holmgren, Julia Lee-Thorp, Gunhild Rosqvist, Frank Rack, Michael Staubwasser, Ralph R. Schneider, and Eric J. Steig. "Holocene Climate Variability." *Quaternary Research* 62 (2004): 243–255. doi: 10.1016/j.yqres.2004.07.001.

McDonald, Diana K. *Serpent Imagery on Ancient Near Eastern Pottery*. Ann Arbor: Columbia University, 1989.

McLachlan, Keith. *The Neglected Garden. The Politics and Ecology of Agriculture in Iran*. London, J.B. Tauris and Co Ltd, 1988.

Mellaart, James. The *Neolithic of the Near East*. London: Thames and Hudson, 1975.

Meyer, Hilbert. *Unterrichtsmethoden. II: Praxisband*, 14th. edition. Berlin: Cornelsen Verlag Scriptor GmbH & Co.KG, 2011.

Micklin, Philip. "The Aral Sea Disaster." *Annual Review of Earth and Planetary Sciences* 35 (2007): 47–72.

Micklin, Philip, Nikolai V. Aladin, and Igor Plotnikov. *The Aral Sea: The Devastation and Partial Rehabilitation of a Great Lake*. Heidelberg: Springer, 2014.

Miller, Naomi F. "The Macrobotanical Evidence for Vegetation in the Near East, c. 18000/16000 BC to 4000 BC." *Paléorient* 23 (1997): 197–207. doi:10.3406/paleo.1997.4661.

Ministère de l'Énergie et des Mines (MEM). "Renewable Energy and Energy Efficiency Program." Accessed May 15, 2017. www.mem-algeria.org/francais/uploads/enr/Programme_ENR_et_efficacite_energetique_en.pdf.

Ministère de l'Energie, des Mines, de l'Eau et de l'Environnement (MEMEE). *Loi 13–09 relative aux énergies renouvelables*. Rabat: MEMEE, 2010.

Ministry of Communications and Information Technologies. "Egyptian Green ICT Strategy: Pillars—Partners—Programs 2010–2018." Accessed October 19, 2017. https://www.epa.gov/sites/production/files/2014-08/documents/egypt-_green_ict.pdf.

Mittendorfer, Max and Michael Nader. "Der Schulgarten als Ort lebensbezogenen Lernens." in *Ernte und Aussaat, Spiritualität und Nachhaltigkeit—Überlegungen und Handlungsimpulse*, vol. 5, edited by Gabriele Hösch-Schagar, Birgit Karre, Erhard Mayerhofer. Wien, Berlin: LIT-Verlag, 2011.

Mittermayer, Catherine. *Die Entwicklung der Tierkopfzeichen. Eine Studie zur syro-mesopotamischen Keilschriftpaläographie des 3. und frühen 2. Jahrtausends v. Chr.* Münster: Ugarit-Verlag, 2005.

Mohanad, H. Ali. "ISIS' Path of Destruction Drains Iraq and Syria's Water Supplies." *Al Arabiya*, June 21, 2014. http://english.alarabiya.net/en/perspective/analysis/2014/06/21/ISIS-path-of-destruction-drains-Iraq-and-Syria-s-water-supplies.html.

Moorey, Peter R.S. "Iran: A Sumerian El-Dorado?" In *Early Mesopotamia and Iran: Contact and Conflict 3500–1600 BC*, edited by John Curtis, 31–43. London: British Museum Press, 1993.

Moortgat-Correns, Ursula. "Die Rosette—ein Schriftzeichen? Die Geburt des Sterns aus dem Geiste der Rosette." *Altorientalische Forschungen* 21 (1994): 359–371.

Morenz, Ludwig D. *Bild-Buchstaben und symbolische Zeichen: Die Herausbildung der Schrift in der hohen Kultur Altägyptens*. Freiburg/Göttingen: Universitätsverlag/Vandenhoeck and Ruprecht, 2004.

Moroccan Agency for Solar Energy (MASEN). *Endless Power for Progress*. Rabat, MASEN, 2016. Accessed May 15, 2017. http://www.masen.ma/media/uploads/documents/MASEN_Brochure_instit_Anglais_finale_FPeQ8Wd.pdf.

Morozova, Galina S. "A Review of Holocene Avulsions of the Tigris and Euphrates Rivers and Possible Effects on the Evolution of Civilizations in Lower Mesopotamia." *Geoarchaeology* 20 (2005): 401–423. doi:10.1002/gea.20057.

Mosammam, Hassan M., Ali M. Mosammam, Mozaffar Sarrafi, and Jamileh Tavakoli Nia. "Analyzing the Potential Impacts of Climate Change on Rainfed Wheat Production in Hamedan Province, Iran, via Generalized Additive Models." *Journal of Water and Climate Change* 7.1 (2016).

Moser, Massimo. "Techno-economic Analysis of Enhanced Dry Cooling for CSP." *Energy Procedia* 49 (2014): 1177–1186. Accessed June 12, 2020. http://www.sciencedirect.com/science/article/pii/S1876610214005815.

Moser, Massimo, Franz Trieb, and Tobias Fichter. "Potential of Concentrating Solar Power Plants for the Combined Production of Water and Electricity in MENA Countries." *Journal of Sustainable Development of Energy, Water and Environment Systems* 1.2 (2013): 122–140. Accessed May 15, 2017. DOI: http://dx.doi.org/10.13044/j.sdewes.2013.01.0009.

Moynihan, Elizabeth B. *Paradise as a Garden in Persia and Mughal India*. London: Scolar Press, 1979.

Muslim, Ḥadīṯ-No. 121, Buch 1, Kapitel 13, Nr.121. Islamische Datenbank. "Riyāḍu ṣ-Ṣāliḥīn," Accessed April 9, 2019, http://islamische-datenbank.de/riyadhu-s-salihin?action=anzeigen&hadithno=121.

Nadji, M. "Karadjis 'Erschließung verborgener Gewässer.' Ein Lehrbuch der Geowissenschaften aus dem 11. Jahrhundert," *Zeitschrift Deutsche Geologische Gesellschaft* 123 (1972): 1–13.

Nasr, Seyyed H. *The Encounter of Man and Nature. The Spiritual Crisis of Modern Man*. Chicago: ABC International Group, 1967.

Nasr, Seyyed H. "Islam and the Environmental Crisis." *al'Ilm* 17 (1997): 1–23.

Nasr, Seyyed H. *Man and Nature. The Spiritual Crisis in Modern Man*. ABC Internation Group Inc., 1997.

Nasr, Seyyed H. "The Islamic Perspective on the Environmental Crisis: Seyyed Hossein Nasr in Conversation with Muzaffar Iqbal." In *Islam and Science* 5.1 (2007).

Nasseef, Abdullah O. *Alliance of Religions and Conservation. The Assisi Declarations: Messages on Humanity and Nature from Buddhism, Christianity, Hinduism, Islam & Judaism*. www.arcworld.org/downloads/THE%20ASSISI%20DECLARATIONS.pdf.

Neef, Reinder. "Plant Remains from Archaeological Sites in Lowland Iraq: Tell El 'Oueili." In *'Oueili—Travaux de 1985*, edited by Jean-Louis Huot, 321–329. Paris: Éditions Recherche sur les Civilisations, 1991.

Nejdawi, Reem, Monia Braham, Jana el-Baba, Susan Razzaz, and Cameron Allen. *Arab Sustainable Development Report*. Beirut: ESCWA, 2015.

Neumann-Lechner, Anton and Florian Preißer. *Umweltpädagogik in Erwachsenenbildung und Gemeinwesenarbeit*, Band 6. Bad Heilbrunn: Verlag Julius Klinkhardt, 1991.

Niang, Isabelle, Oliver C. Ruppel, Mohamed Abdrabo, and Essel Ama. "Africa." In *Intergovernmental Panel on Climate Change (IPCC)—Climate Change 2014: Impacts, Adaptation, and Vulnerability. Part B: Regional Aspects. Contribution of Working Group II to the Fifth Assessment Report of the Intergovernmental Panel on Climate Change*. Cambridge, New York: Cambridge University Press, 2014.

Nissen, Hans J. "*The early history of the Ancient Near East, 9000–2000 B.C.*" Chicago/London: Chicago University Press, 1988.

Nissen, Hans J. *Geschichte Altvorderasiens*. Munich: R. Oldenbourg Verlag, 1999.

NourEnergy. "NourEnergy." Accessed August 15, 2018. http://www.nour-energy.com/.

Ohlson, Leif, ed. *Hydropolitics. Conflicts over Water as a Development Constraint*. Dhaka: University Press, London, New Jersey: ZED Books, 1995.

Okazaki, Shoko. "The Japanese Qanat (Mambo): Its Technology and Origin. Beaumont." In *Qanat, Kariz and Khattara: Traditional Water Systems in the Middle East and North Africa*, edited by Peter Beaumont, Michael Bonine, and Keith McLachlan, 267–279. London: University of London, Centre for Middle eastern Studies, MENAS Press 1989.

Organisation for Economic Co-Operation and Development (OECD). "OECD Country Risk Classification." Last accessed June, 12, 2020. http://www.oecd.org/tad/xcred/crc.htm.

Ospina, Angelica V. and Richard Heeks. *Unveiling the Links between ICTs & Climate Change in Developing Countries: A Scoping Study*. Manchester, Institute for Development Policy and Management, SED, 2010.

Ospina, Angelica V., and Richard Heeks. "Linking ICTs and Climate Change Adaptation: A Conceptual Framework for e-Resilience and e-Adaptation." In *ICTs, Climate Change and Development: Themes and Strategic Actions*, edited by Richard Heeks and Angelica V. Ospina. Manchester: University of Manchester Press, 2012.

Österreicher, Herbert. *Natur- und Umweltpädagogik für sozialpädagogische Berufe*. Köln: Bildungsverlag EINS, 2014.

Otto, Adelheid. *Die Entstehung und Entwicklung der Klassisch-Syrischen Glyptik*. Berlin: Walter de Gruyter, 2000.

Ouis, Seyyed P. "Islamic Ecotheology Based on the Qur'ān." *Islamic Studies* 37.2 (1998): 151–181.

Özdemir, Ibrahim. *Science and the Environment: Is Science Responsible for Environmental Crisis?* JESS/CSBD, 1996.

Özdemir, Ibrahim. "Towards an Understanding of Environmental Ethics from a Qur'anic Perspective." In *Islam and Ecology*, edited by Richard C. Foltz, Frederick M. Denny, and Azizan Baharuddin, 3–38. Cambridge: Harvard University Press, 2003.

Özdemir, Ibrahim. "Turkey." In *Environmentalism Environmentalism in the Muslim World*, edited by Richard C. Foltz. New York: Nova Science Publishers, Inc., 2005.

Pearcy, George E. *The Middle East—An Indefinable Region*. Washington: Department of State Publication, 1964.

Pedrick, Clare. "A Future for Farming in the Dry Lands: Smart Solutions for Climate Change." Accessed July 28, 2017. https://ccafs.cgiar.org/blog/future-farming-dry-lands-smart-solutions-climate-change#.WXtwb7ZpxLM.

Peritore, N. Patrick. "Iran: From Revolution to Ecological Collapse." In *Third World Environmentalism: Case Studies from the Global South*. Gainsville: University Press Florida, 1999.

Petruccioli, Attilio. *Der islamische Garten. Architektur—Natur—Lanfschaft*. Stuttgart: DVA. 1995.

Pientka-Hinz, Rosel. "Schlange. A. In Mesopotamien." in *Reallexikon der Assyriologie und Vorderasiatischen Archäologie*, edited by Michael P. Streck. Berlin: Walter de Gruyter, 2009.

Pientka-Hinz, Rosel. "Bunte Kühe? Zu den frühesten Farbbezeichnungen im Alten Orient." In *The Empirical Dimension of Ancient Near Eastern Studies: Die empirische Dimension altorientalischer Forschungen*, edited by Gebhard J. Selz and Klaus Wagensonner. Wien: LIT Verlag, 2011.

Pientka-Hinz, Rosel. "Skorpion." In *Reallexikon der Assyriologie und Vorderasiatischen Archäologie*, edited by Michael P. Streck. Berlin: Walter de Gruyter, 2011.

Pilgrim. "Pilgrim aktuell." Accessed August 13, 2018. http://pilgrim.at/aktuell.html.

Pint, Anna, Max Engel, Sandra Melzer, Peter Frenzel, Birgit Plessen, and Helmut Brückner. "How to Discriminate Athalassic and Marginal Marine Microfaunas? Foraminifera and other Fossils from an Early Holocene Continental Lake in Northern Saudi Arabia." *Journal of Foraminiferal Research* 47 (2017): 175–187. doi:10.2113/gsjfr.47.2.175.

Polanyi, Karl. *The Great Transformation. The Political and Economic Origins of our Time*. New York: Farrar and Rinehart, 1944.

Pollock, Susan. *Ancient Mesopotamia*. Cambridge: Cambridge University Press, 1999.

Pongratz-Leisten, Beate. "Sintflut (flood). A. In schriftlichen Quellen. In *Reallexikon der Assyriologie und Vorderasiatischen Archäologie*, vol. 12, edited by Michael P. Streck, 525–526. Berlin: Walter de Gruyter, 2011.

Popp, Herbert and A. Kassah. *Les ksour du sud Tunisien. Atlas illustré d'un patrimoine culturel*. Bayreuth; Naturwissenschaftliche Gesellschaft, 2010.

Popp, Herbert, M. Aït Hamfa, and Brahim El Fasskaoui. *Les agadirs de l'Anti-Atlas occidental. Atlas illustré d'un patrimoine culturel du Sud marocain*. Bayreuth: Naturwissenschaftliche Gesellschaft, 2011.

Popp, Herbert and Brahim El Fasskaoui. *Die Straße der Kasbahs—einst und jetzt. Le route des Kasbahs—jadis et maintenant—The Road of the Kasbahs—Past and Present*. Bayreuth: Michael Imhof Verlag, 2017.

Pournelle, Jennifer R. "Marshland of Cities: Deltaic Landscapes and the Evolution of Early Mesopotamian Civilization". PhD diss., University of California, San Diego, 2003. doi:10.6067/XCV8J67GK3.

Pournelle, Jennifer R. "Physical Geography." In *The Sumerian World*, edited by Harriet Crawford, 13–32. London: Routledge, 2013. doi:10.4324/9780203096604.ch1.

Posch, Peter. "The Study: 'Environment and School Initiatives': Phase One." In *Evaluating Innovation in Environmental Education*. Paris: OECD documents, 1994.

Postgate, John N. *Early Mesopotamia: Society and Economy at the Dawn of History*. London: Routledge, 1992/1994.

Progressive Muslims Network (PMN). "Progressive Islam: A Definition and Declaration." Last accessed via the Wayback Machine at Archive.org, March 25, 2019. http://web.archive.org/web/20020403150253/www.progressivemuslims.com/index2.html.

Qantara. "Environmental Protection through the Koran?" Last accessed January 28, 2019. https://en.qantara.de/content/islam-and-development-cooperation-environmental-protection-through-the-koran.

Qantara. "Islam, Glaube und Klimawandel." Accessed January 26, 2019. https://de.qantara.de/inhalt/oeko-islam-islam-glaube-und-klimawandel.

Rahsad, Mahmoud. *Die Entwicklung der vor- und frühgeschichtlichen Stempelsiegel in Iran: im Vergleich mit Mesopotamien, Syrien und Kleinasien—von ihren Anfängen bis zum Beginn des 3. Jahrtausends v.Chr.* Berlin: Dietrich Reimer Verlag, 1990.

Raikes, R.L. "The Physical Evidence for Noah's Flood." *Iraq* 28 (1966): 52–63. doi:10.2307/4199795.

Ramadan, Tariq. *Der Islam. Gott-Mensch-Welt. Begegnung mit dem Islam* Berlin: Greenpalace, 2002.

Ramezani, Mehrian, Majid Raul P. Hernandez, Ahmad Reza Yavari, Shahrzad Faryadi, and Esmaeil Salehi. "Investigating the Causality of Changes in the Landscape Pattern of Lake Urmia Basin, Iran, Using Remote Sensing and Time Series Analysis." *Environmental Monitoring and Assessment* 188 (2016).

Raz, Simcha. *A Tzaddik in our Time. The Life of Rabbi Aryeh Levin*. Jerusalem: Feldheim, 2008.

Riehl, Simone, Konstantin Pustovoytov, Aron Dornauer, and Walther Sallaberger. "Mid-to-Late Holocene Agricultural System Transformations in the Northern Fertile Crescent: A Review of the Archaeobotanical, Geoarchaeological, and Philological Evidence." In *Climates, Landscapes, and Civilizations*, edited by Liviu Giosan, Dorian Q. Fuller, Kathleen Nicoll, Rowan K. Flad, and Peter D. Clift, 115–136. Washington: American Geophysical Union, 2012. doi:10.1029/2012GM001221.

Riehl, Simone, Konstantin Pustovoytov, Heike Weippert, Stefan Klett, and Frank Hole, "Drought Stress Variability in Ancient Near Eastern Agricultural Systems Evidenced by $\delta^{13}C$ in Barley Grain." *Proceedings of the National Academy of Sciences of the USA* 111 (2014): 12348–12353. doi: 10.1073/pnas.1409516111.

Riehl, Simone, Mohsen Zeidi, and Nicholas J. Conard. "Emergence of Agriculture in the Foothills of the Zagros Mountains of Iran." *Science* 341 (2013): 65–67. doi:10.1126/science.1236743.

Rivas-Martinez, Salvador and Salvador Rivas-Saenz. "Worldwide Bioclimatic Classification System, 1996–2017. Phytosociological Research Center, University Complutense of Madrid, Spain." Last modified January, 2017, accessed July 14, 2017. http://www.pendientemigracion.ucm.es/info/cif/plot/diagram.htm.

Roberts, Neil, Warren J. Eastwood, Catherine Kuzucuoğlu, Girolamo Fiorentino, and Valentina Caracuta. "Climatic, Vegetation and Cultural Change in the Eastern Mediterranean during the Mid-Holocene Environmental Transition." *The Holocene* 21 (2011): 147–162. doi:10.1177/0959683610386819.

Rollefson, Gary, Daniel Schyle, Laurens Thissen, Henrieta Todorova, and Christoph Zielhofer. "The Impact of Rapid Climate Change on Prehistoric Societies during the Holocene in the Eastern Mediterranean." *Documenta Praehistorica* XXXVI (2009): 7–59. doi:10.4312/dp.36.2.

Root, Margaret C. "Animals in the Art of Ancient Iran." In *A History of the Animal World in the Ancient Near East*, edited by Billie J. Collins, 169–209. Leiden: Brill, 2002.

Rosen, Arlene M. *Civilizing Climate. Social Responses to Climate Change in the Ancient Near East*. Lanham: Altamira Press, 2007.

Rosenberger, Michael. "'Wenn ich wüsste, dass morgen die Welt untergeht …'. Christliche Schöpfungsspiritualität zwischen Angst und Hoffnung." In *Gärten in der Wüste, Schöpfungsethik zwischen Wunsch und Wirklichkeit*, edited by Severin J. Lederhilger, 39–49. Regensburg: Verlag Friedrich Pustet, 2018.

Rossignol-Strick, Martine. "The Holocene Climatic Optimum and Pollen Records of Sapropel 1 in the Eastern Mediterranean, 9000–6000 BP." *Quaternary Science Reviews* 18 (1999): 515–530. doi: 10.1016/S0277-3791(98)00093-6.

Roth, Martha T. *Law Collections from Mesopotamia and Asia Minor*. Atlanta: Scholars Press, 1997.

Rova, Elena. "The Eagle and the Snake. Remarks on the Iconography of Some Archaic Seals." *KASKAL* 3 (2006): 1–29.

Ruchser, Matthias. *Desertec or the Mediterranean Solar Plan—Whose Sun is Shining Brighter?* Bonn: Deutsches Institut für Entwicklungspolitik (DIE), 2010.

Rudorf, Wilhelm. *Zur Geschichte und Geographie alteuropäischer Kulturpflanzen. Eine Übersicht aufgrund archäologischer Quellen*. Berlin, Hamburg: Verlag Paul Parey, 1969.

Safei-Eldin, A. Hamed. "Egypt." In *Environmentalism in the Muslim World*, edited by Richard C. Foltz. New York: Nova Science Publishers, Inc., 2005.

Safi-Nejad, Javad. *Talebabad. Case Study of a Village. Tehran*. Teheran: University of Tehran: Institute of Social Research, 1966.

Said, Edward W. *Orientalism. Western Conceptions of the Orient*. New York: Routledge 1978, reprint in London: Penguin Books 1991.

Saidi, Abbas. "Agrarwirtschaftlicher Wandel in Iran unter besonderer Berücksichtigung der Bewässerungswirtschaft. Am Beispiel des Gebietes Assad'abad (West Iran)." Unpublished Ph.D. diss., Universität Marburg.

Sallaberger, Walther. *Wenn Du mein Bruder bist …' Interaktion und Textgestaltung in altbabylonischen Alltagsbriefen*. Groningen: STYX Publications, 1999.

Sanlaville, Paul. "Considérations sur l'évolution de la Basse Mésopotamie au cours des derniers millénaires." *Paléorient* 15 (1989): 5–27. doi:10.3406/paleo.1989.4506.

Santarius, Tilman. "Digitalization, Efficiency and the Rebound Effect." Accessed July 26. 2017, https://www.degrowth.info/en/2017/02/digitalization-efficiency-and-the-rebound-effect/.

Saqqaf, Abdulaziz Y. *The Middle East City. Ancient Traditions Confront a Modern World*. New York: Paragon House Publishers, 1987.

Sardar, Ziauddin. "Towards an Islamic Theory of the Environment." In *Islamic Futures: The Shape of Ideas to Come*, edited by Ziauddin Sardar. London, New York, 1985.

Sarnthein, Michael. "Sediments and History of the Postglacial Transgression in the Persian Gulf and Northwest Gulf of Oman." *Marine Geology* 12 (1972): 245–266. doi:10.1016/0025-3227(72)90002-3.

Sarvati, Mohammad R. *Geomorphologische Untersuchungen in der Region Kashmar NE/Iran*. Würzburg: Würzburger geographische Arbeiten, 1986.

Schär, Christoph. "Climate Extremes: The Worst Heat Waves to Come." *Nature Climate Change* 6 (2016): 128–129. doi:10.1038/nclimate2864.

Schatzschneider, Isabel. *Animal Welfare in Ḥalāl Market Standards. European Conceptions Coined in Islamic Terms*, Bonner Islamwissenschaftliche Hefte, Heft 42. Berlin: EB, 2018.

Schenkel, Wolfgang. *Die Bewässerungsrevolution im Alten Ägypten*. Mainz: Philipp von Zabern: Deutsches Archäologisches Institut, Abteilung Kairo, 1978.

Schimmel, Annemarie. *Yunus Emre. Ausgewählte Gedichte*. Köln: Önel-Verlag, 1991.

Schimmel, Annemarie. *Die Zeichen Gottes. Die religiöse Welt des Islams*. München: C. H. Beck, 1995.

Schmidt, Matthias. "Wasserkrise am Urmiasee im Iran: Eine Umwelt- und Sozialkatastrophe des Anthropozäns." *Geographische Rundschau* 70.1–2 (2018).

Schmitt, Thomas M. "Protection du patrimoine culturel et transformation socioculturelle: la « vallée du M'zab ». Un site classé patrimoine mondial en Algérie." In *Les pays du Maghreb. Contributions de la géographie humaine allemande*, edited by Herbert Popp, 38–49. Bayreuh: University of Bayreuth, 2009.

Scholz, Fred. *Muscat—Sultanat Oman. Geographische Skizze einer einzigartigen arabischen Stadt*, 2 vols. Berlin, Das Arabische Buch, 1990.

Schwab, Markus J. Frank Neumann, Thomas Litt, Jörg F.W. Negendank, and Mordechai Stein. "Holocene Palaeoecology of the Golan Heights (Near East): Investigation of Lacustrine Sediments from Birkat Ram Crater Lake." *Quaternary Science Reviews* 23 (2004): 1723–1731. doi:10.1016/j.quascirev.2004.05.001.

Schwartz, Richard. *Judaism and Vegetarianism*. New York: Lantern, 2012.

Schwarzschild, Steven. "The Unnatural Jew." *Environmental Ethics* 6 (1984): 347–362.

Schwencke, Anne Marieke. *Globalized Eco-Islam. A Survey of Global Islamic Environmentalism*. Leiden: Leiden University, 2012.

SEDAC—Socioeconomic Data and Applications Centre. "Gridded Population of the World: Future Estimates." Accessed May 18, 2018. http://sedac.ciesin.columbia.edu/mapping/popest/gpw-v4/.

Seidl, Ursula. "Das Flut-Ungeheuer abūbu." *Zeitschrift für Assyriologie und Vorderasiatische Archäologie* 88 (1998): 100–113.

Seidl, Ursula and Manfred Krebernik. "Rosette." In *Reallexikon der Assyriologie und Vorderasiatischen Archäologie*, edited by Michael P. Streck, 442–447. Berlin: Walter de Gruyter, 2007.

Sekem. "Ecology." Accessed November 1, 2019. https://www.sekem.com/en/ecology/.

Sekem. "Index." Accessed November 1, 2019. https://www.sekem.com/en/index/.

Selz, Gebhard J. "'Elam' und 'Sumer'—Skizze einer Nachbarschaft nach inschriftlichen Quellen der vorsargonischen Zeit." In *Mesopotamie et Elam*, ed. Léon. de Meyer and Hermann. Gasche, 23–43. Ghent: University of Ghent, 1991.

Selz, Gebhard J. "Über Mesopotamische Herrschaftskonzepte: Zu den Ursprüngen mesopotamischer Herrscherideologie im 3. Jahrtausend." In *dubsar anta-men: Studien zur Altorientalistik. Festschrift für Willem H.Ph. Römer zur Vollendung seines 70. Lebensjahres mit Beiträgen von Freunden, Schülern und Kollegen*, edited by Manfried Dietrich and Oswald Loretz, 281–344. Münster: Ugarit-Verlag, 1998.

Selz, Gebhard J. "'Wirtschaftskrise—Legitimationskrise—Staatskrise': Zur Genese mesopotamischer Rechtsvorstellungen zwischen Planwirtschaft und Eigentumsverfassung." *Archiv für Orientforschung* 46/47 (1999/2000): 1–44.

Selz, Gebhard J. "Schrifterfindung als Ausformung eines reflexiven Zeichensystems." *Wiener Zeitschrift für die Kunde des Morgenlandes* 90 (2000): 169–200.

Semple, Ellen C. *Influences of Geographic Environment*. New York: 1911.

Semsar, Yazdi, Ali Asghar, and Majid Labbaf Khaneiki. *Qanat Knowledge: Construction and Maintenance*. Dordrecht: Springer, 2016.

Serjeant, Robert B. and Ronald Lewcock. *Sana. An Arabian Islamic City*. Cambridge: World of Islam Fetival Trust, 1983.

Shadkam, Somayeh, Fulco Ludwig Michelle van Vliet, and Amandine V. Pastor. "Preserving the World Second Largest Hypersaline Lake under Future Irrigation and Climate Change." *Science of the Total Environment* 559 (2016): 317–325.

Shadkam, Somayeh, Pieter Oel, Pavel Kabat, and Fulco Ludwig. "Water-saving Interventions Assessment Framework: An Application for the Urmia Lake Restoration Program." *Geophysical Research Abstracts* 19 (2017).

Shakir, Amena and Said Topalovic. *Islamstunde 7. Wie werde ich stark? Religionsbuch für die Sekundarstufe I*. Wien: Oldenburg-Verlag, 2015.

Sieferle, Rolf P. "The Ends of Nature." In *Environment across Cultures. Wissenschaftsethik und Technikfolgenbeurteilung*, ed. Eckart Ehlers, and Carl F. Gethmann. Berlin: Springer, 2003.

Sima, Somayeh and Mohammad Tajrishy. "Using Satellite Data to Extract Volume—Area—Elevation Relationships for Urmia Lake, Iran." *Journal of Great Lakes Research* 39.1 (2013): 90–99.

Simenowitz, Shmuel. "Water Conservation and Halacha: An Unorthodox Approach." In *Compendium of Sources in Halacha and the Environment*, edited by Ora Shenson and Shai Spetgant. Jerusalem: Canfei Nesharim, 2005.

Staubwasser, Michael and Harvey Weiss. "Holocene Climate and Cultural Evolution in late Prehistoric-Early Historic West Asia." *Quaternary Research* 66 (2006): 371–387. doi:10.1016/j.yqres.2006.09.001.

Stein, Ilana. "Le'ovda Uleshomra: Judaism and the Environmental Ethic." In *Compendium of Sources in Halacha and the Environment*, edited by Ora Shenson and Shai Spetgant. Jerusalem: Canfei Nesharim, 2005.

Steinkeller, Piotr. "On Rulers, Priests and Sacred Marriage: Tracing the Evolution of Early Sumerian Kingship." in *Priests and Officials in the Ancient Near East*, edited by Kazuko Watanabe, 103–137. Heidelberg: Universitätsverlag C. Winter, 1999.

Stern, Nicholas. *The Economics of Climate Change. The Stern Review*. Cambridge: Cambridge University Press, 2007.

Stetter, Daniel. "Enhancement of the REMix Energy Model—Global Renewable Energy Potentials Optimized Power Plant Siting and Scenario Validation". PhD diss., University Stuttgart and DLR, 2012.

Stevens, Lora A. Emi Ito, Antje Schwalb, and Herbert E. Wright Jr. "Timing of Atmospheric Precipitation in the Zagros Mountains Inferred from a Multi-proxy Record from Lake Mirabad, Iran." *Quaternary Research* 66 (2006): 498–499. doi:10.1016/j.yqres.2006.06.008.

Stevens, Karel G. "Eine ikonographische Untersuchung der Schlange im vorgeschichtlichen Mesoptamien." In *Archaeologia Iranica et orientalis: miscellanea in honorem Louis Vanden Berghe*, edited by Léon de Meyer and Ernie Haerinck. Gent: Peters, 1989.

Stiftung Umwelt und Entwicklung NRW. "Faire Moschee." Accessed March 26, 2019. www.sue-nrw.de/2016/02/29/faire-moschee/www.sue-nrw.de/2016/02/29/faire-moschee/.

Stöber, Georg. "Nahost—Mittelost—Vorderer Orient: Wer ist was?" In *Der Islamische Orient. Grundlagen zur Länderkunde eines Kulturraums*, edited by Eckart Ehlers and Abdoldjavad Falaturi. Köln: Studien zum Islam in Interkulturellen Wechselbeziehungen. Schriftenreihe der Islamischen Wissenschaftlichen Akademie Bd. 1, 1990.

Stol, Marten. "Kanal(isation). A. Philologisch." In *Reallexikon der Assyriologie und Vorderasiatischen Archäologie vol. 5*, edited by Dietz O. Edzard. Berlin: Walter de Gruyter, 1976–1980. 355–365.

Svensson, Birgit. "Flood Disaster in Iraq: Water as an Instrument of War." *Qantara.de*, June 13, 2014, https://en.qantara.de/content/flood-disaster-in-iraq-water-as-an-instrument-of-war.

Swain, Ashkok and Anders Jägerskog, *Emerging Security Threats in the Middle East. The Impact of Climate Change and Globalisation*. Latham MD: Rowman and Littlefield, 2016.

Swain, Ashkok, Anders Jägerskog, and Michael Schulz, ed. *Handbook of Middle East Security*. Latham MD: Rowman and Littlefield 2019.

Tahmasebi, Asghar. "Indigenous Knowledge for Water Management in Iran's Dry Land—Siraf." *International Journal of Environmental Studies* 66.3 (2009): 317–325.

Tahmasebi, Asghar, Eckart Ehlers, and Conrad Schetter. "Climate Change and Mountain Pastoralism—The Shahsevan of Northwest Iran." *Erdkunde* 67.4 (2013): 309–323.

Taravat, Alireza, Masih Rajaei, Iraj Emadodin, Hamidreza Hasheminejad, Rahman Mousavian, and Ehsan Biniyaz. "A Spaceborne Multisensory, Multitemporal Approach to Monitor Water Level and Storage Variations of Lakes." *Water* 8.11 (2016): 478.

Teller, Jim T., Kenneth W. Glennie, Nicholas Lancaster, and Ashok K. Singhvi. "Calcareous Dunes of the United Arab Emirates and Noah's Flood: The Postglacial Reflooding of the Persian (Arabian) Gulf." *Quaternary International* 68–71 (2000): 297–308. doi:10.1016/S1040-6182(00)00052-5.

Thalen, Derk C.P. *Ecology and Utilization of Desert Shrub Rangelands in Iraq*. The Hague: Dr. W. Junk Publishers, 1979). doi:10.1007/978-94-009-9622-9.

The Jakarta Post of June 5, 2018. Accessed March 26, 2019. www.thejakartapost.com/news/2018/06/05/greenpeace-indonesia-holds-eco-iftar-in-pondok-indah.html.

Thomas, Elizabeth R., Eric W. Wolff, Robert Mulvaney, Jorgen P. Steffensen, Sigfus J. Johnsen, Carol Arrowsmith, James W.C. White, Bruce Vaughn, and Trevor Popp. "The 8.2 ka Event from Greenland Ice Cores," *Quaternary Science Reviews* 26 (2007): 70–81. doi:10.1016/j.quascirev.2006.07.017.

Tirosh-Samuelson, Hava. "Jewish Environmentalism: Bridging Scholarship, Faith and Activism." In *Jewish Thought and Jewish Belief*, edited by Daniel Lasker, 65–117. Beer Sheva: Ben Gurion University Press, 2012.

Tirosh-Samuelson, Hava. "Judaism and the Science of Ecology." In *The Routledge Companion to Religion and Science*, edited by James W. Haag, George R. Peterson, Michael L. Spezio, 345–355. London: Routledge, 2011.

Topalovic, Said, Claudia Ausweger, Hanna Hodzic, Ursula F. Kowanda-Yassin, Sanela Mahmutovic, Mevlida Mesanovic, Amena Shakir, Azra Tajar, Aida Tuhcic, Fatima Tupo, Yasin Ulu. *Islamstunde 8. Die Anderen respektieren. Religionsbuch für die Sekundarstufe I*. Oldenburg: VERITAS.

Tourianet, Mohammad J., B. Devaraju, N. Sneeuw, O. Elmi, and Q. Chen, and S. Roohi. "A Spaceborne Multisensor Approach to Monitor the Desiccation of Lake Urmia in Iran." *Remote Sensing of Environment* 156 (2015): 349–360.

Torabian, Elham. "Exploring Social Vulnerability and Environmental Migration in Urmia Lake of Iran: Comparative Insights from the Aral Sea." In *The State of Environmental Migration 2014*, edited by François Gemenne, Pauline Brücker, and Dina Ionesco. Geneva: IOM, 2014.

Trieb, Franz. *Trans-Mediterranean Interconnection for Concentrating Solar Power (TRANS-CSP)*. Stuttgart, German Aerospace Center (DLR), 2006.

Trieb, Franz and Hans Müller-Steinhagen. "Europe—Middle East—North Africa Cooperation for Sustainable Electricity and Water." *Sustainability Science* 2.2 (2007): 205–219.

Trieb, Franz, Marlene O'Sullivan, Thomas Pregger, Christoph Schillings, Wolfram Krewitt. *Characterization of Solar Electricity Import Corridors from MENA to Europe—Export Potentials, Infrastructures and Costs*. Stuttgart: DLR, 2009.

Trieb, Franz. "Solar Electricity Imports from the Middle East and North Africa to Europe." *Energy Policy* 42 (2012): 341–353.

Trieb, Franz. Tobias Fichter, and Massimo Moser. "Concentrating Solar Power in a Sustainable Future Electricity Mix." *Sustainability Science* 9.1 (2014): 47–60. DOI 10.1007/s11625-013-0229-1.

Trieb, Franz, Cristina De la Rúa, N. Caldés, and Marta Santamaria. *Bringing Europe and Third Countries Closer Together through Renewable Energies—WP3 North Africa Case Study*. Project Report 2015.

Trieb, Franz, Juergen Kern, Natàlia Caldés, Cristina de la Rua, Dorian Frieden, Andreas Tuerk. "Rescuing the Concept of Solar Electricity Transfers from North Africa to Europe." *International Journal of Energy Sector Management* 10.3 (2016): 448–473. Accessed May 15, 2017. www.emeraldinsight.com/1750–6220.htm.

Tropp, Håkan and Anders Jägerskog. *Water Scarcity Challenges in the Middle East and North Africa (MENA)*. UNDP, 2006.

Uchupi, Elazar, Stephen. A. Swift, and David A. Ross. "Late Quaternary Stratigraphy, Paleoclimate and Neotectonism of the Persian (Arabian) Gulf Region." *Marine Geology* 160 (1999): 1–23. doi:10.1016/S0025-3227(99)00011-0.

Ucko, Peter J. and G.W. Dimpleby. *The Domestication and Exploitation of Plants and Animals. Proceedings of a Meeting of the Research Seminar in Archaeology and Related Subjects Held at the Institute of Archaeology*. London: London University, 1969.

Uehlinger, Christoph. "Drachen und Drachenkämpfe im alten Vorderen Orient." In *Auf Drachenspuren: Ein Buch zum Drachenprojekt des Hamburgischen Museums für Völkerkunde*, edited by Bernd Schmelz und Rüdiger Vossen. Bonn: Holos, 1995.

UNESCO. *Water for People, Water for Life*. The United Nations World Water Development Report. UNESCO, 2003.

Union for the Mediterranean (UfM), Secretariat of the Union for the Mediterranean. "The UfM: An Action-Driven Organisation with a Common Ambition." Accessed May 15, 2017. http://ufmsecretariat.org/wp-content/uploads/2017/10/UfM-Roadmap-for-action-2017.pdf.

United Nations Department of Economic and Social Affairs (UN DESA). *World Population Prospects 2017*. New York: United Nations Department of Economic and Social Affairs, Population Division, 2017.

United Nations Development Programme (UNDP). *Arab Human Development Report 2016: Youth and the Prospects for Human Development in a Changing Reality*. New York: United Nations Development Programme Regional Bureau for Arab States, 2016.

United Nations Development Programme (UNDP)–GEAS. "The Drying of Iran's Lake Urmia and its Environmental Consequences," *Environmental Development* 2 (2012).

United Nations World Water Assessment Programme (WWAP)/UN-Water. *The United Nations World Water Development Report 2018: Nature-Based Solutions for Water*. Paris: UNESCO, 2018.

United Nations World Water Assessment Programme (WWAP). *The United Nations World Water Development Report 2016: Water and Jobs*. Paris: UNESCO, 2016.

Urmia Lake Restoration Program (ULRP). "Urmia Lake Restoration Program: Brief Report and Projects Outline." Last modified October 2, 2015, accessed May 18, 2018. https://usu.instructure.com/files/61057414.

van Ess, Margarete. "Uruk—Verortung in Raum und Zeit." In *Uruk—5000 Jahre Megacity. Begleitband zur Ausstellung „Uruk—5000 Jahre Megacity" im Pergamonmuseum. Curt-Engelhorn-Stiftung für die Reiss-Engelhorn-Museen Mannheim, dem Deutschen Archäologischen Institut—Orient-Abteilung, der Deutschen Orient-Gesellschaft e. V. und dem Vorderasiatischen Museum—Staatliche Museen zu Berlin*, ed. Nicole Crüsemann, Margarete van Ess, Markus Hilgert, and Beate Salje, 39-45, Petersberg: Imhof Verlag, 2013.

van Ess, Margarete. "Die topographische Entwicklung der Stadt Uruk im 4. und 3. Jt. v. Chr." In: *Uruk. Altorientalische Metropole und Kulturzentrum*, edited by Margarete van Ess. Wiesbaden: Harrassowitz-Verlag, in press.

Verheyden, Sophie, Fadi H. Nader, Hai J. Cheng, Lawrence R. Edwards, and Rudy Swennen. "Paleoclimate Reconstruction in the Levant Region from the Geochemistry of a Holocene Stalagmite from the Jeita Cave, Lebanon." *Quaternary Research* 70 (2008): 368–381. doi:10.1016/j.yqres.2008.05.004-.

Verhoeven, Kristiaan. "Changing Watercourses in Babylonia." In *Geomorphological Research in the Mesopotamian Flood Plain Changing Watercourses in Babylonia. Towards a Reconstruction of the Ancient Environment in Lower Mesopotamia, Vol. I*, edited by Hermann Gasche and Michel Tanret, 159–245. Chicago: University of Chicago Press, 1998.

Volk, Konrad. *Inanna und Šukaletuda: Zur historisch-politischen Deutung eines sumerischen Literaturwerkes*. Wiesbaden: Harrassowitz Verlag, 1995.

von der Osten-Sacken, Elisabeth. *Der Ziegen-'Dämon': 'Obed- und Uruk-zeitliche Götterdarstellungen*. Kevelaer: Butzon and Bercker, 1992.

von der Osten-Sacken, Elisabeth. "Schlange. C. In der Bildkunst." In *Reallexikon der Assyriologie und Vorderasiatischen Archäologie*, edited by Michael P. Streck. Berlin: Walter de Gruyter, 2009.

von Lossow, Tobias. "Klimawandel und Wassermangel in Afrika: Möglichkeiten der Kooperation am Beispiel Nil," in *Klimawandel und Sicherheit. Herausforderungen, Reaktionen und Handlungsmöglichkeiten*, edited by Steffen Angenendt, Susanne Dröge, and Jörn Richert. Baden-Baden: Nomos, 2011.

von Lossow, Tobias. "The Rebirth of Water as a Weapon: IS in Syria and Iraq." *The International Spectator* 51.3 (2016): 82–99.

von Lossow, Tobias. "Blaues Wunder. Wasser war für Israel stets von existenzieller Bedeutung—gesellschaftlich, politisch, wirtschaftlich." *Internationale Politik (IP)* 2 (2016): 53.

von Lossow, Tobias. "The Multiple Crisis. Perspectives on Water Scarcity in the Euphrates and Tigris Basin." *ORIENT* 58 (2017): 45–53.

von Lossow, Tobias. *More than Infrastructures: Water Challenges in Iraq*. Den Haag: Clingendael, July 2018.

von Lossow, Tobias. "Weaponization of Water in the Middle East—'Lessons Learned' from IS." In *Water and Conflict in the Middle East*, edited by King, Marcus D. Hurst Publishers, 2019.

von Thun, Friedemann Schulz, Kathrin Zach, and Karen Zoller. *Miteinander reden von A-Z. Lexikon der Kommunikationspsychologie*. Reinbek bei Hamburg: Rowohlt-Verlag GmbH, 2012.

von Wickede, Alwo. *Prähistorische Stempelglyptik in Vorderasien*. München: Profil Verlag, 1990.

Wagstaff, Malcolm T. ed. *Landscape and Culture. Geographical Archaeological Perspectives* (Oxford, New York; Blackwell, 1987.

Waissbein, Oliver, Yannik Glemarec, Hande Bayractar and Tobias S. Schmidt. *De-risking Renewable Energy Investment—A Framework to Support Policymakers in Selecting Public Instruments to Promote Renewable Energy Investments in Developing Countries*. New York, United Nations Development Programme (UNDP), 2013.

Wals, Arjen E.J. *Social Learning towards a Sustainable World. Principles, Perspectives, and Praxis*. Wageningen: Academic Publishers 2007.

Ward, Christopher. Sabine Beddies, and Sandra Ruckstuhl. *Managing Water Scarcity to Reduce Conflict Risks in the Middle East and North Africa*. Washington, World Bank, 2011.

Ward, Christopher and S. Ruckstuhl, ed. *Water Scarcity, Climate Change and Conflict in the Middle East*. London: Tauris, 2017.

Ward, W.A. "Relations Between Egypt and Mesopotamia from Prehistoric Times to the End of the Middle Kingdom." *Journal of the Economic and Social History of the Orient* 7 (1964).

Waskow, Arthur. "What Is Eco-Kosher?" In *This Sacred Earth: Religion, Nature, Environment*, edited by Gottlieb, Roger S. New York: Routledge, 1996.

Watanabe, Chikako E. *Animal Symbolism in Mesopotamia. A Contextual Approach*. Vienna: Institut für Orientalistik der Universität Wien, 2002.

Weatherall, Pauline, Karen M. Marks, Martin Jakobsson, Thierry Schmitt, Shin Tani, Jan E. Arndt, Marzia Rovere, Dale Chayes, Vicki Ferrini, and Rochelle Wigley. "A New Digital Bathymetric Model of the World's Oceans." *Earth and Space Science* 2 (2015): 331–345. doi:10.1002/2015EA000107.

Webster, Frank. *Theories of the Information Society*. London, New York: Routledge, 2007.

Weisberger, Atara. "An Exploration of Tza'ar Ba'alei Chaim." In *Compendium of Sources in Halacha and the Environment*, edited by Ora Shensonand Shai Spetgant. Jerusalem: Canfei Nesharim, 2005.

Weiss, Harvey, Marie-Agnès Courty, Wilma Wetterstrom, François Guichard, Luis Senior, Richard H. Meadow, and Andrew Curnew. "The Genesis and Collapse of 3rd Millennium North Mesopotamian Civilization." *Science* 261 (1993): 995–1004. doi:10.1126/science.261.5124.995.

Weninger, Bernhard, Lee Clare, Eelco J. Rohling, Ofer Bar-Yosef, Utz Böhner, Mihael Budja, Manfred Bundschuh, Angelica Feurdean, Hans-Georg Gebel, Olaf Jöris, Jörg Linstädter, Paul Mayewski, Tobias Mühlenbruch, Agathe Reingruber, Gary Rollefson, Daniel Schyle, Laurens Thissen, Henrieta Todorova, and Christoph Zielhofer. "The impact of rapid climate change on prehistoric societies during the Holocene in the eastern Mediterranean." *Documenta Praehistorica* XXXVI (2009): 7–59. doi:10.4312/dp.36.2.

White, Lynn jr. "Die historischen Ursachen unserer ökologischen Krise." In *Gefährdete Zukunft. Prognosen angolamerikanischer Wissenschaftler*, edited by Michael Lohmann. München, Wien: Hanser, 1970.

Wick, Lucia. Gerry Lemcke, and Michael Sturm. "Evidence of Lateglacial and Holocene Climatic Change and Human Impact in Eastern Anatolia: High-resolution Pollen, Charcoal, Isotopic and Geochemical Records from the Laminated Sediments of Lake Van, Turkey." *The Holocene* 13 (2003): 665–675. doi:10.1191/0959683603hl653rp.

Wiggermann, Franz A.M. "Transtigridian Snake Gods." In *Sumerian Gods and their Representations*, edited by Irvine L. Finkel and Markham J. Geller. Groningen: Styx Publications, 1997.

Wiggermann, Frans A.M. "Niraḫ, Irḫan." In *Reallexikon der Assyriologie und Vorderasiatischen Archäologie*, edited by Dietz O. Edzard. Berlin: Walter de Gruyter, vol. 9, 1998–2001.

Wiggermann, Frans A. M. "The Four Winds and the Origins of Pazuzu." In *Das geistige Erfassen der Welt im Alten Orient*, ed. Claus Wilcke. Wiesbaden: Harrassowitz Verlag, 2007.

Wilcke, Claus. "Weltuntergang als Anfang." In *Weltende*, edited by Adam Jones. Wiesbaden: Harrassowitz-Verlag, 1999.

Wilkinson, Tony J. "Early Channels and Landscape Development around Abu Salabikh, a Preliminary Report." *Iraq* 52 (1990): 75–83. doi:10.2307/4200320.

Wilkinson, Tony J. "Regional Approaches to Mesopotamian Archaeology: The Contribution of Archaeological Surveys." *Journal of Archaeological Research* 8 (2000): 219–267. doi:10.1023/A:1009487620969.

Wilkinson, Tony J. *Archaeological Landscapes of the Near East*. University of Arizona Press, 2003.

Wilkinson, Tony J., Graham Philip, Jennie Bradbury, Robert Dunford, Danny Donoghue, Nikolaos Galiatsatos, Dan Lawrence, Andrea Ricci, and S. L. Smith. "Contextualizing Early Urbanization: Settlement Cores, Early States and Agro-pastoral Strategies in the Fertile Crescent During the Fourth and Third Millennia BC." *Journal of World Prehistory* 27 (2014). doi:10.1007/s10963-014-9072-2.

Wirth, Eugen. "Einleitung: Der Orient—Versuch einer Definition und Abgrenzung." In *Nordafrika und Vorderasien*, edited by Horst Mensching und Eugen Wirth, 323–332. Frankfurt a.M.: Fischer, 1973.

Wirth, Eugen. "Die Beziehungen der orientalischen Stadt zum umgebenden Land." In *Erdkuntliches Wissen*. Stuttgart: Steiner Verlag, 1973.

Wirth, Eugen. *Die orientalische Stadt im islamischen Vorderasien und Nordafrika. Städtische Bausubstanz und räumliche Ordnung, Wirtschaftsleben und soziale Organisation*. Mainz: Verlag Philipp von Zabern, 2000.

Wisdom in Nature. "Permaculture and Sufism. A Return to the Natural Patterning." Accessed November 01, 2018. http://wisdominnature.org/category/sufism.

Wissenschaftlicher Beirat der Bundesregierung Globale Umweltveränderungen. *Welt im Wandel: Herausforderungen für die deutsche Wissenschaft*. Jahresgutachten 1996. Bremerhaven: WBGU, 1996.

Wissenschaftlicher Beirat der Bundesregierung Globale Umweltveränderungen (WBGU). *Sicherheitsrisiko Klimawandel*. Berlin, Heidelberg: Springer, 2008.

Wissenschaftlicher Beirat der Bundesregierung Globale Umweltveränderungen (WBGU). *Welt im Wandel. Gesellschaftsvertrag für eine Große Transformation*. Berlin: WBGU, 2011.

Wissenschaftlicher Beirat der Bundesregierung Globale Umweltveränderungen (WBGU). *Der Umzug der Menschheit: Die transformative Kraft der Städte*. Berlin: WBGU, 2016.

Wittfogel, Karl A. "The Hydraulic Civilizations." In *Man's Role in Changing the Face of the Earth*, edited by William L. Thomas Jr., 152–164. Chicago: The University of Chicago Press, 1956.

Woertz, Eckart. "Agriculture and Development in the Wake of the Arab Spring." *International Development Policy / Revue internationale de politique de développement* 7.7 (2017).

Wolf, Aaron T. "Conflict and Cooperation along International Waterways." *Water Policy* 1 (1998): 251–265.

Woolley, C. Leonard. "Excavations at Ur of the Chaldees." *The Antiquaries Journal* 3 (1923): 311–333. doi:10.1017/S0003581500014980.

World Bank. *Beyond Scarcity: Water Security in the Middle East and North Africa*. Washington, DC. World Bank, 2017.

Word Bank. "World Bank Indicators." Accessed December 20, 2018. https://data.worldbank.org/indicator/SP.POP.GROW?end=2017&locations=ZQ&start=2016.

World Bank. "Middle East and North Africa. Integration of Electricity Networks in the Arab World: Regional Market Structure and Design." Accessed July 28, 2017. https://openknowledge.worldbank.org/bitstream/handle/10986/19329/ACS71240ESW0WH0I0and0II000Final0PDF.pdf?sequence=1&isAllowed=y, 2.

World Bank and African Development Bank. *ICTs for Agriculture in Africa*. World Bank, 2012. Accessed July 28, 2017. http://siteresources.worldbank.org/EXTINFORMATIONANDCOMMUNICATIONANDTECHNOLOGIES/Resources/282822-1346223280837/Agriculture.pdf.

World Bank Group. *Agriculture and Rural Development in MENA*. World Bank Group, 2008.

World Economic Forum. "How Islam can Help Combat Climate Change." Accessed January 26, 2019. https://www.weforum.org/agenda/2015/09/how-islam-can-help-combat-climate-change/.

Wulff, Hans E. *The Traditional Crafts of Persia. Their Development, Technology, and Influence on Eastern and Western Civilizations*. Cambridge: MIT Press, 1966.

Wulff, Hans E. "The Qanats of Iran." *Scientific American* 218 (1968): 94–105.

Zafarnejad, Fatemeh. "The Contribution of Dams to Iran's Desertification." *International Journal of Environmental Studies* 66.3 (2009): 327–341.

Zand, Kamran V. "Zu den Schreibungen des Anzud-Vogels in der Fāra-Zeit." In *Von Göttern und Menschen. Beiträge zu Literatur und Geschichte des Alten Orients. Festschrift für Brigitte Groneberg*, edited by Dahlia Shehata, Frauke Weiershäuser and Kamran V. Zand, 415–442. Leiden: Brill, 2010.

Zarghami, Mahdi. "Effective Watershed Management. Case Study of Urmia Lake, Iran." *Lake and Reservoir Management* 27.1 (2011): 87–94.

Zarghami, Mahdi, Amin Abdi, Iman Babaeian, and Yousef Hassanzadeh. "Impacts of Climate Change on Runoffs in East Azerbaijan, Iran." *Global and Planetary Change* 78 (2011): 137–146.

Zarrineh, Nina and Mohammad Azari Najaf Abad. "Integrated Water Resources Management in Iran: Environmental, Socio-economic and Political Review of Drought in Lake Urmia." *International Journal of Water Resources and Environmental Engineering* 6.1 (2014).

Zawahri, Neda A. *Water Security in the Middle East and North Africa*. Durham: Duke University, 2017.

Zbidi, Monika. "*The Call to Eco-Jihad. The Islamic Environmental Movement.*" Art&Thought, 99 (2013).

Zeier, Hans. "Umwelt, Angst und Stress." In *Umweltpädagogik. Ansätze, Analysen, Ausblicke*, edited by Felix von Cube Volker and Storch. Heidelberg: Edition Schindele, 1988.

Zeinoddini, Mostafa, Arash Bakhtiari, and Majid Ehteshami. "Long-term Impacts from Damming and Water Level Manipulation on Flow and Salinity Regimes in Lake Urmia, Iran." *Water and Environment Journal* 29 (2015).

Zeit Online. "*Tote bei Protesten im Irak.*" 23 July, 2018. https://www.zeit.de/politik/ausland/2018-07/irak-sueden-demonstrationen-korruption-arbeitslosigkeit-staatliche-dienstleistungen.

Zeitoun, Mark. *Power and Water in the Middle East. The Hidden Politics of the Palestinian-Israeli Water Conflict*. London, New York: Tauris, 2008, Paperback 2011.

Zickfeld, Florian. *DESERT Power 2050—Perspectives on a Sustainable Power System for EUMENA*. Munich: Dii, 2012.

Ziebertz, Hans-G. and Ulrich Riegel. "Eigenverantwortliches Lernen im Religionsunterricht: Projekt und Freiarbeit." In *Religionsdidaktik. Ein Leitfaden für Studium, Ausbildung und Beruf*, ed. Georg Hilger, Stephan Leimgruber, and Hans G. Ziebertz. München: Kösel-Verlag, 2010.

Ziebertz, Hans-G. "Wozu religiöses Lernen?—Religionsunterricht als Hilfe zur Identitätsbildung." In *Religionsdidaktik. Ein Leitfaden für Studium, Ausbildung und Beruf*, edited by Georg Hilger, Stephan Leimgruber, and Hans G. Ziebertz, 142–151. München: Kösel-Verlag, 2010.

Zinkin, John. "Islam and CSR: A Study of the Compatibility between the Tenets of Islam, the UN Global Compact and the Development of Social, Human and Natural Capital. *Corporate Social Responsibility and Environmental Management* 14.4 (2007): 206–218.

Zou'bi Moneef R. "The Arab States." in *UNESCO Science Report: Towards 2030*, edited by UNESCO. Paris: UNESCO, 2015.

Index